BIOMEDICAL ELECTRONICS

BIOMEDICAL ELECTRONICS

by Howard M. Yanof, Ph.D.

Assistant Professor of Physiology
Saint Louis University School of Medicine

F. A. DAVIS COMPANY, Philadelphia

The circuit diagrams included in this manual are included for illustration of typical electronic applications and are not intended as constructional information. For this reason, component values, wattage ratings of resistors, and voltage ratings of capacitors are not necessarily given. Similarly, shielding techniques and alignment methods which may be necessary in some circuit layouts are not indicated. Although reasonable care has been taken in their preparation to insure their technical correctness, no responsibility is assumed by the author, publisher, or any of the various contributors for any consequences of their use.

The vacuum tube and semiconductor devices and arrangements disclosed herein may be covered by patents held by the contributors or others. The disclosure of any information herein does not convey any license under patent claims combinations of these devices with other devices or elements. In the absence of an express written agreement to the contrary, neither the author nor the publisher assumes any liability for patent infringement arising out of any use of any circuit, any device, or any element by any purchaser or reader of this book.

Printed in the United States of America

LIBRARY OF CONGRESS CATALOG CARD NUMBER 65–21881

This book is dedicated to
my wife, Esther, and my children,
Denise, Jeffrey, and Michelle,
who in their own way provided
cooperation which allowed
this book to become a reality.

Preface

This book was designed to serve as a textbook of biomedical electronic instrumentation for two separate populations. Section 1 is needed by biological students who have very little or no background in electronics. Section 2, in addition to introducing the student with a background in medicine or one of the biological sciences to examples of instruments presently in use, will also serve the electronic engineer as a guide to some of the basic problems which confront the biomedical research worker.

I wish to thank Mr. Louis Rosso, Beckman Instrument Co., Mr. Huntly Millar, E & M Instrument Co., and Mr. C. A. Meyer of the Radio Corporation of America as well as the Department of the U. S. Army for the many illustrations they provided for use in this book. I wish also to thank the many other individuals and companies who provided material for the book.

Special thanks are due to Ray W. Kenworthy, Ph.D., of the Department of Physics at the University of Washington, who so kindly permitted me to use numerous illustrations from his textbook, College Physics.

I would also like to thank Mrs. Joseph R. Klein, Jr., Miss Marian Parmley, Mrs. Bruce Wolfe, and my wife for their help in preparing the manuscript, Mrs. Helen Smith for doing some of the art work, and Mr. Grant Nugent for proofreading the pages.

Lastly, I would like to express gratitude to Dr. Alrick B. Hertzman, other members of the Physiology Department, and fellow faculty members for their encouragement and assistance.

A special note of thanks is due to Mr. Paul Salz, the electronic engineer with whom I worked at the E. O. Lawrence Radiation Laboratory, University of California, Berkeley. His patience, insight, and skill have succeeded in teaching me most of what I know about biomedical instrumentation. I am also indebted to Mr. William Blissard and to Mr. Nick Dana for teaching me a variety of electronic fabrication techniques.

HOWARD M. YANOF

Table of Contents

INTRODUCTION xi

SECTION 1—Fundamentals of Electronics 1

Chapter 1 Basic Physical Concepts 3
Chapter 2 Introduction to AC Circuit Theory . . . 25
Chapter 3 The Measurement of Voltage and Current. 49
Chapter 4 The Rectifier and the Diode 69
Chapter 5 Amplification 83
Chapter 6 The Oscillator 129
Chapter 7 The Power Supply 145

SECTION 2—Introduction to Biomedical Instruments 159

Chapter 8 General Remarks 161
Chapter 9 The Transducer 171
Chapter 10 The Output Signal and What to Do with It 215
Chapter 11 Some Examples of Biomedical Instrumentation 251
Chapter 12 Noise and Its Elimination 307

APPENDICES 317

I Some Physical Units and Conversion Factors . . . 319
II Conventional Electronic Symbols 321
III Identification of Components (Color Codes, etc.) . . 323
IV Typical Vacuum-Tube Characteristic Curves . . . 327
V Typical Transistor Characteristic Curves 341
VI Operational Amplifier Applications 345

INDEX 349

Introduction

In the past decade it has become increasingly necessary for the biological investigator to use precise instruments in order to carry out his work. Similarly, it has also become necessary for the physician to make use of complicated devices both in diagnosis and treatment. Even the person untrained as a scientist has become acutely aware of the technological upheaval which surrounds him in the form of electrocardiograms, computers, and Telstar.

The result of all this is an ever growing need for scientists not only trained in one or more of the biological or medical sciences, but also in the field of electronic instrumentation. It is difficult enough for them to choose a proper instrument for a particular purpose; they must also know enough about the instrument to use it, and often to repair it as well.

This book is intended to serve as a basic textbook for those individuals who seek a beginning knowledge in the field of biological and medical instrumentation. It is not intended to train design engineers or even competent electronic technicians. It is intended to provide a basic course of instruction, a ground work for future study.

The first section of the book is devoted to introductory electricity and electronics. The student should try to grasp the fundamental concepts and terminology, and whenever possible he should attempt to build the circuits described and examine them closely. The second section presents important aspects of instrumentation with particular emphasis on biological and medical instruments. The physical and electronic principles involved in the design of each instrument are discussed. The inclusion of an instrument in this book does not imply that it is the instrument of choice, but rather that the particular instrument is typical or demonstrates a particular point. The reader is encouraged to insist that the manufacturer provide a usable circuit diagram so that the user cannot only better understand the instrument but can more easily repair it when it fails.

SECTION 1

Fundamentals of Electronics

1

Basic Physical Concepts

charge
voltage
current
the conduction of electrical current
semiconductors
resistance
resistors in parallel and series combination
Ohm's law
electrical power
capacitance
capacitors in parallel and series combination
the magnetic field
inductance
inductors in parallel and series combination
transformers
references

This section of the book is intended to familiarize the student with the behavior of electrons as they flow in a closed circuit. The circuit will contain any number of components intended to modify that flow of electrons in such a way as to accomplish a specific purpose. It will be necessary that the student develop a facility for predicting the effect of a wide variety of active and passive components with the same facility with which he might be expected to predict how water would flow through a given circuit of plastic pipes, elbows, tees, and valves. It is also important to develop a technical vocabulary. The student will have to understand such terms as voltage, current, resistance, capacitance, and inductance before he can appreciate the function of the tube, transistor, and tunnel diode.

CHARGE

Two small light-weight pith balls are suspended with silk threads. If a glass rod is rubbed with a piece of silk and then momentarily placed against the pith balls, the balls repel each other and may be seen to separate as in Figure 1-1(a). This will occur because each ball has had a small positive charge deposited on its surface. Were one of the pith balls touched with a hard rubber rod which had been rubbed with cat fur, the balls would attract as shown in Figure 1-1(b).

The electroscope, Figure 1-2, is another device which demonstrates the presence of charge. The ball on the top of the electroscope has been touched with a glass rod made negative by rubbing with silk. The electroscope will also accumulate charge in the presence of ionizing radiation; a specially constructed electroscope, in which the movable thin gold leaf is attached to a pointer which can be viewed against a scale through its optical system, is used to determine the amount

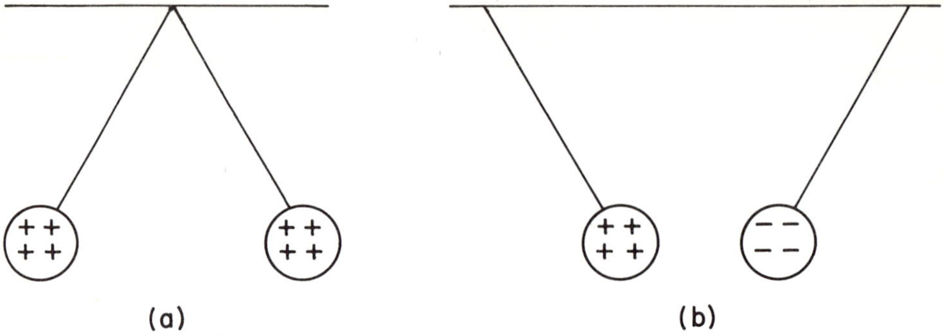

Figure 1-1. Like charges repel and unlike charges attract one another. (Courtesy of R. W. Kenworthy, College Physics, F. A. Davis Co.)

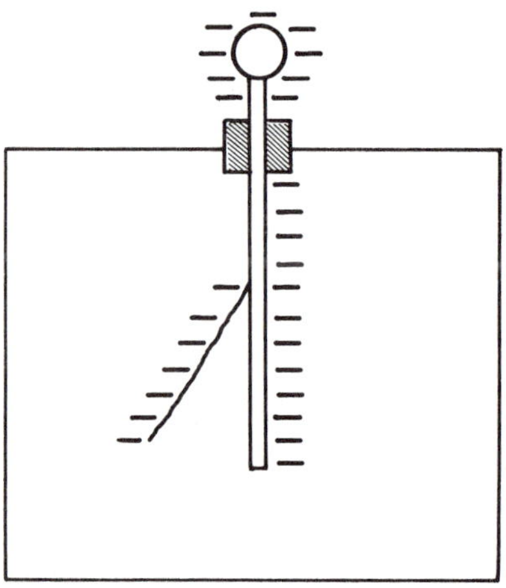

Figure 1-2. A metal foil electroscope. (Courtesy of R. W. Kenworthy, College Physics, F. A. Davis Co.)

of radiation in a particular place. Such an electroscope is often called a pocket dosimeter.

All matter is composed of molecules which are, in turn, made of atoms. The atom itself consists of a nucleus surrounded by an orderly array of particles of considerably less mass called electrons. The nucleus contains one or more positively charged particles called protons and may also contain one or more particles of equal mass but no charge called neutrons. Each atom species has a particular structure, the details of which depend on a wide variety of atomic and nuclear factors, *e.g.,* the atomic weight.

We are particularly interested at this time in the electrons which form the outer layer of the electron cloud surrounding the nucleus. In some materials, such as hard rubber, atomic forces exist which tend not only to hold these outer electrons tightly but tend to be more stable in the presence of additional electrons. In other materials such as cat fur, the atom is more stable when one or more electrons are given up. Thus by rubbing the cat fur on the hard rubber rod, hence bringing the atoms of the cat fur in close association with the atoms of the hard rubber, the hard rubber is left with an excess of electrons. These electrons may then be transferred to the pith balls by momentary physical contact. The electrons tend to flow to the pith balls because there is no excess of electrons on the pith balls, and during the momentary contact the excess electrons will evenly coat the two pith balls; in other words, an equilibrium will be established.

The more electrons transferred to the pith balls, the further they tend to separate, that is, the pith balls exert a repulsive force on each other. The property of electrons which causes them to repel each other with a force gives rise to Coulomb's law which says that the charge (q) necessary to repel a similar charge (q') with a force of 1 dyne (F) when placed at a distance (s) of 1 cm in a vacuum is 1 *esu* (electrostatic unit in the centimeter-gram-second system*). k is a constant of proportionality and is equal to 1 in the cgs system. Coulomb's law also provides us with a definition of charge.

$$F = k\frac{q\,q'}{s^2} \qquad 1.1$$

In the meter-kilogram-second system* the unit of charge is called the coulomb, which is equal to 3×10^9 *esu*. Each electron has been found to have a charge of 1.59×10^{-19} coulomb. The proton has a charge of the same magnitude but opposite sign.

If an electron could be suspended in a convenient place and if the force on another charged particle, called a test charge, could be measured, and the force at each point in space plotted, it would be found that the force would decrease as the square of the distance from the electron. The volume surrounding the electron is said to be an electric field. The intensity of the electric field (E) is equal to the force (F) imparted to the test charge divided by the magnitude of the charge. It must be noted that the electric field is a vector, that is, it has direction as well as magnitude. Figure 1-3 illustrates the electrostatic lines of force around positively and negatively charged bodies. Each charged particle contributes to the field, and thus in real physical situations the electric field intensity at any point is the sum of contributions from many charged particles.

VOLTAGE

The movement of a charged particle in an electric field necessitates the expenditure of work on that particle. The total energy of the charged particle is the sum of its kinetic energy (equal to $1/2\ mv^2$) and its positional or potential energy. When work is done on a particle to move that particle in an electric field, the potential energy is changed. In mechanics, all changes in potential energy are referred to some arbitrary height above sea level. The electrical potential, however, is referred to a position at infinity. A charged particle which is infinitely distant from another charged particle experiences no Coulomb's force and is said to have *zero potential.* The absolute potential of a particle brought from infinity to point A is given by:

$$V_A = \frac{W}{q'} \qquad 1.2$$

*There are three systems of units in general use in the physical sciences. The centimeter-gram-second system is used in most scientific fields with the exception of engineering, which prefers the meter-kilogram-second system of units. In addition to these two systems, there is the English system (foot-pound-second). See Appendix I.

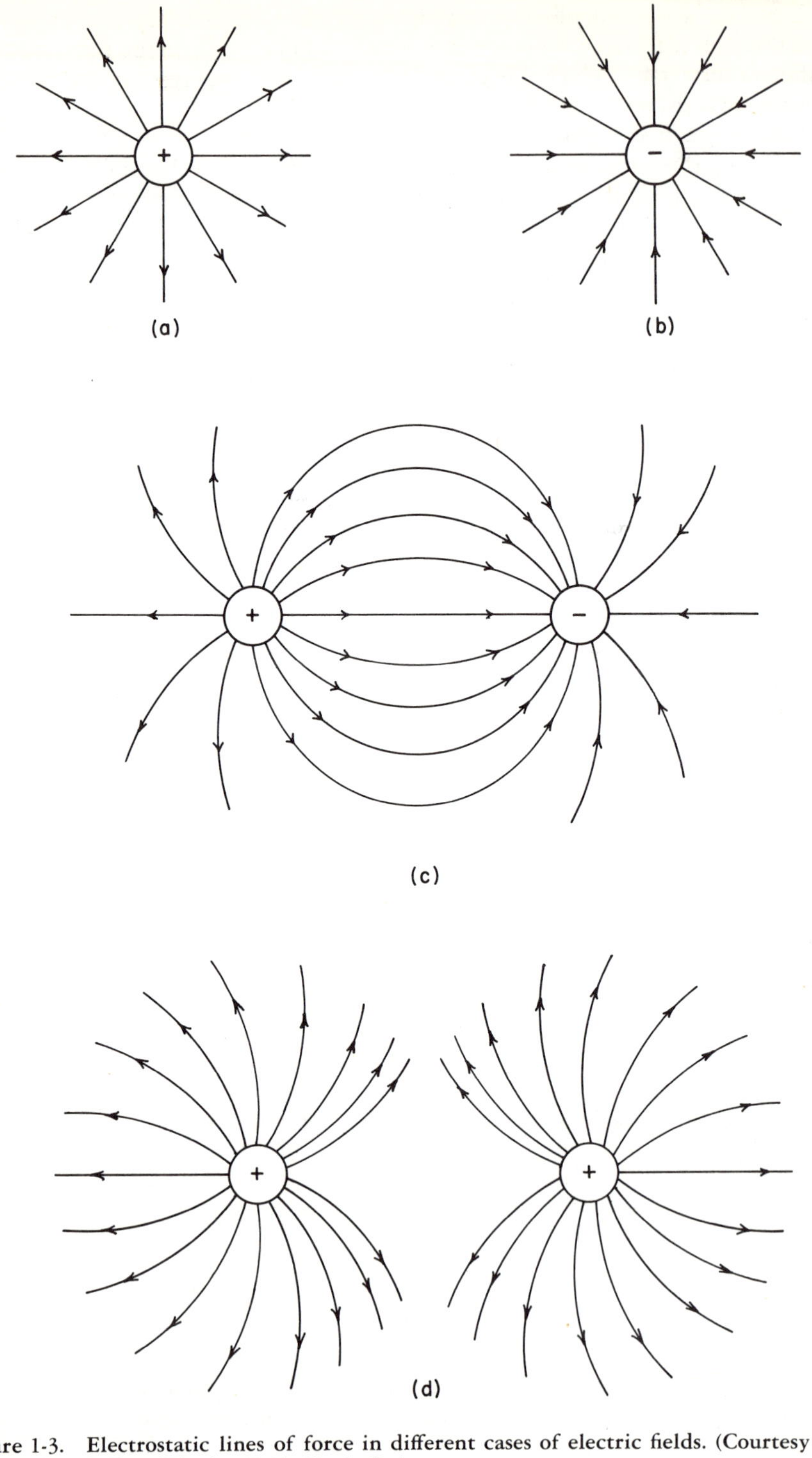

Figure 1-3. Electrostatic lines of force in different cases of electric fields. (Courtesy of R. W. Kenworthy, College Physics, F. A. Davis Co.)

For practical reasons, the absolute potential is not used. It is much more convenient to measure the amount of work (W), per unit of charge (q'), required to move a charged particle from a real point A to another point B in an electric field. The potential difference is also the change in potential energy per unit of charge when a charged particle is moved from point A to point B in an electric field. The units of potential and potential difference are called *volts* (named after the Italian scientist who first described them, Allesandro Volta). A volt is equivalent to 1 joule of work per coulomb of charge in the mks system or 1 erg per statcoulomb in the cgs system.

A standard 12-volt automobile battery has one terminal marked positive and the other negative. In terms of what has been said above, this would imply that the positive terminal has an absolute potential which is 12 volts higher than the negative terminal, and that if a test positive charge were to be moved from the negative terminal to the positive terminal, 12 joules of energy would have to be expended. The electrostatic lines of force are the same as shown in Figure 1-3(c).

In electronic circuitry the potential difference is almost always stated with respect to *ground,* symbolized by ⏚. That is to say, the earth or ground is so large that it is considered to have an infinite supply of atoms and can fill deficits or absorb an excessive number of electrons and still maintain, for all practical purposes, a charge neutrality.

CURRENT

The number of charges moving from one place to another per unit of time is referred to as the current (I):

$$I = \frac{q}{t} \qquad 1.3$$

The practical unit of current is the *ampere,* which is the movement of 1 coulomb of charge per second. It is important that the student comprehend the difference between the concepts of voltage and current. Let us compare the concepts of voltage and current to a hydraulic system such as a swimming pool with a continuously operating filter. It is necessary to have a pump to push the water through the filter and back to the swimming pool. The pressure developed by the pump is analogous to the voltage; the volume of water pumped per second is analogous to the current. It would take much greater pump pressure to push a given amount of water through a filter of finer porosity than through one made of more coarse material. The filter is said to resist the flow of water; it is analogous to the electrical resistance to be discussed below.

The Conduction of Electrical Current

According to its ability to conduct electricity, all matter can be classified into three broad categories: conductors, insulators, and semiconductors. Electrical current is usually understood to involve the passage of electrons through a conducting material such as a copper wire. In fact, the movement of 6.24×10^{18} electrons per second is 1 ampere of current. The electron which enters a length of copper wire is probably not one of the electrons which leaves the copper wire. It will be remembered that each atom in the copper wire consists of a nucleus surrounded by an electron cloud. The electrons in the outer orbit are not tightly bound to the atom and are free to drift from atom to atom. This drifting is random, like

Brownian motion, and although the electrons are traveling with thermal velocities of about 250,000 miles per hour, the net current in any one direction is zero. If one end of the wire is at a potential which is not equal to the potential at the other end, a current will tend to flow. The amount of current that actually does flow is a function not only of the potential difference, but of the material and the geometry of the conductor through which it is flowing. If current flows readily through a material, the material is referred to as a *conductor*; if it is highly resistant to the flow of current, it is called an *insulator.* The concept of resistance is more thoroughly discussed below.

The movement of electrons through a material may be likened to the following situation: Consider one row of seats in a movie theater with people sitting in all but one seat. That vacant seat is on the right aisle. A person now approaches that row from the left aisle. There are two ways that he can get a seat in that row without walking around. He can go past all the people in that row, or each person in the row can move over one seat and he can sit on the left aisle. The movement of each person to the next seat is analogous to the movement of electrons in a conductor. In the above analogy, we have compared the electrons to people sitting in a theater. At the same time that the people were moving to the right, the vacant seat was moving to the left. In the atom, the vacant seat is comparable to the hole left by the electron moving in the opposite direction.*

SEMICONDUCTORS

Semiconductors are materials which lie between conductors and insulators in their properties of conduction. Presently, silicon and germanium are used in the construction of semiconductor devices such as diodes and transistors. Both silicon and germanium atoms have four electrons which would be free to move from atom to atom were they not involved in the covalent bonding between atoms which is necessary to form a silicon or germanium crystal. Methods have been devised, however, to add a controlled amount of either boron or antimony impurity to the silicon or germanium crystal. If boron is added, the silicon or germanium crystal will have a *hole* in each location where a boron atom has been incorporated into its lattice in place of a germanium or silicon atom. This type of semiconductor is called P type or acceptor type because it has an excess of holes. Should one of the germanium or silicon atoms lose one of its loosely held electrons, it would be snatched up by the boron impurity, and then another electron would fill in for the first one leaving another hole. In this way the hole moves through the semiconductor. If antimony is used as the impurity rather than boron, the semiconductor is of the N type or donor type. Antimony, when incorporated into the crystal lattice of germanium, provides an extra electron which is free to move. Figure 1-4 illustrates N-type and P-type conduction when a voltage is applied.

RESISTANCE

The amount of current which flows in a conductor as a result of the applied voltage is a function of the number of electrons in each of its atoms that is free to drift or move through the conductor, as previously discussed. The current is also a function of a variety of other factors such as the temperature of the conductor. It is generally true that a good conductor of electricity is also a good conduc-

---

*In electrical terminology, current is said to move from the positive terminal to the negative terminal, but electrons are actually moving from negative to positive.

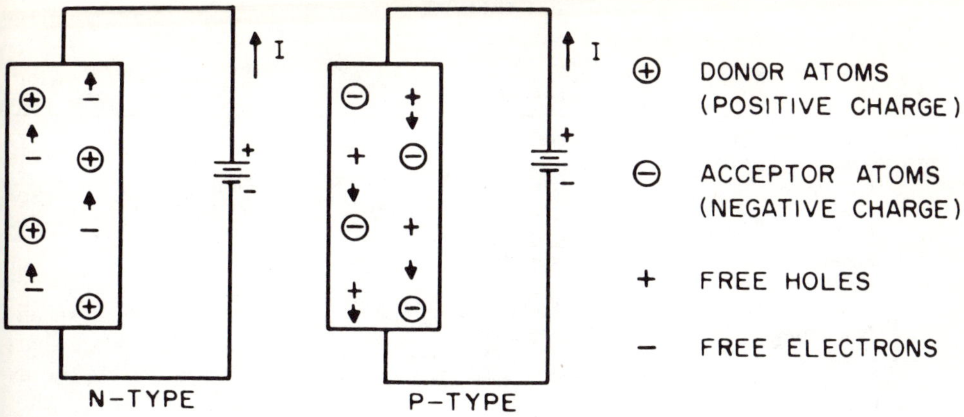

Figure 1-4. Conduction in N-type and P-type semiconductors. (Courtesy of G.E. Transistor Manual, 6th Ed.)

tor of heat. The property of the conductor which describes its ability to conduct quantitatively is called its *conductance.* It is more practical, however, to use a term which is the reciprocal of conductance. This term is *resistance.* The resistance of a conductor to the flow of current is a function of the length of the conductor, its cross-sectional area, and the nature of the conductor itself. The unit of resistance is the *ohm*, which is named after the physicist who first described it, George Simon Ohm. The resistance of a 1-cm length of a conductor whose cross-sectional area is 1 cm^2 is called its *resistivity.* The resistivity of some materials classed as conductors and some materials which are classed as insulators is listed in Table 1-1. The actual resistance (R) of a material may be determined by using the following relationship:

$$R = \rho \times \frac{\text{length}}{\text{cross-sectional area}} \tag{1.4}$$

where ρ is the resistivity.

The resistance will change with temperature, as follows:

$$R_t = R_o\,(1 + \alpha t) \tag{1.5}$$

where R_t is the resistance at $t°$ C
R_o is the resistance at 0° C
α is the temperature coefficient in Table 1.1.

The resistance does not change very much as a function of temperature in most commonly used materials. It does, however, change enough to become an important factor in some circuits in which the resistance must be held extremely constant for a particular application. Metal film resistors have been made commercially available which change less than 5 parts per million per degree centigrade. The resistivity of metals usually increases with increasing temperature. If, however, a metal is cooled to temperatures near absolute zero (−273° C), the metal will have, for all practical purposes, zero resistivity. The conduction of current under these conditions is called *super-conductivity.* Non-metals will generally have temperature

TABLE 1-1

Resistivities and Temperature Coefficients at Room Temperature

Material	*Resistivity* (ρ) *ohm-cm*	*Temperature Coefficient* (α) $°C^{-1}$
Aluminum	2.63×10^{-6}	0.0039
Brass	$6\text{-}8. \times 10^{-6}$	0.0020
Carbon	$3500. \times 10^{-6}$	−0.0005
Copper	1.72×10^{-6}	0.0039
Iron	$10. \times 10^{-6}$	0.0050
Silver	1.47×10^{-6}	0.0038
Amber	5×10^{14}	
Bakelite	$2 \times 10^{5} - 2 \times 10^{14}$	
Glass	$10^{10} - 10^{14}$	
Hard Rubber	$10^{13} - 10^{16}$	
Mica	$10^{11} - 10^{15}$	
Quartz	7.5×10^{17}	
Wood	$10^{8} - 10^{11}$	

coefficients which decrease with increasing temperature. Solutions of electrolytes behave as non-metals in this respect.

Some materials change resistance as a function of mechanical stress or a change in hydrostatic pressure. Other materials, such as bismuth, experience a change of resistance when a magnetic field is applied; still others, such as selenium, change resistance when illuminated. Each of these properties may be used to transduce or convert changes in the physical environment to changes in resistance which can be then measured with appropriate electronic circuitry.

The symbol  represents the component used in building electronic circuits which offers a specific resistance. It is sometimes convenient to have a resistance in the circuit which is adjustable. In electrical terminology this is called a rheostat or variable resistor and is identified by the symbol 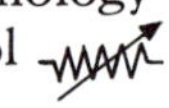.

RESISTORS IN PARALLEL AND SERIES COMBINATION

There are two basic ways that a component can be connected to a similar component in a circuit. One way is to connect the resistors together in an end to end fashion, known as in series. The other way is to connect the two or more resistors in such a way that all the left ends are connected together and all the right ends are connected together, known as in parallel (Fig. 1-5).

The total resistance of resistors in series is simply:

$$R_{\text{total}} = R_1 + R_2 + R_3 + \cdots\cdots R_n \qquad 1.6$$

The total resistance of resistors connected in parallel is:

$$\frac{1}{R_{\text{total}}} = \frac{1}{R_1} + \frac{1}{R_2} + \frac{1}{R_3} + \cdots\cdots \frac{1}{R_n} \qquad 1.7$$

The total resistance for two resistors in parallel is:

$$R_{\text{total}} = \frac{1}{\dfrac{1}{R_1} + \dfrac{1}{R_2}} = \frac{R_1 R_2}{R_1 + R_2} \qquad 1.8$$

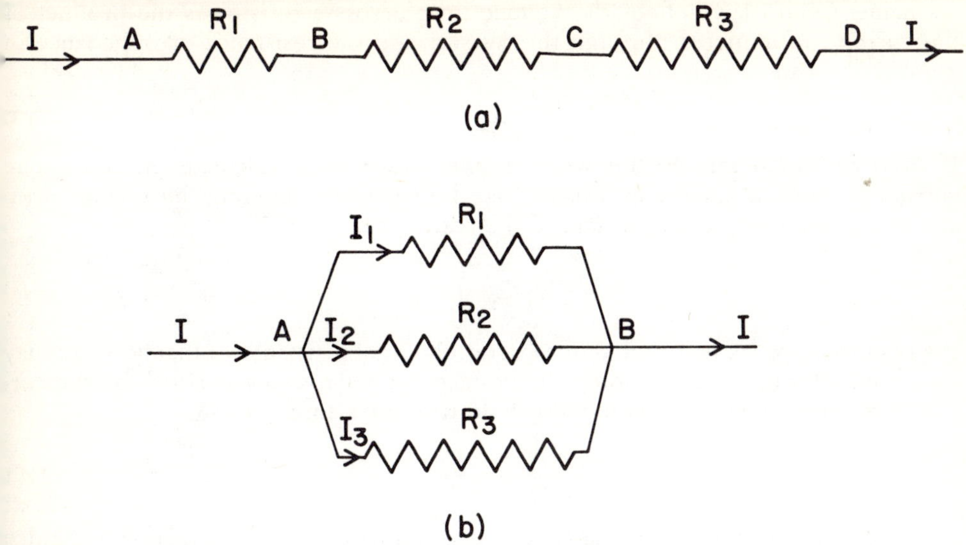

Figure 1-5. Resistors in series and in parallel. (Courtesy of R. W. Kenworthy, College Physics, F. A. Davis Co.)

When a circuit is composed of resistors arranged in a complicated manner, the student should draw a box around convenient groups of resistors and determine the resistance of each box. He should then redraw the circuit and proceed to simplify with further boxes. Eventually the circuit will be resolved into a simple equivalent series or parallel arrangement. Figure 1-6 shows such a circuit which the student should reduce to a single resistance from *A* to *B*.

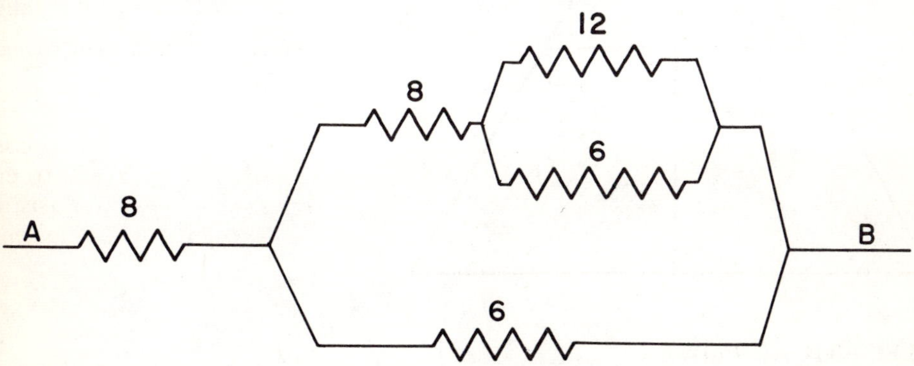

Figure 1-6. Courtesy of R. W. Kenworthy, (College Physics, F. A. Davis Co.)

OHM'S LAW

In every real closed circuit there is a definite relationship between voltage, current, and resistance.* This relationship was first described by Ohm and bears

*In the discussion of circuits energized by alternating current, it will be necessary to include a term called *reactance* with resistance to form impedance (*z*). (See Chapter 3.) Reactance is the contribution of the capacitive and inductive elements of the circuit to the total impedance.

his name. Ohm's law defines the voltage drop across a resistor as the product of the current, in amperes, through the resistor and the resistance or impedance in ohms.

$$V = I\,R \tag{1.9}$$

If any two parameters are known, it is then possible to calculate the third. The current through a particular resistor may be found by dividing the voltage drop across a resistor by the magnitude of the resistance.

$$I = \frac{V}{R} \tag{1.10}$$

If a resistor network is included in a circuit, the total resistance of that network may be easily determined by dividing the voltage drop across the network by the current through it. The same would also hold true for a single resistor.

$$R = \frac{V}{I} \tag{1.11}$$

Ohm's law may be more easily remembered with the aid of Figure 1-7. Notice that if one term, *R* for example, is blocked out with a finger, the other two are arranged as in equation 1.10 (V/I).

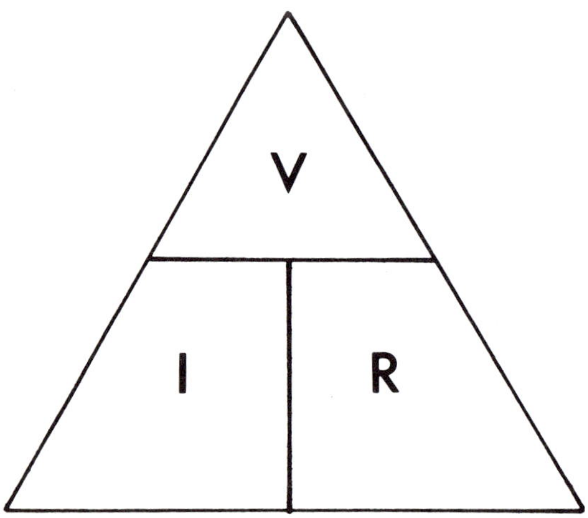

Figure 1-7. Ohm's law triangle.

ELECTRICAL POWER

Power is defined as the amount of work done per unit of time. Reference to page 5 will show that voltage is an expression of the amount of work per unit of charge.

$$\text{Voltage} = \frac{\text{Work}}{\text{Charge}} \tag{1.12}$$

If this term is then multiplied by the current, which is the amount of charge transferred per unit of time, the result will be electrical power. (It should be remembered that *power* is a vector quantity and so has direction as well as magnitude.)

$$\text{Power} = \frac{\text{work}}{\text{charge}} \times \frac{\text{charge}}{\text{time}} = \frac{\text{work}}{\text{time}} \qquad 1.13$$

$$\text{Power} = V \times I \qquad 1.14$$

When voltage is expressed in volts and current is expressed as amperes, electrical power is given in watts. When large numbers of watts are involved, it is easier to use the term kilowatts (1 kilowatt = 1000 watts). Since Ohm's law has both a voltage and a current term, the expression for electrical power can be stated in terms of resistance and current or resistance and voltage as well as voltage and current.

$$P = VI = I^2R = \frac{V^2}{R} \qquad 1.15$$

The amount of heat liberated from a resistor (or any electronic component having resistance) is proportional to the power. This is a very important point as electronic components, resistors for example, are made to withstand only a certain definite amount of heat. A resistor may have a value of 100 ohms (Ω) but if it is a 1/2-watt resistor, it can withstand only 70 ma (.070 amperes). Certain components such as zener diodes and transistors, especially when made with germanium, are very sensitive to heat and can easily be ruined if the power exceeds the specifications of the manufacturer.* Wire can be obtained in many diameters; it is important not to exceed the current requirements of the size chosen for use. Table 1-2 lists wire by gauge size, diameter, and current capacity.

CAPACITANCE

The capacitance of a conductor may be defined as the ratio of its charge (q) to its potential (V). The amount of charge that a given object can hold depends to a large extent on its size and shape. A sphere can hold more charge than a rod, and a rod can hold less charge if it is sharpened at one or both ends. The potential of the conductor depends on its position. Therefore, it can be seen that the capacitance of a conductor depends on a great many factors.

The capacitor used in electronic circuitry is represented by the symbol $\dashv\vdash$ or $\dashv\vdash^{+}$. This *capacitor* consists of a conductor, usually in the form of two plates or other pairs of similar shapes with a large surface area, separated by a non-conducting material, called an insulator or, in the case of a capacitor, a *dielectric.*

When the capacitor is made a part of a closed circuit, having a source of voltage, at the time the circuit is initially closed, one plate will accumulate a certain amount of positive charge while the other plate will accumulate an equal amount of negative charge. The charge is then said to have been transferred from one plate to the other; in reality, however, no charge has crossed the dielectric separator. While the charge is building up on the plate, the capacitor is said to be in the process of charging. The initial charging process takes a small but finite amount of time which is characteristic of the capacitor and other factors. The time that it takes the capacitor to reach 63 per cent of its final charge is called its *time constant.* The very rapid charging is also referred to as the transient response of the capacitor, which will be examined in greater detail in Chapter 3.

---

*A note of caution is in order here. One should always leave himself a margin of safety as some manufacturers tend to state power ratings which are above what the component can withstand.

TABLE 1-2

Wire Size A.W.G.	*Diameter 0.001 inch*	*Circular† Mil Area*	*Ohms per 1000 ft. 25° C.*	*Current-Carrying Capacity**
1	289.3	83690	0.1264	119.6
2	257.6	66370	0.1593	94.8
3	229.4	52640	0.2009	75.2
4	204.3	41740	0.2533	59.6
5	181.9	33100	0.3195	47.3
6	162.0	26250	0.4028	37.5
7	144.3	20820	0.5080	29.7
8	128.5	16510	0.6405	23.6
9	114.4	13090	0.8077	18.7
10	101.9	10380	1.018	14.8
11	90.74	8234	1.284	11.8
12	80.81	6530	1.619	9.33
13	71.96	5178	2.042	7.40
14	64.08	4107	2.575	5.87
15	57.07	3257	3.247	4.65
16	50.82	2583	4.094	3.69
17	45.26	2048	5.163	2.93
18	40.30	1624	6.510	2.32
19	35.89	1288	8.210	1.34
20	31.96	1022	10.35	1.46
21	28.46	810.1	13.05	1.16
22	25.35	642.4	16.46	0.918
23	22.57	509.5	20.76	0.728
24	20.10	404.0	26.17	0.577
25	17.90	320.4	33.00	0.458
26	15.94	254.1	41.62	0.363
27	14.20	201.5	52.48	0.288
28	12.64	159.8	66.17	0.228
29	11.26	126.7	83.44	0.181
30	10.03	100.5	105.2	0.144
31	8.928	79.70	132.7	0.114
32	7.950	63.21	167.3	0.090
33	7.080	50.13	211.0	0.072
34	6.305	39.75	266.0	0.057
35	5.615	31.52	335.0	0.045
36	5.000	25.00	423.0	0.036
37	4.453	19.83	533.4	0.028
38	3.965	15.72	672.6	0.022
39	3.531	12.47	848.1	0.018
40	3.145	9.88	1069.0	0.014

* at 700 circular mils per ampere

†The area of the wire in circular mils is calculated as $1/4\pi d^2$ with d, the diameter, in units of mils (0.001 inch).

When the capacitor is charging, current is flowing in the circuit. Electrons are accumulating on one plate while the other plate is supplying electrons to the circuit. The capacitance of a given capacitor is the amount of charge that is apparently transferred from one plate to the other divided by the voltage drop across the conducting plates (Fig. 1-8).

$$C = \frac{q}{V_{ab}} \qquad 1.16$$

(V_{ab} is the potential difference, or voltage, between points a and b)

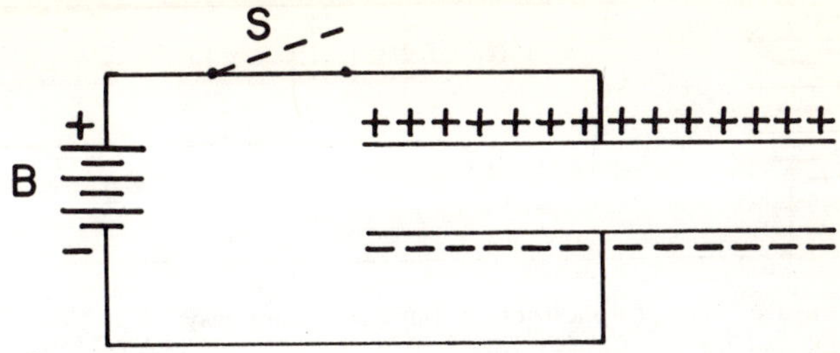

Figure 1-8. A parallel plate capacitor. (Courtesy of R. W. Kenworthy, College Physics, F. A. Davis Co.)

The capacitance is 1 *farad* (named after Sir Michael Faraday) when 1 coulomb of charge is transferred and the voltage drop (from *a* to *b* above) is 1 volt. The farad, however, is a very large unit, and values such as the microfarad (μf, 1×10^{-6} farad) and the picofarad (pf or $\mu\mu$f, 1×10^{-12} farad) are more commonly used.

CAPACITORS IN PARALLEL AND SERIES COMBINATION

The total capacitance of several capacitors attached to each other in series, as shown in Figure 1-9, can be calculated as follows:

$$\frac{1}{C_{\text{total}}} = \frac{1}{C_1} + \frac{1}{C_2} + \frac{1}{C_3} + \cdots\cdots\cdots + \frac{1}{C_n} \qquad 1.17$$

or for two capacitors in series:

$$C_{\text{total}} = \frac{C_1 C_2}{C_1 + C_2} \qquad 1.18$$

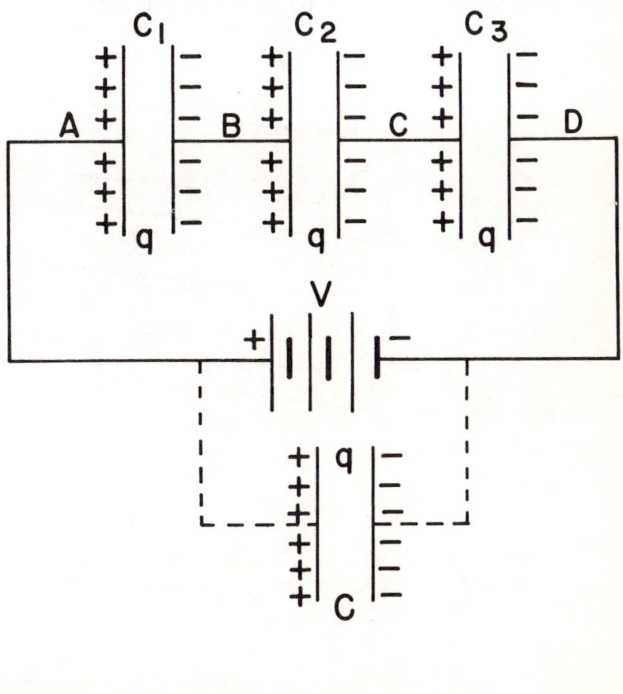

Figure 1-9. Capacitors in series. (Courtesy of R. W. Kenworthy, College Physics, F. A. Davis Co.)

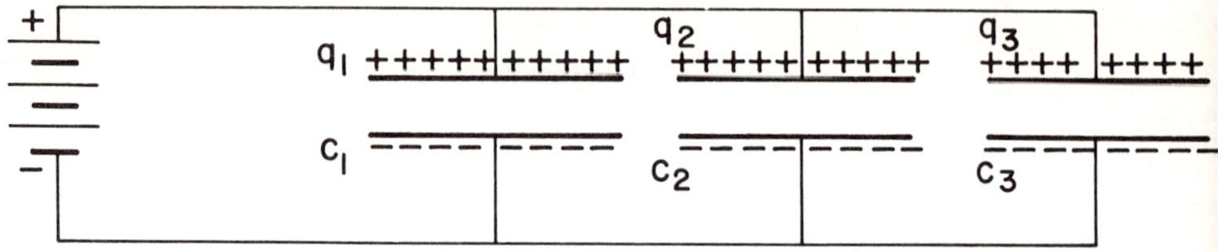

Figure 1-10. Capacitors in parallel. (Courtesy of R. W. Kenworthy, College Physics, F. A. Davis Co.)

The capacity of two or more capacitors in parallel (Fig. 1-10) can be determined by simply adding their respective values. Notice that the procedure for determining the total capacitance of capacitors arranged in parallel is identical to that for calculating the total resistance of resistors in series.

THE MAGNETIC FIELD

All of us have observed that certain materials, namely iron and some ferrous compounds and alloys, exhibit properties which are referred to as *magnetic properties*. Two magnetic bars may be seen to attract if one end is brought near the proper end of the other. The magnets may also be seen to repel if one of the magnetic bars is rotated and the other end is presented to the same end of the first bar (Fig. 1-11). The earth itself appears to be a large magnet, and thus all magnets, when properly suspended, will orient themselves with respect to the *north* and *south poles* of the earth. It is this principle that is used to construct the compass. Thus, the ends of the magnetic bars may be labeled "north pole" and "south pole" respectively. Experimentation will quickly verify that *unlike* poles attract (north pole of one and the south pole of the other) and *like* poles repel (north of one to north of the other or south of one to south of the other).

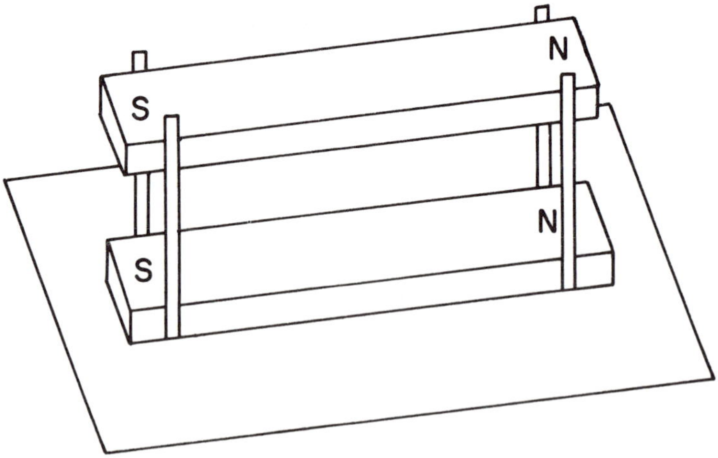

Figure 1-11. The force of repulsion between the two magnets is equal to the weight of the upper magnet. (Courtesy of R. W. Kenworthy, College Physics, F. A. Davis Co.)

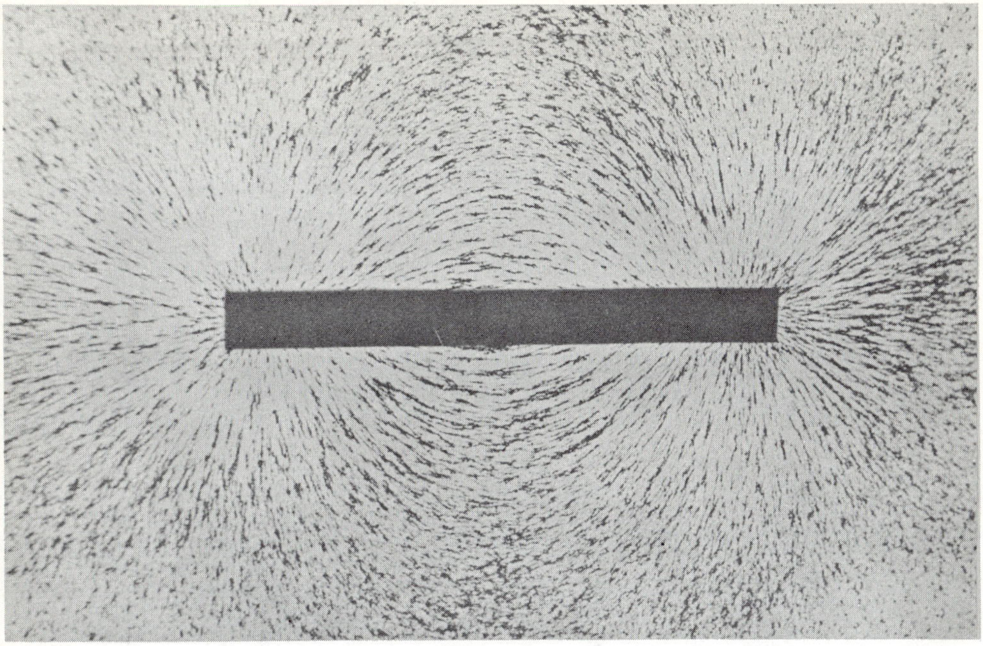

Fig. 1-12A

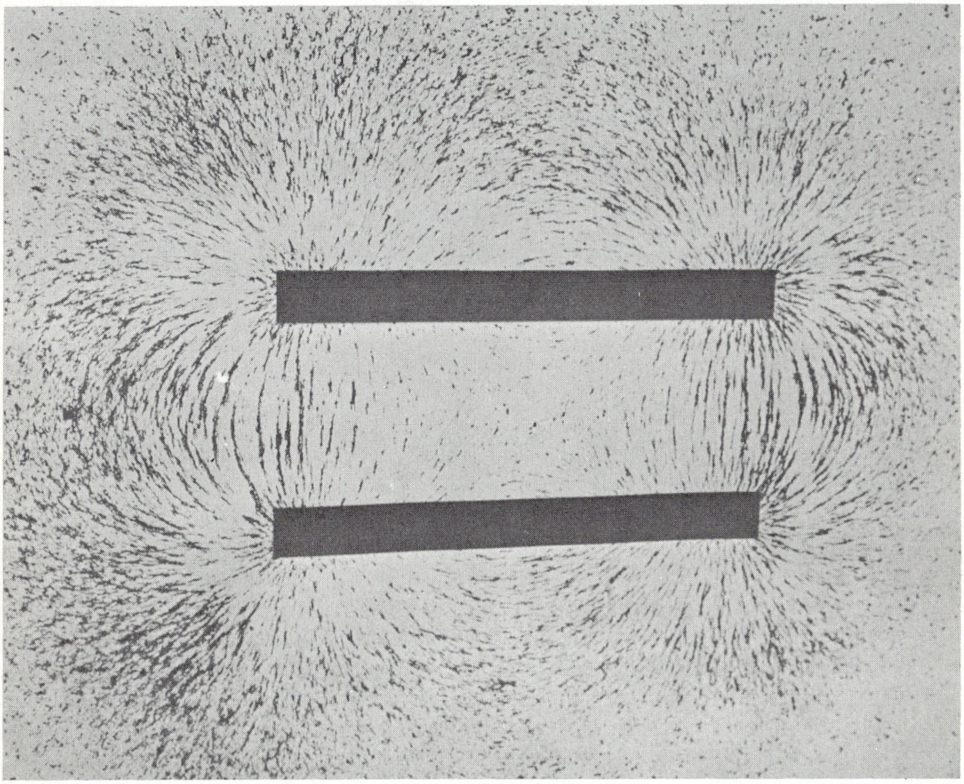

Fig. 1-12B

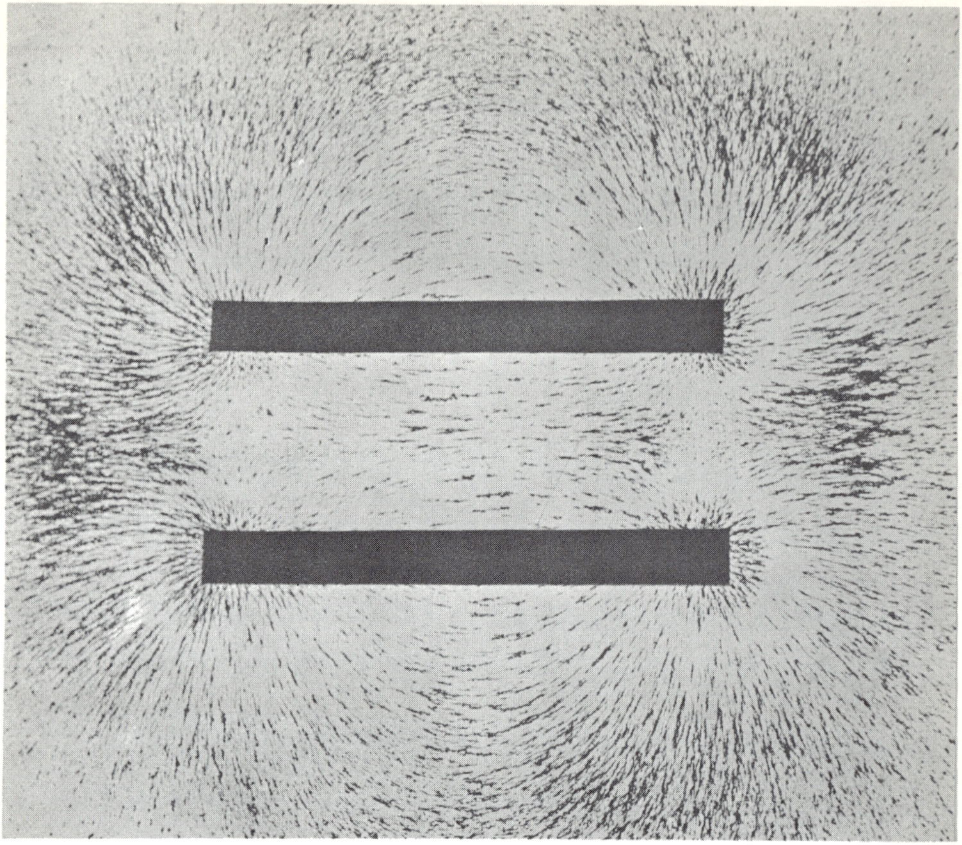

Fig. 1-12C

Figure 1-12. Lines of force around a bar magnet. (Courtesy of R. W. Kenworthy, College Physics, F. A. Davis Co.)

If a small test magnet is brought near a magnet and the force of attraction or repulsion is measured, it will be found that there is a definite magnetic field which surrounds the magnet. The forces thus determined will follow a pattern as shown in Figure 1-12. The field strength is expressed as *lines per unit area.* The number of lines per unit area at right angles to their direction is called the *flux density* and is measured in *webers per square meter* in the meter-kilogram-second (mks) system of units and *gauss* in the centimeter-gram-second system (cgs). The magnetic field intensity is given as *ampere-turns per meter* in the mks system and *oersted* in the cgs system.

An electron moving through a wire will generate a magnetic field, and assuming that the wire is non-ideal, in other words, it has some resistance, it will also have an electric field, as shown in Figure 1-13. If that wire is formed into a spiral or solenoid, it will generate a field through its center as shown in Figure 1-14; if the solenoid is wound in the opposite sense, the field will be directed in the opposite direction.

The magnetic field that can be generated in the solenoid can be increased by a factor of 100 to 1000 times by filling the space around the solenoid with a *core* of *magnetic material* (Fig. 1-15) such as iron or an alloy of nickel and iron. The ability

Figure 1-13. Lines of force around a long straight conductor. (Courtesy of R. W. Kenworthy, College Physics, F. A. Davis Co.)

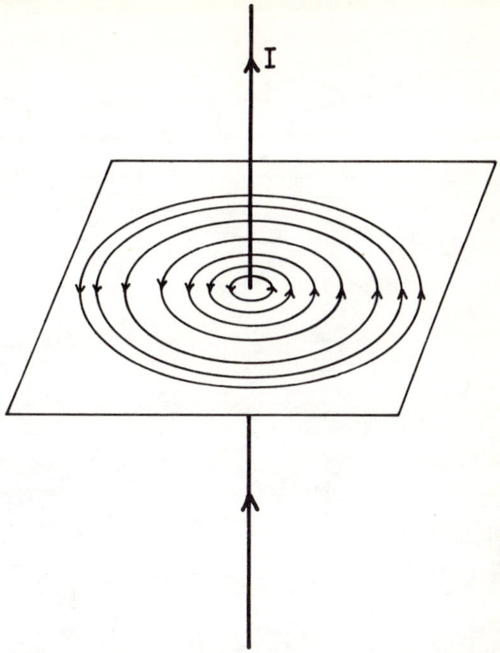

Fig. 1-13A

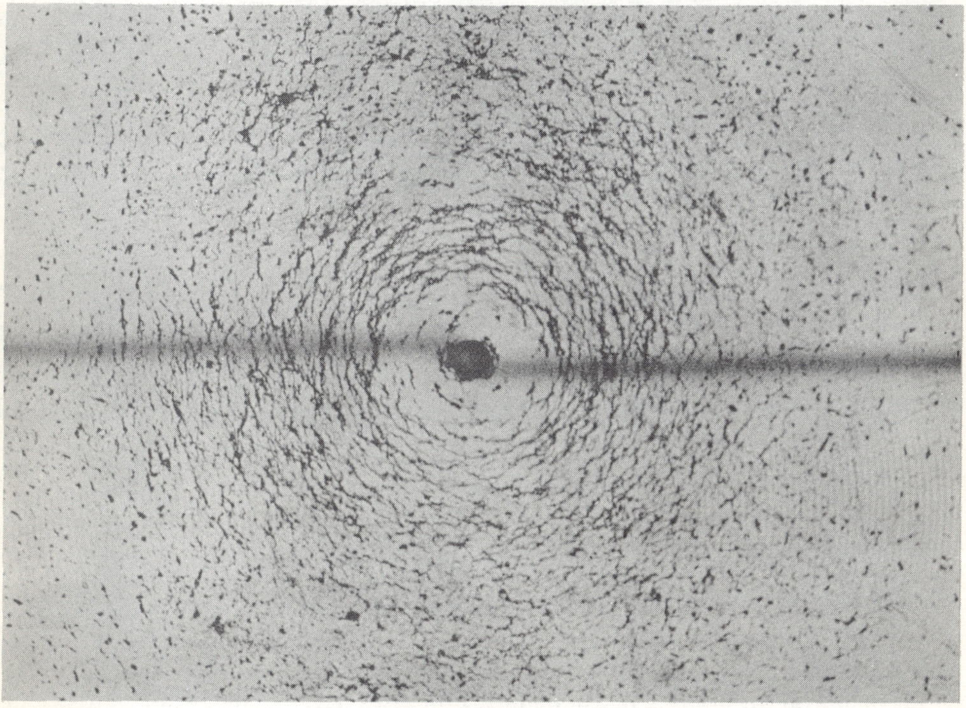

Fig. 1-13B

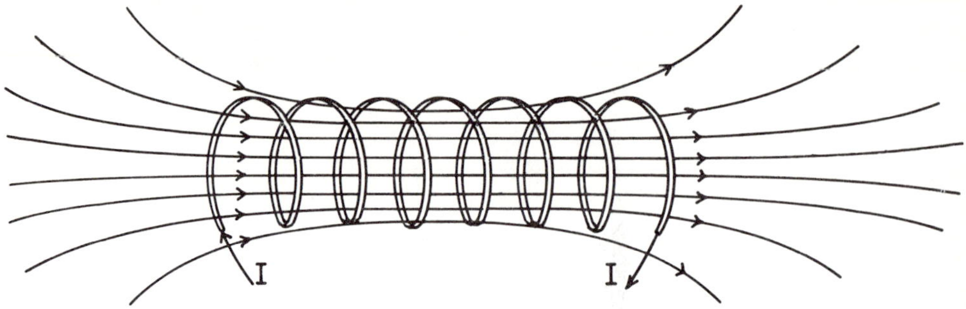

Fig. 1-14A

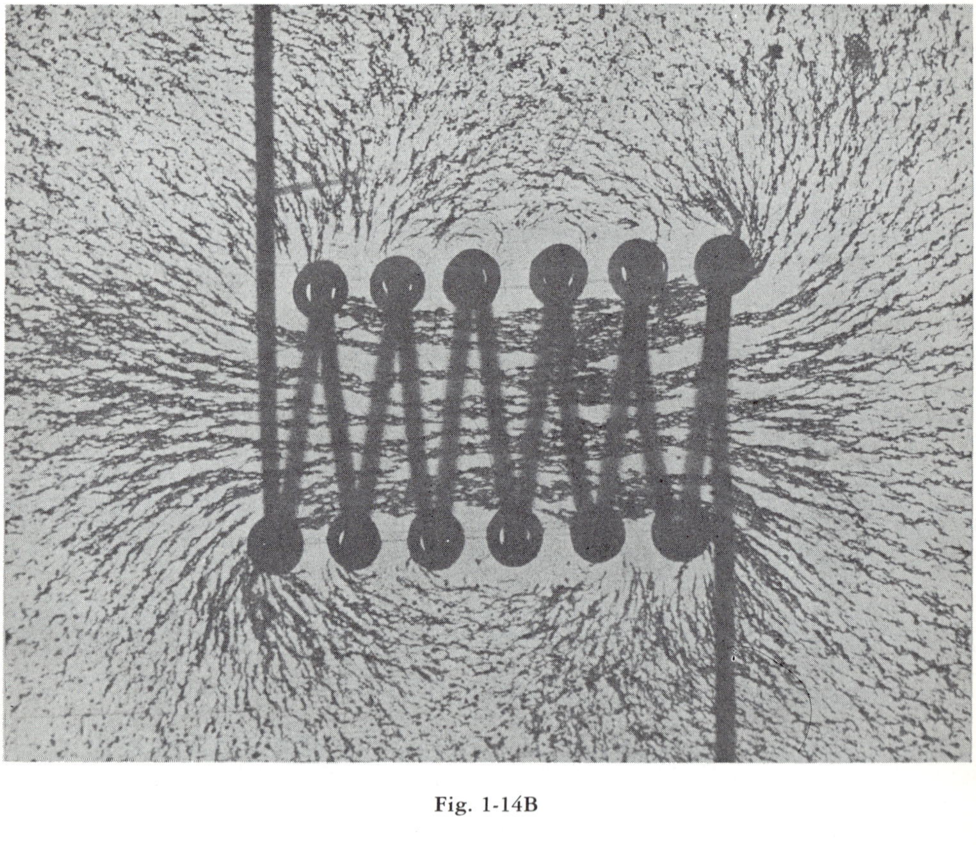

Fig. 1-14B

Figure 1-14. The magnetic field produced by a long solenoidal coil in which there is a current I. (Courtesy of R. W. Kenworthy, College Physics, F. A. Davis Co.)

of the core material to hold a magnetic field is a function of the *permeability* of the material. Some core materials have been developed that are thousands of times better than iron for holding magnetic flux. Processes have been developed to orient the atoms to improve the permeance of the core even further.

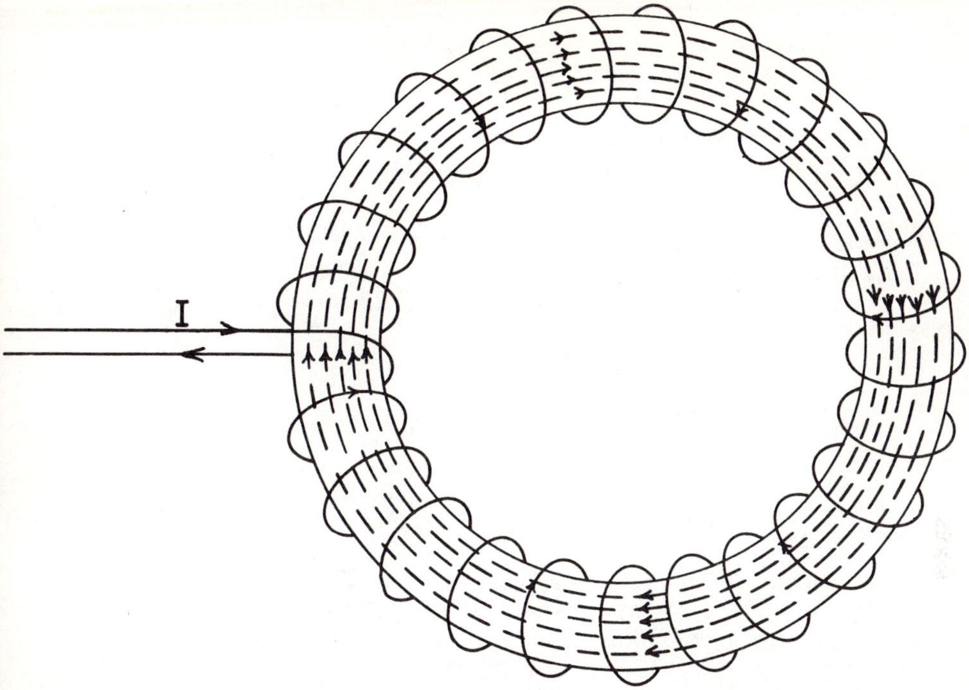

Figure 1-15. The magnetic field produced by a toroidal coil in which there is a current I. (Courtesy of R. W. Kenworthy, College Physics, F. A. Davis Co.)

INDUCTANCE

The magnetic field generated by the solenoid shown in Figure 1-15 will tend to decelerate the electrons traveling through the wire. This phenomenon is called *inductance.* A solenoid or coil of wire will have an inductance which is expressed in units, named after Joseph Henry, called henrys. The inductance of a coil is equal to:

$$L = \frac{\mu_0 A N^2}{\ell} \times 10^{-8} \qquad 1.19$$

where

L = the inductance in henrys (cgs system)
μ_o = the coefficient of permeability of the core
A = the cross-sectional area of the coil (cm^2)
N = the number of turns
ℓ = the length of the coil (cm)

Such a coil will generate a voltage or electromotive force (emf) opposite to the one driving the current. This is called a *back-emf* (ϵ).

$$\epsilon = -L\,\frac{\text{change in current}}{\text{change in time}} \qquad 1.20$$

or, in terms of the calculus:

$$\epsilon = -L\frac{di}{dt} \qquad 1.21$$

The result of this back-emf is to cause the current to be delayed. The effects of the emf on the flow of current in the circuit will become more clear in the next chapter.

INDUCTORS IN PARALLEL AND SERIES COMBINATION

Inductors, represented in electronic circuitry as , behave in parallel and series combinations in the same way as resistors; in parallel arrangement:

$$\frac{1}{L_{\text{total}}} = \frac{1}{L_1} + \frac{1}{L_2} + \frac{1}{L_3} + \cdots\cdots\cdots + \frac{1}{L_n} \qquad 1.22$$

and in series:

$$L_{\text{total}} = L_1 + L_2 + L_3 + \cdots\cdots\cdots + L_n \qquad 1.23$$

TRANSFORMERS

In the section in which the magnetic field was discussed, it was pointed out that acceleration of the electron through a wire generates a magnetic field. When the wire is bent to form a loop, or many loops as in the case of the toroid shown in Figure 1-16(b), the magnetic field is directed through its center; the amount of magnetic flux can be increased when a ferrous core is used. If a loop can generate a magnetic field, then it follows that if a magnetic field passes through another loop, it can generate (or induce) a current in that second loop. The current that is induced in the second loop will flow in the opposite direction to that of the first loop. This is known as Lenz's law, and is shown in Figure 1-16(a).

A *transformer* is a component in which one or more continuous loops of wire are wrapped around a ferrous core, forming what is known as the *primary coil.* Elsewhere on the magnetic core another coil of one or more continuous loops is wrapped around the core. This second coil is referred to as the *secondary coil* (Figure 1-17). There are several useful relationships which will further clarify the function of the transformer in electronic circuitry. The first of these relationships compares the voltage to the number of turns:

$$\frac{\text{voltage drop across the primary}}{\text{voltage drop across the secondary}} = \frac{\text{number of turns in primary}}{\text{number of turns in secondary}} \qquad 1.24$$

$$\frac{V_p}{V_s} = \frac{N_p}{N_s} \qquad 1.25$$

Equation 1.24 tells us that if 110 volts are applied to the primary coil of a transformer and we wish to have 12.6 volts across the secondary coil which contains 126 turns, then there must be 1100 turns in the primary coil. In other words, this ratio must be 1100/126 or about 8.7 to 1.

It is clear from the conservation of energy concept that the electrical power going into the primary must equal the electrical power available from the secondary (with a small adjustment for heat loss in the core itself). This is expressed as

$$V_pI_p = V_sI_s \qquad 1.26$$

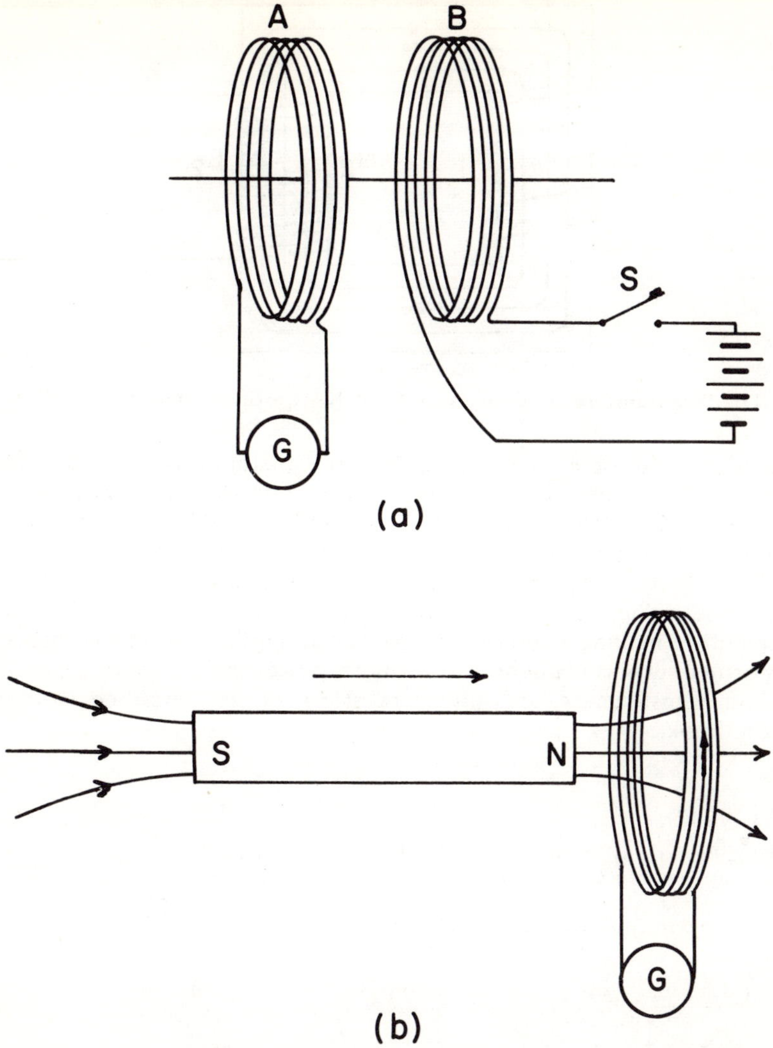

Figure 1-16. Methods of producing an induced emf in a coil. (Courtesy of R. W. Kenworthy, College Physics, F. A. Davis Co.)

Combining equations 1.25 and 1.26

$$\frac{I_p}{I_s} = \frac{V_s}{V_p} = \frac{N_s}{N_p} \qquad 1.27$$

Many different varieties of transformers are used in electronic circuitry. They are used to convert one voltage into another in some instances, and in others they are used merely to isolate one circuit from another. (No current traveling through the primary actually travels through the secondary and, hence, the primary and secondary circuits are isolated from each other.)

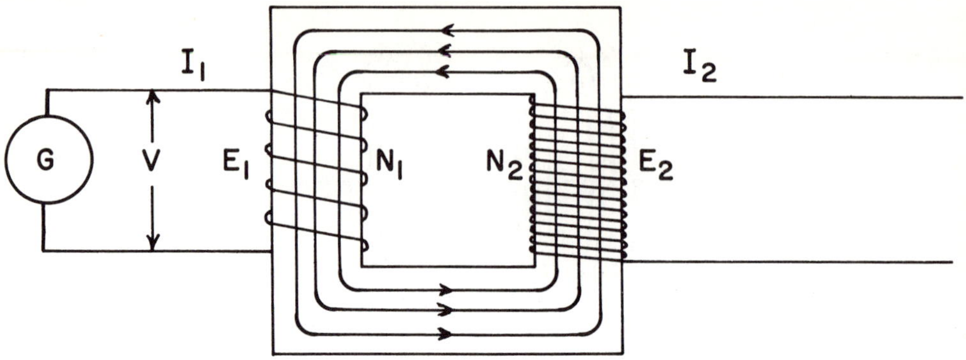

Figure 1-17. The transformer. (Courtesy of R. W. Kenworthy, College Physics, F. A. Davis Co.)

There are almost as many types of core material as there are varieties of transformers. Cores may be fabricated in a variety of shapes, *i.e.*, E's, rectangles, circular, or toroidal. The core is often laminated (made of thin sheets separated with insulator) to prevent *eddy* currents or currents that cross the core and upset the magnetic field. It is important that the correct transformer be picked for each use that arises. The student will have to study the manufacturer's specifications carefully before making a choice for a particular application. The common varieties of transformer such as filament transformers, power transformers, input, interstage, and output transformers, and pulse transformers are described throughout the rest of the book.

REFERENCES:

College Physics, Ray W. Kenworthy (F.A. Davis Co., Philadelphia, 1961).

Electricity and Magnetism, Francis W. Sears (Addison-Wesley Publishing Co., Reading, Mass., 1951).

G. E. Transistor Manual, J. F. Cleary, Ed. (General Electric Co., 1962).

2

Introduction to Alternating Current Circuit Theory

basic concepts of alternating current
phase
the concept of impedance
the resistive branch
the RC circuit
the RL circuit
the RLC circuit
the transient response in the RLC circuit
series resonance and Q
steady state response in a series RLC circuit
Kirchhoff's laws
Kirchhoff's first law
Kirchhoff's second law
special filtering networks
the low-pass RC network
the high-pass RC network
the twin-tee network

The student should study this chapter carefully. In it many necessary concepts are described. The biologically or medically trained student will feel, however, that he is not prepared to understand the material. This is not so. The material has been carefully presented with this in mind, but if some difficulty is noted, the student should proceed to the chapters that follow and return to this one later.

Aside from the obvious applications to the study of electronics, the biologist will find that many of these concepts have analogies in the study of hydraulics, mechanics, acoustics, etc. These concepts may be applied directly or indirectly to the study of biological systems or to the construction of models of biological systems.

BASIC CONCEPTS OF ALTERNATING CURRENT

There are two broad categories of electronic circuits. In one category, the voltage is supplied by an electrochemical cell or battery. An ideal battery is a source of voltage which does not vary with time. The amount of current which the voltage provided by a battery causes to flow in a resistive circuit is constant, and gives rise to the term used to describe one broad category of electronic circuitry—the *direct current* or *DC* circuit. The other category is called the *alternating current* or *AC* circuit. In this type of circuitry, the voltage source is not constant, and both the voltage and the current vary with time. This variation or oscillation as a function of time is often sinusoidal since the sinusoidal alternating current is easily generated, but it could be a square wave, saw-tooth, or triangular oscillation. These waveforms are shown in Figure 2-14.

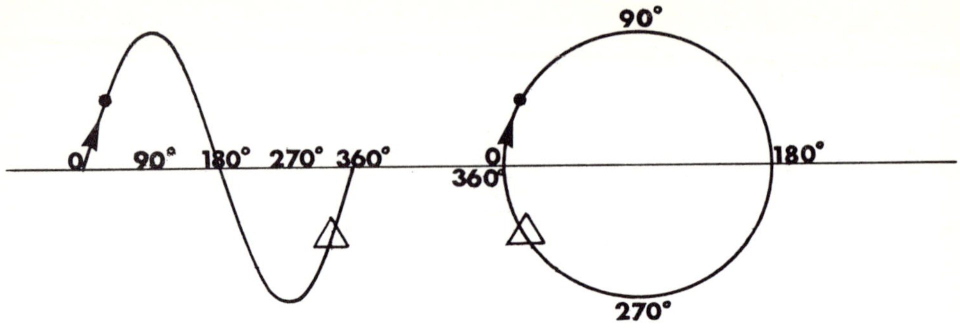

Figure 2-1. Generating a sine wave.

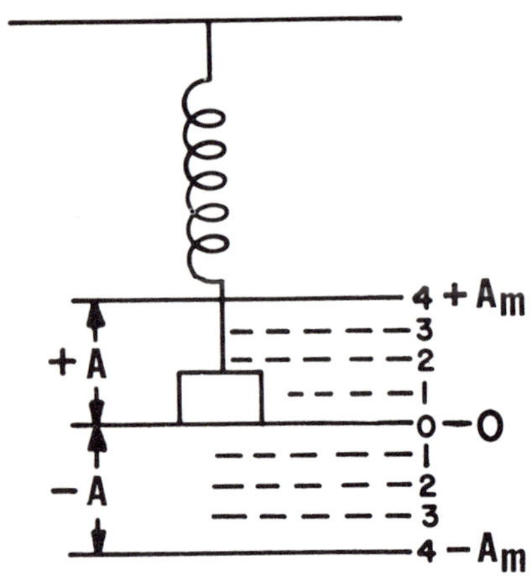

Figure 2-2. A mass suspended by a spring generates simple harmonic motion. (Courtesy of R. W. Kenworthy, College Physics, F. A. Davis Co.)

The motion of the ball around the circle on the right is represented as a graph in Figure 2-1. The motion of a mass suspended from a spring, Figure 2-2, is also periodic, and its motion may be plotted as shown in Figure 2-1. This curve, called a sine wave, is periodic, that is, it repeats itself over and over with the passage of time. Notice that time is represented along the abscissa in terms of the circular unit, the degree. Each time the wave repeats itself, it completes *one cycle.* The number of times the wave repeats itself per unit of time is referred to as its *frequency* and is usually expressed as the number of *cycles per second* (*cps*), although it may be expressed in degrees or radians/sec. The peak amplitude of a sinusoidal wave is defined as one half of the measured peak-to-peak distance. The amplitude, at any instant, is described mathematically as:

$$A = A_m \sin 2\pi ft = A_m \sin \omega t \qquad 2.1$$

where

A_m = maximum amplitude
f = frequency in cps
t = time in seconds
ω = frequency in radians/second
(2π radians = 360° of a circle)

PHASE

If two events, both described by sinusoidal waves of equal amplitude and frequency, occur at the same time, the two events are said to be *in phase.* Should they occur at slightly different times, they would be *out of phase* with each other. In this case one event would have to lag behind the other event. The amount of *lag* (or in the opposite sense, *lead*) is measured in degrees (fractions of a cycle).

Figure 2-3 is a graph representing the voltage and the current in a given circuit. In this circuit the voltage leads the current by 30°, or the voltage and current are 30° out of phase.

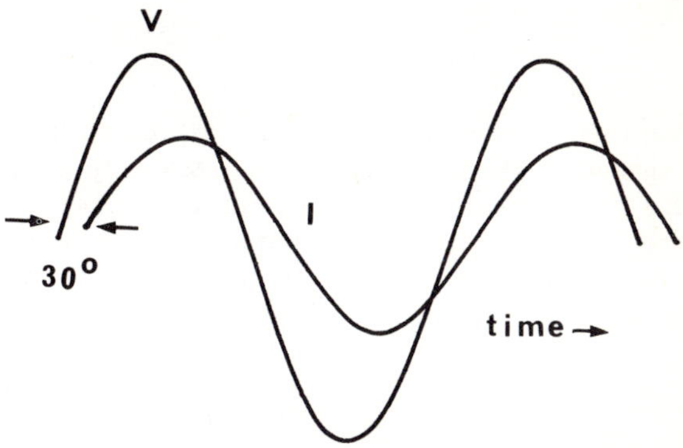

Figure 2-3. Voltage leading the current by 30°.

The voltage, labeled V in Figure 2-3, may be expressed mathematically as:

$$V = V_m \sin(2\pi ft + \phi) \qquad 2.2$$

and the current, labeled I in Figure 2-3, would then be expressed as:

$$I = I_m \sin(2\pi ft) \qquad 2.3$$

The Greek letter ϕ usually represents phase in electronics. V_m and I_m are maximum voltage and current respectively.

THE CONCEPT OF IMPEDANCE

Ohm's law was defined in the last chapter as:

$$V = IR \qquad 2.4$$

This relationship, as written, applies to a circuit containing only resistive elements. When the circuit contains capacitive and/or inductive elements, it is necessary to define a term to replace resistance (R) in equation 2.4. This new more general term is called *impedance* and is symbolized by the letter Z.

Impedance consists of two parts, namely, *resistance* and *reactance.* The resistance portion of the impedance is the same resistance that was discussed in the last chapter. The reactance, however, is the contribution of the inductive and the capacitive elements to the total impedance. Those elements of the circuit which are purely resistive contribute only to the resistance portion of the impedance. Both capacitive and inductive elements contribute to the reactance portion of the total impedance. Real capacitive and inductive elements almost always have some resistance associated with them, and thus these components will also contribute to the resistive portion of the impedance as well. A coil generating an inductance will have some resistance because it is made with wire. Resistors may also have some inherent capacitance or inductance in them.

The contribution of both capacitive and inductive elements to the impedance depends on the frequency of the sinusoidal oscillation of the voltage and current in the circuit. The symbol X_L is generally used to represent the reactance of an inductive element and may be calculated as follows:

$$X_L = 2\pi f L = \omega L \qquad 2.5$$

The reactance of a capacitive element X_C is given by,

$$X_C = -\frac{1}{2\pi f C} = -\frac{1}{\omega C} \qquad 2.6$$

The total impedance of a circuit, expressed in *ohms*, is a geometrical summation of resistance and reactance; that is to say, if the impedance represents the hypotenuse of a right triangle, then the resistance represents one leg of the triangle and the total reactance (inductive and capacitive) is represented by the other leg. The geometrical relationship among these terms is represented by the Pythagorean theorem (Fig. 2-4).

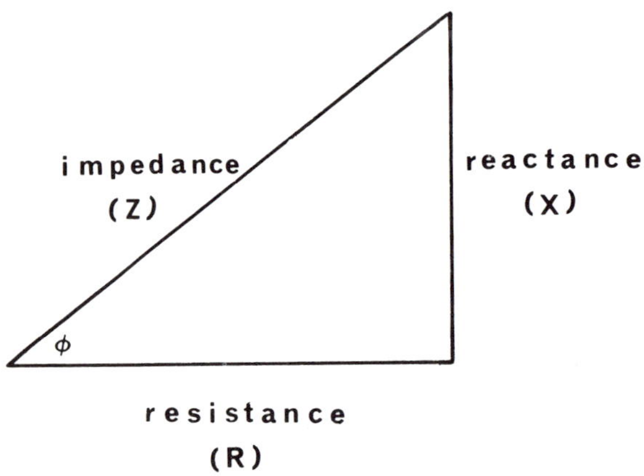

Figure 2-4. Vector diagram for the impedance Z in an alternating current circuit.

Mathematically the geometrical relationship in Figure 2-4 may be expressed:

$$Z^2 = R^2 + [X_L - X_C]^2 \qquad 2.7$$

The impedance of a circuit also has a phase angle associated with it. This may be more easily understood if the impedance is stated as Ohm's law.

$$V = IZ \qquad 2.8$$

in which case

$$Z = \frac{V}{I} \qquad 2.9$$

In terms of equations 2.2 and 2.3, the impedance (Z) equals,

$$Z = \frac{V_m \sin(2\pi ft + \phi)}{I_m \sin 2\pi ft} \tag{2.10}$$

which equals,

$$Z = \frac{V_m}{I_m} \left\{ \frac{\sin(2\pi ft + \phi)}{\sin 2\pi ft} \right\} \tag{2.11}$$

The maximum voltage (V_m) and the maximum current (I_m) in equation 2.11 represent, in each case, the peak amplitude of the sinusoidal waveform. These are not, however, the voltage and current that would be measured with an ordinary voltmeter or ammeter. These meters read an *average value.* Since sinusoidal waves are symmetrical about the zero axis, the word *average*, when applied to a meter movement, means the square root of the average of the squares of the current or voltage over the whole cycle. This type of average is known as the *RMS* or *root-mean-square* value. It is calculated as follows:

$$I_{RMS} = \sqrt{\frac{I_1^2 + I_2^2 + \cdot\cdot\cdot\cdot\cdot\, I_n^2}{n}} = .707\, I_{max} \tag{2.12}$$

where n is the number of samples taken of a cycle. It will be most accurate when n approaches infinity. If, however, the average is taken over only one half of the cycle it would be equal to only .636 I_m. This will be important in the discussion of rectification (Chapter 4).

If a biologist were measuring sinusoidal blood pressure in an animal, using a mercury manometer, and sinusoidal blood flow, using a flowmeter which could only measure mean blood flow, then he would not be able to determine the instantaneous impedance or the phase difference between pressure (analogous to voltage) and flow (analogous to current). He would have to choose a better instrument.

The remainder of this chapter will be devoted to the study of the relationship among voltage, current, and impedance in circuits containing resistance, resistance and capacitance, resistance and inductance, and a circuit containing all three: resistance, capacitance, and inductance.

Both transient and steady state responses will be examined. The *transient* response is the description of what happens to the voltage and the current when the circuit is first energized, *e.g.*, closing a switch in a circuit in which a battery is the source of voltage. The *steady state* response occurs after the circuit has been energized for some time. The time necessary for the circuit to reach the steady state varies with the nature of the components in the circuit and with the frequency of the sinusoidal voltage supplying the circuit.

THE RESISTIVE BRANCH

Figure 2-5 shows a resistance circuit. Actually, it is impossible to have a purely resistive component. The actual components will always have some inherent inductance and/or capacitance. A voltage drop equal to the voltage supplied by the generator will appear across the resistor at the instant the switch is closed. A current, due to the voltage drop across the resistor, will flow without any delay. In other words, the voltage and current will be *in phase.* Were there two resistors of equal value in this circuit, the voltage supplied by the generator would appear as a voltage drop equal to one half of the supply voltage across each resistor, *i.e.*, the sum of the

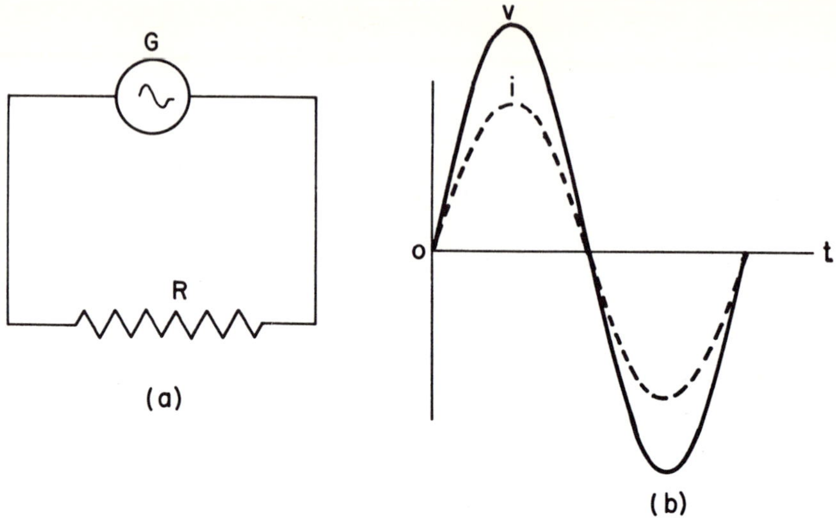

Figure 2-5. The current and voltage are in phase when there is an alternating current through a pure resistance. (Courtesy of R. W. Kenworthy, College Physics, F. A. Davis Co.)

voltage drops across any number of resistors must equal the voltage supplied by the generator.

Figure 2-5 (b) shows the phase relationship in an AC circuit with only a resistive branch. When the switch is closed the voltage and current will be *in phase.* This is a characteristic of the resistive branch. The voltage and the current in a capacitive or inductive branch will not be in phase. The transient and the steady state responses are identical in a purely resistive circuit.

THE RC CIRCUIT

The circuit in Figure 2-6 contains a capacitive element and a resistor as well as a voltage source and a switch.

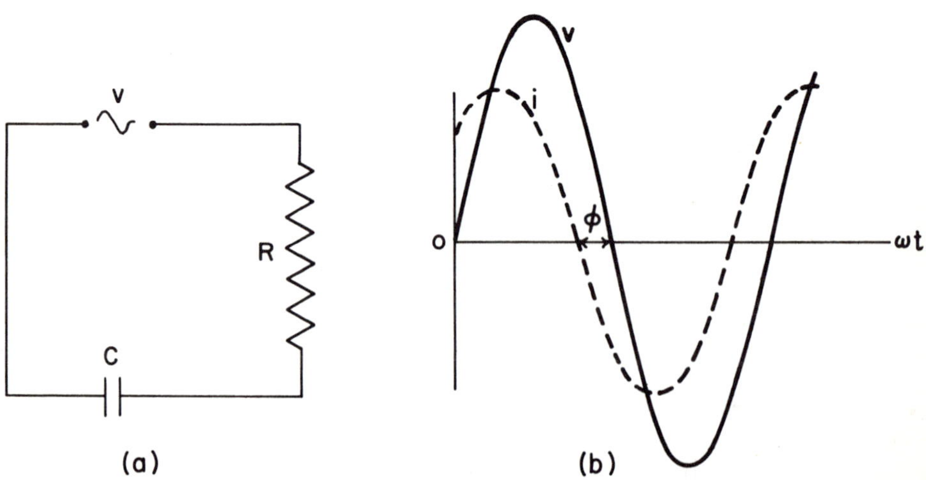

Figure 2-6. The current is a maximum at an earlier time than the voltage in a circuit containing a resistance and a capacitance. (Courtesy of R. W. Kenworthy, College Physics, F. A. Davis Co.)

If the generator in Figure 2-6 (a) is suddenly turned on, the total voltage drop (equal to the generator voltage) will occur initially across the resistor. The phase relation between the voltage and current in this circuit is shown in Figure 2-6 (b). Current will then flow in the circuit, accumulating negative charge on one side of the capacitor. As the charge accumulates, the voltage across the capacitor increases and the current in the circuit decreases. Since the sum of the voltage across the resistor and the capacitor must equal the generator voltage, when the voltage across the capacitor increases, the voltage across the resistor must decrease. Notice that the current in the circuit is equal to the voltage across the resistor divided by the resistance. Thus, if the voltage across the resistor decreases, one would expect the current through the circuit to decrease. The mathematical expression for the transient response of the current in the circuit shown in Figure 2-6 (a) may be derived from the fact that the voltage supplied by the generator is equal to the voltage across the resistor plus the voltage across the capacitor at any time following the closing of the switch:

$$V_{\text{generator}} = [RI]_{\text{resistor}} + \left[\frac{Q}{C}\right]_{\text{capacitor}} \qquad 2.13$$

where Q is the charge on the capacitor.
Equation 2.13 can be differentiated* with respect to time

$$0 = R\frac{dI}{dt} + \frac{1}{C}\frac{dQ}{dt} \qquad 2.14$$

but

$$I = \frac{dQ}{dt} \qquad 2.15$$

so

$$0 = R\frac{dI}{dt} + \frac{I}{C} \qquad 2.16$$

or

$$\frac{dI}{dt} + \frac{I}{RC} = 0 \qquad 2.17$$

Equation 2.16 can be solved for I as follows:

$$I = \frac{V}{R}e^{-\frac{t}{RC}} \qquad 2.18$$

from which it can be seen that when $t = RC$, the current has decreased exponentially from its original value of V/R at $t = 0$ to $(V/R - .707\ V/R)$ or about 29.3 per cent of V/R. The time at which $t = RC$ in equation 2.18 is known as the *time constant* of the circuit. The time constant is a very important parameter of any circuit. It tells the instrumentalist how fast the circuit can respond to changes of voltage or current.

---

*The term *differentiated* is a mathematical term which describes how the slope of the graphic curve of I (in this case) varies as a function of time rather than how the *curve* of I itself varies as a function of time. The calculus provides methods for doing this mathematically without resorting to the graphic display of I.

The time constant is usually represented by the Greek letter τ and in an RC circuit it is equal to:

$$\tau = RC \qquad 2.19$$

where

τ = time constant in seconds
R = resistance in ohms
C = capacitance in farads

The transient response of the RC circuit shown in Figure 2-6 (a) is illustrated graphically in Figure 2-7.

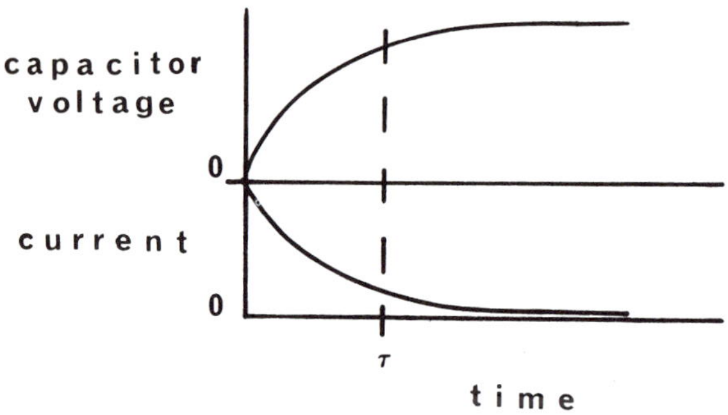

Figure 2-7. Transient response of an RC circuit.

When the generator in Figure 2-6 (a) has been on for some time (at least ten times the time constant), it can be said that the circuit is in the *steady state.* Figure 2-6 (b) is a graph of the voltage and current across the capacitor in Figure 2-6 (a).

At the instant the voltage across the capacitor is zero (capacitor uncharged), maximum current will flow in the circuit. When the voltage reaches its maximum one fourth of a cycle later, the capacitor will be fully charged and no current will flow in the circuit. Then, as the voltage decreases, the capacitor will discharge current into the circuit until the voltage reaches zero. The voltage then continues to go in a negative direction, and the opposite plate of the capacitor charges. Then as the voltage climbs back toward zero, the capacitor discharges into the circuit in the opposite direction. Thus, we see that the current always leads the capacitor voltage in a capacitive circuit. This may be expressed mathematically as:

$$V = V_{\text{max}} \sin 2\pi ft \qquad 2.20$$

$$I = I_{\text{max}} \sin (2\pi ft + 90°) = I_{\text{max}} \cos 2\pi ft \qquad 2.21$$

The impedance of an RC circuit is given by

$$Z = \sqrt{R^2 + \left(\frac{1}{2\pi fC}\right)^2} \qquad \phi = \arctan \frac{-\dfrac{1}{2\pi fC}}{R} \qquad 2.22$$

THE RL CIRCUIT

A series RL circuit (Fig. 2-8 (a)) is suddenly energized by an alternating voltage by closing a switch. As the current surges through the circuit it generates a voltage

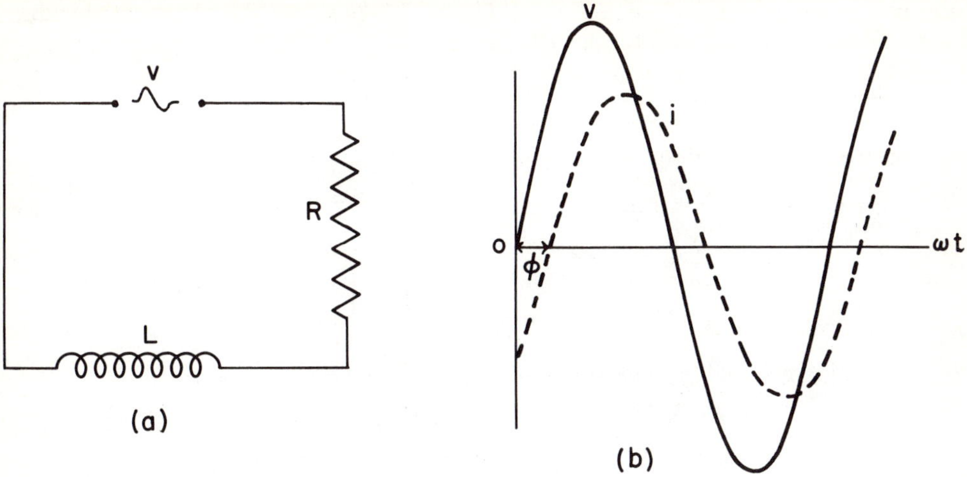

Figure 2-8. The current is a maximum at a later time than the voltage in a circuit containing an inductance. (Courtesy of R. W. Kenworthy, College Physics, F. A. Davis Co.)

opposite in sign to that supplied by the generator. This is called the back-emf and was discussed on page 21. The current through the circuit is given by:

$$I=\frac{V}{R}-\frac{V}{R}e^{-\frac{Rt}{L}} \qquad 2.23$$

The first term V/R is the current that will result in the steady state. During the transient state the current will be decreased by (V/R) exponent $-Rt/L$. The time that it takes for the current to rise to .707 of its steady state value (V/R) is called the time constant of the RL circuit, just as it was in the RC circuit, except that the time constant (τ) of an RL circuit is equal to L/R. Figure 2-9 illustrates the transient behavior of the current through an RL circuit.

Example: The time constant of an RL circuit in which a 10-ohm resistor is in series with a 0.2-henry inductor is

$$\tau=\frac{L}{R}=\frac{0.2}{10}=0.02 \text{ seconds}$$

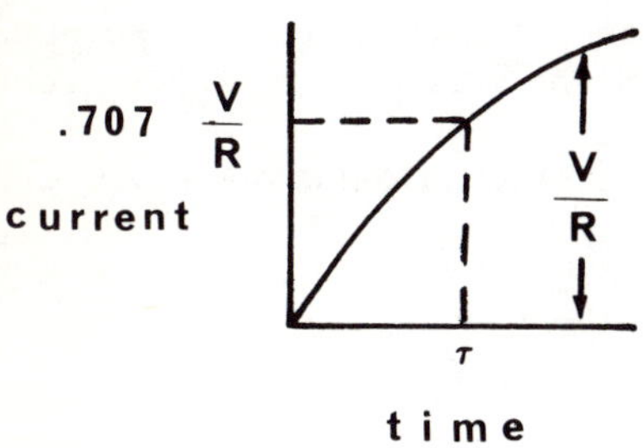

Figure 2-9. The transient behavior of an RL circuit.

The inductive circuit tends to resist change. A sudden increase in voltage will transfer energy to the magnetic field generating the back-emf. The back-emf will decrease the voltage and hence tend to return the voltage across the inductor to the voltage which existed before the increase. Inductors are extremely useful in smoothing out ripples in a DC power supply.

The steady state response of the RL circuit shown in Figure 2-8(a) can be predicted from the behavior of the resistor and the inductor considered as separate entities. The voltage crop across each component of a circuit, such as the one shown in Figure 2-8(a), can be added to yield the voltage supplied to the circuit by the generator. Figure 2-8(b) shows the phase relationship between voltage and current in an RL circuit.

The voltage drop across a resistor may be found by applying Ohm's law:

$$V_{\text{resistor}} = I R \qquad 2.24$$

The voltage across the inductor is equal to the inductance (L) multiplied by the rate of change of current with time, written as $\frac{dI}{dt}$ using the calculus (see footnote, page 31).

$$V_{\text{inductor}} = L \frac{dI}{dt} \qquad 2.25$$

So the total sinusoidal voltage supplied by the generator is equal to:

$$V_{\text{generator}} = (RI)_{\text{resistor}} + \left(L \frac{dI}{dt}\right)_{\text{inductor}} \qquad 2.26$$

Solving equation 2.26 involves a knowledge of differential equations not presupposed for readers of this book. The student should nonetheless study what follows as I have written it with this in mind.

If the generator voltage is $V_{\max} \sin 2\pi ft$, then the solution of equation 2.26 is:

$$\underbrace{V_{\max} \sin 2\pi ft}_{\text{generator}} = \underbrace{RI_{\max} \sin 2\pi ft}_{\text{resistor}} + \underbrace{2\pi fLI_{\max} \cos 2\pi ft}_{\text{inductor}} \qquad 2.27$$

Notice that the voltage across the inductor is 90° out of phase with the applied voltage. In an RL circuit the current leads the voltage across the inductor by 90°.

The impedance and phase difference ϕ between the voltage and the current of the circuit shown in Figure 2-8(b) and discussed above may be calculated as follows:

$$Z = \sqrt{R^2 + (2\pi fL)^2} \qquad \text{and} \qquad \phi = \arctan \frac{2\pi fL}{R} \qquad 2.28$$

Notice that equation 2.28 may be represented by a right triangle, in which the impedance (Z) is the hypotenuse, and the resistance (R) and the reactance ($2\pi fL$) are the sides (see Fig. 2-4). Thus, equation 2.28 may be seen as the application of the Pythagorean theorem. The Greek letter ϕ represents the phase difference between the sinusoidal voltage and the sinusoidal current in the RL circuit, as will be seen below.

Now using Ohm's law to derive the expression for the current in the RL circuit, Figure 2-8(a):

$$I_{RL} = \frac{R}{\sqrt{R^2 + (2\pi fL)^2}} I_{\max} \sin 2\pi ft + \frac{2\pi fL}{\sqrt{R^2 + (2\pi fL)^2}} I_{\max} \cos \omega t \qquad 2.29$$

from Figure 2-8 (b):

$$\sin \phi = \frac{2\pi fL}{Z} = \frac{2\pi fL}{\sqrt{R^2 + (2\pi fL)^2}} \qquad 2.30$$

$$\cos\phi = \frac{R}{Z} = \frac{R}{\sqrt{R^2 + (2\pi fL)^2}} \tag{2.31}$$

Sin ϕ and cos ϕ can be substituted in equation 2.29 to give:

$$I = I_{max}\,[\sin 2\pi ft \cos\phi + \sin\phi \cos 2\pi ft] \tag{2.32}$$

which can be simplified, using a trigonometric identity, to:

$$I = I_{max} \sin(2\pi ft + \phi) \tag{2.33}$$

and the voltage would then be given by:

$$V = V_{max} \sin 2\pi ft \tag{2.34}$$

The impedance has already been defined for the RL circuit in equation 2.28. In a purely inductive circuit, that is, with no resistance and no capacitance, the phase difference would equal 90° and the current would lag the voltage by 90°.

THE RLC CIRCUIT

Consider now a circuit which has a resistive element, a capacitive element, and an inductive element all in series, as shown in Figure 2-10.

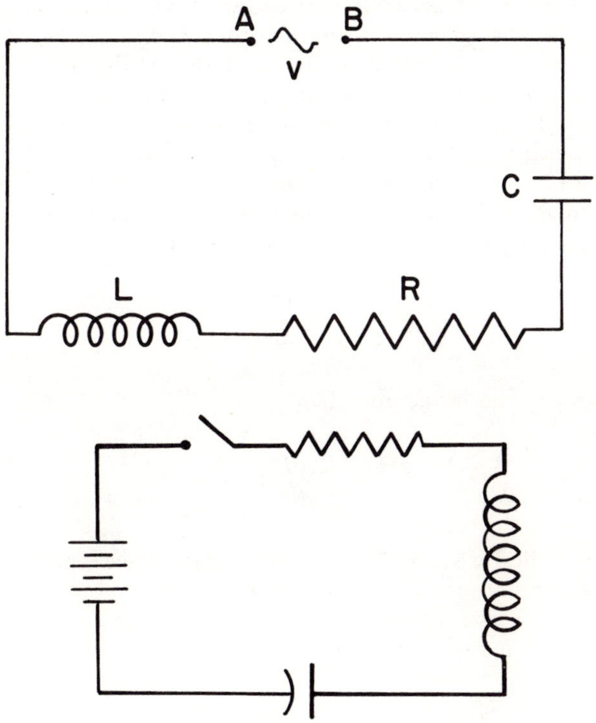

Figure 2-10. A network containing resistance, capacitance, and inductance. (Courtesy of R. W. Kenworthy, College Physics, F. A. Davis Co.)

If no current is flowing in the circuit in Figure 2-10 and the switch is suddenly closed (at $t = o$), the voltage will be equal to the inductance (L) *times* the rate of change of the current $\left(\frac{dI}{dt}\right)$, *plus* the resistance (R) *times* the current (I) *plus* the amount of current as a function of time (expressed in the language of the calculus as the integral of the current ($\int Idt$), divided by the capacitance. All of this is stated mathematically as:

$$L\frac{dI}{dt} + RI + \frac{1}{C}\int Idt = V_{\text{generator}} \qquad 2.35$$

Equation 2.35 is called a *differential equation*, and its mathematical solution is beyond the scope of this book. Nevertheless, it is instructive to show how it is solved and to discuss its solution in graphic terms which are within the scope of the student's presupposed level of knowledge.

If the current were plotted as a function of time then dI/dt would be a plot of the slope of I plotted as a function of time. This is called the *first derivative* of I with respect to time (t). An equation containing only first derivatives is called a first-order differential equation. If the slope of the previous curve (dI/dt) is then determined and is also plotted as a function of time, this new curve would be called the *second derivative* of I with respect to time (t) and is written $\left(\frac{d^2I}{dt^2}\right)$. Equations containing second derivatives as well as first derivatives are called second-order differential equations. The calculus is, among other things, the study of the procedure for determining the derivatives directly from the mathematical expressions. The derivative of the $\int Idt$ term, in equation 2.35, as a function of time would merely equal the current I. The *integral* and the *derivative* may be considered opposite mathematical operations. The above is, of course, an oversimplification, and the student is urged to study a textbook of mathematics for greater detail.

To solve equation 2.35, it is first necessary to determine the derivative of each term either graphically or using the calculus. The equation which results is:

$$L\frac{d^2I}{dt^2} + R\frac{dI}{dt} + \frac{I}{C} = 0 \qquad 2.36$$

Dividing through by the inductance (L):

$$\frac{d^2I}{dt^2} + \frac{R}{L}\frac{dI}{dt} + \frac{I}{LC} = 0 \qquad 2.37$$

Equation 2.37 is called a second-order differential equation. Notice that the term R/L, which is a time constant, appears in equation 2.37. The technique used to solve equation 2.37 is as follows:
First make the following substitutions:

$$D = \frac{dI}{dt} \qquad 2.38$$

$$D^2 = \frac{d^2I}{dt^2} \qquad 2.39$$

Then substitute D and D^2 in equation 2.36:

$$LD^2 + RD + \frac{1}{C} = 0 \qquad 2.40$$

Equation 2.40 is an algebraic quadratic equation whose solutions are:

$$D = \frac{-R \pm \sqrt{R^2 - 4L/C}}{2L} = -\frac{R}{2L} \pm \sqrt{\frac{R^2}{4L^2} - \frac{1}{LC}} \qquad 2.41$$

We see from equation 2.38 that the symbol D is actually a *derivative.* Hence to solve equation 2.41, it is necessary to perform an integration. The final solution for the current I which results from suddenly closing the switch in the circuit shown in Figure 2-10 is:

$$I = \frac{CV - Q_o}{\sqrt{R^2C^2 - 4LC}}\, e^{-\frac{R}{2L} + \sqrt{\frac{R^2}{4L^2} - \frac{1}{LC}}} - e^{-\frac{R}{2L} - \sqrt{\frac{R^2}{4L^2} - \frac{1}{LC}}} \qquad 2.42$$

Q_o, in equation 2.42, represents the initial charge on the capacitive element *before* the switch was suddenly closed. Equation 2.42 looks complicated indeed, but it is only the algebraic difference between two exponential curves. Almost every *real* electronic circuit has resistance, capacitance, and inductance, and most circuits are much more complicated than a simple series RLC circuit. Fortunately, the current and the voltage may be easily measured with the proper instruments. Instruments may also be used to measure directly the phase difference between voltage and current. There are a few pitfalls involved in the measurements, but these are discussed in detail in the next chapter.

THE TRANSIENT RESPONSE IN THE RLC CIRCUIT

The transient build-up of current which results when a voltage is suddenly applied to an RLC circuit may be divided into four classes: oscillatory, underdamped, critically damped, and overdamped responses. The oscillatory response occurs when $R^2/4L$ is less than $1/LC$ (damping $\ll 1.0$). Notice that in Figure 2-11 the current shoots up and oscillates, but the peak-to-peak amplitude of the oscillations becomes smaller and smaller until the current settles down to a steady state

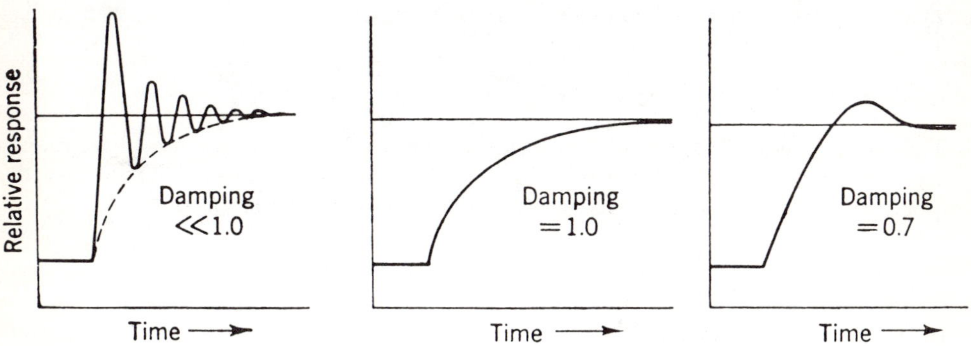

Figure 2-11. Transient response of second-order instruments. (From Biological and Medical Electronics, by Ralph Stacy, Copyright © 1960. McGraw-Hill Book Company. Used by permission.)

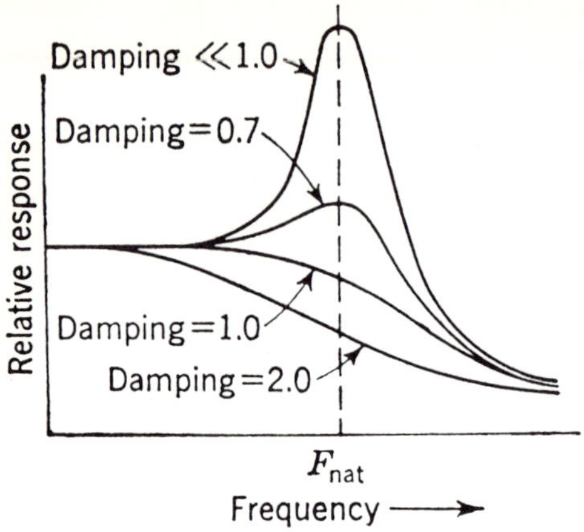

Figure 2-12. Effect of damping on the frequency response of an RLC circuit. (From Biological and Medical Electronics, by Ralph Stacy. Copyright © 1960. McGraw-Hill Book Company. Used by permission.)

value. When the current shoots up to a maximum and then settles down exponentially approaching its steady state value, the response is called an *underdamped response* (Fig. 2-11; damping = 1). When the square root in the exponential term in equation 2.42 equals 0, which occurs when $R^2/4L$ equals $1/LC$, the response is called a *critical response.* This is shown graphically in Figure 2-11 (damping = 1.0). The *overdamped* response is a non-oscillatory response which occurs when $R^2/4L$ is greater than $1/LC$. This is illustrated in Figure 2-11 (damping = 0.7). As the difference between $R^2/4L$ and $1/LC$ becomes more and more pronounced, the current falls more and more rapidly and the response is said to be more damped. The amount of damping is expressed as the coefficient of damping and is related to the *time constant* (τ) of the circuit (Fig. 2-12).

SERIES RESONANCE AND Q

If in equation 2.40 the inductive reactance is equal to the capacitive reactance, the total reactance will equal zero. The current through a series RLC current, such as is shown in Figure 2-10, will equal V/R. This is the maximum current that can flow through the circuit. Both inductive and capacitive reactance are dependent on the frequency of the oscillation. Therefore, if one sets the inductive reactance equal to the capacitive reactance and solves for the frequency, the result is the frequency f_R at which the current through the circuit is maximum.

$$f_R = \frac{1}{2\pi\sqrt{2C}} \qquad 2.43$$

The current in an RLC circuit is plotted in Figure 2-13 as a function of frequency. The narrowness of the curve in Figure 2-13 is called the *Q of the circuit* and is a function of the resistance, capacitance, and inductance of the circuit.

$$Q = \frac{f_R}{f_2 - f_1} \qquad 2.44$$

In the series RLC circuit

$$Q = \frac{1}{R}\sqrt{\frac{L}{C}} \qquad 2.45$$

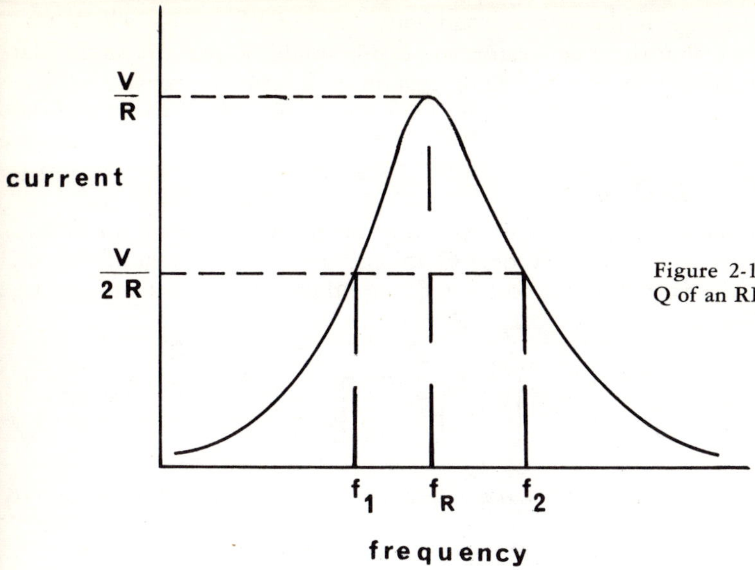

Figure 2-13. Graph showing Q of an RLC circuit.

STEADY STATE RESPONSE IN A SERIES RLC CIRCUIT

The current which results some time after the switch is closed in the DC circuit is zero. The steady state current which results when a sinusoidal generator is left on for some time is a sinusoid. The voltage, current, and impedance will then be defined as previously described for the RL and RC circuits except that the phase angle will depend on the ratio of the inductance to the capacitance as well as the frequency of oscillation. The current is given by:

$$I = \frac{V}{Z} \qquad 2.46$$

The voltage across the inductance is

$$V_L = \frac{V\,(2\pi f L)}{Z} \qquad 2.47$$

and the voltage across the capacitor is

$$V_C = \frac{V\left(\dfrac{1}{2\pi f C}\right)}{Z} \qquad 2.48$$

where Z is the impedance and in a series RLC circuit is given by:

$$Z = \sqrt{R^2 + (2\pi f L - 1/2\pi f C)^2} \qquad 2.49$$

The phase difference ϕ, which represents the separation of the sinusoidal voltage and current in time in fractions of a cycle, is equal to:

$$\phi = \arctan \frac{\omega L - \dfrac{1}{\omega C}}{R} \qquad 2.50$$

Were a DC voltage to be applied suddenly to an RLC circuit which was already supplied by a sinusoidal generator, the result would be that the sinusoidal current would be added to the transient response described previously. The resulting current would be the graphic sum of the steady state AC current and the transient DC current.

THE NON-SINUSOIDAL WAVE

All of the mathematical theory described thus far in this chapter has referred to either DC voltage from a battery or AC voltage from a sinusoidal generator. Other waveforms, however, can and do exist. Some other common waveforms are shown in Figure 2-14.

Figure 2-14. Examples of common non-sinusoidal waves.

Mathematical techniques do exist for expressing the waveforms shown in Figure 2-14 as a number of sine waves which when added together may be superimposed on the original non-sinusoidal waveform. This technique is called *Fourier analysis*. The sine terms which result from a Fourier analysis form what is called a *Fourier series* when the terms are added together as shown below:

$$A = A_0 + A_1 \sin x + B_1 \cos x + A_2 \sin 2x + B_2 \cos 2x + A_3 \sin 3x + B_3 \cos 3x + \cdots + A_n \sin nx + B_n \cos nx$$

where A is the amplitude at any point in a cycle; A_0 is the amplitude at $t = 0$ and is a function of the mean value or superimposed DC value; A_1 and B_1 are called the fundamental terms; A_2 and B_2 are the second harmonic terms, etc.

Figure 2-15 illustrates a rectangular wave and its Fourier harmonics.

Some waves require more harmonic terms to describe them than others. In general, the steeper the rise and fall of the wave, the more harmonic terms it will take to describe it. A square or rectangular wave would thus require an infinite number of terms to describe it completely.

There are no (or almost no) biological phenomena which are represented by pure sinusoidal waves. Yet the same mathematical tools described in this chapter can be applied to biological phenomena when certain assumptions are made, and they may therefore be useful biological tools. There are, however, some mathematical drawbacks to the indiscriminate use of the Fourier analysis in biological systems. These mathematical details are beyond the scope of this book, and the biologist interested in applying them should consult a competent biophysicist or biomathematician. Fourier analyses can be used more freely in most electronic circuit analyses.

KIRCHHOFF'S LAWS

There are some circuit arrangements which because of their complexity cannot be reduced to an equivalent circuit element as described thus far in this chapter. Before application of procedures based on Kirchhoff's laws, it is necessary to define two terms, namely, the *branch point* (often called a *node*) and the *loop*. A

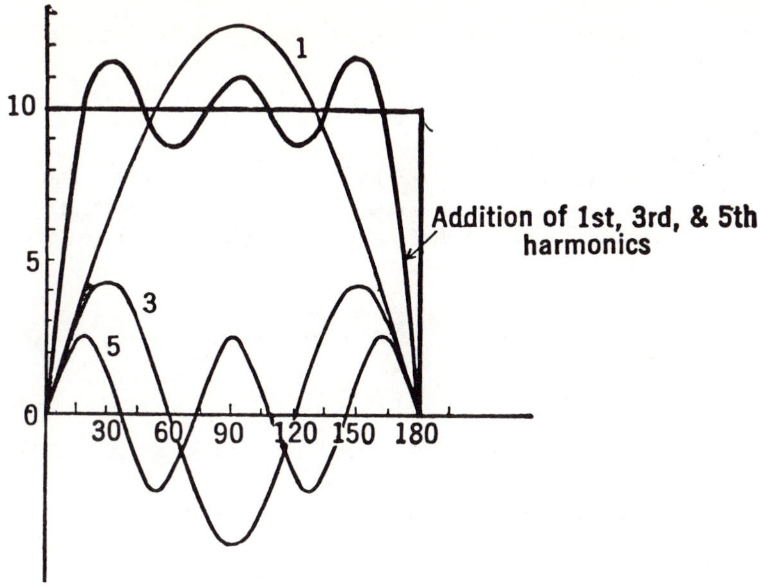

Figure 2-15. The Fourier analysis of a rectangular wave.

branch point is a point in a network, such as the one shown in Figure 2-16, where three or more conductors are joined. Points *a,b,c,* and *d* in Figure 2-16 are branch points. A *loop* is a closed circuit, starting at a branch point or any other convenient point, continuing around the network along *any* path, and finally returning to the same point. In other words, it is a closed path such as *abda, abcda, abcVa,* and *adcVa* in Figure 2-16.

Figure 2-16. A resistive network showing branch points.

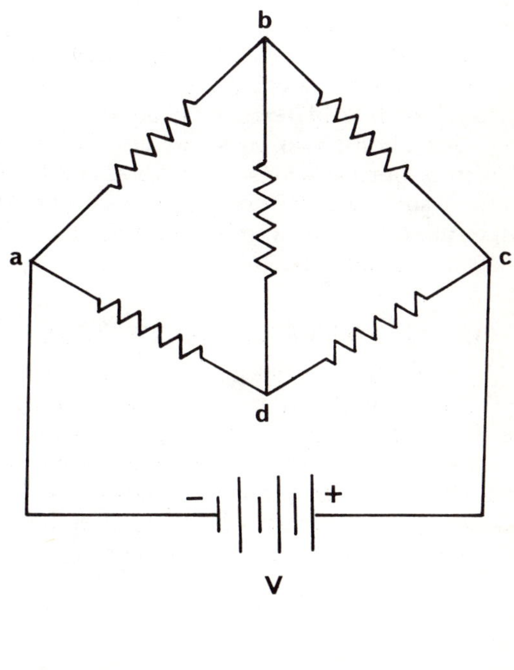

Kirchhoff's First Law

The algebraic sum of the currents into and out of each branch point in a given closed network is equal to zero.

In other words, electrons do not accumulate at the intersection of three or more wires in a network.

Example: The resistance from a to b in Figure 2-16 is 5 ohms, and the voltage drop across it is 10 volts. The resistance from a to d is 2 ohms, and the voltage drop from a to d is 20 volts. What current is supplied by the battery?

$$I_{\text{battery}} + \frac{V_{ab}}{R_{ab}} + \frac{V_{ad}}{R_{ad}} = 0$$

$$I_{\text{battery}} + \frac{5}{10} + \frac{2}{20} = 0$$

$$I_{\text{battery}} = -0.6 \text{ amperes}$$

This means that the battery supplies 0.6 amperes and it is flowing into the branch point, whereas the current flowing from a to b and from a to d is flowing away from the branch point, hence the minus sign.

Kirchhoff's Second Law

The voltages produced as a result of current(s) flowing in a closed path of a network are equal to the sum of all internal, applied, and induced voltages in the same path.

If the absolute potential at point a in Figure 2-16 is V_a, then if we add all the voltage differences, i.e., V_{ab}, V_{bc}, V_{cd}, V_{da}, around a given path, we should still conclude that the absolute potential at a is V_a. In order to use Kirchhoff's second law it is necessary to determine the direction and magnitude of the current through each branch of the circuit. The voltage drop across each component is then calculated. The correct algebraic sign must be given to each voltage. The sum of the voltages across the components must then equal the sum of voltages from voltage sources such as batteries or generators.

Kirchhoff's two laws may also be applied to AC circuits containing not only resistors but capacitors, inductors, and even active circuit elements such as transistors and vacuum tubes. In the case of AC circuits, however, it is important to determine the correct phasic relationships between voltages and currents in each part of the network.

SPECIAL FILTERING NETWORKS

It is often necessary to utilize the frequency-dependent characteristics of a passive network. (A *passive* network is a group of resistors, capacitors, inductors, and, possibly, diodes. The opposite is an active network which contains not only passive components but also one or more stages of a vacuum tube or transistor amplifier.) A signal which is allowed to enter a given network is called its *input* (V_i or I_i), and the signal which, after being modified by the network, leaves the network is called the output (V_o or I_o).

The Low-Pass RC Network

If sinusoidal voltages of different frequencies but equal amplitudes are applied to the input terminals of the network in Figure 2-17, only the low frequencies pass through the network undiminished in amplitude (unattenuated). At some frequency the amplitude will begin to become attenuated, and as the frequency is increased still further, the amplitude at the output will fall off steeply.

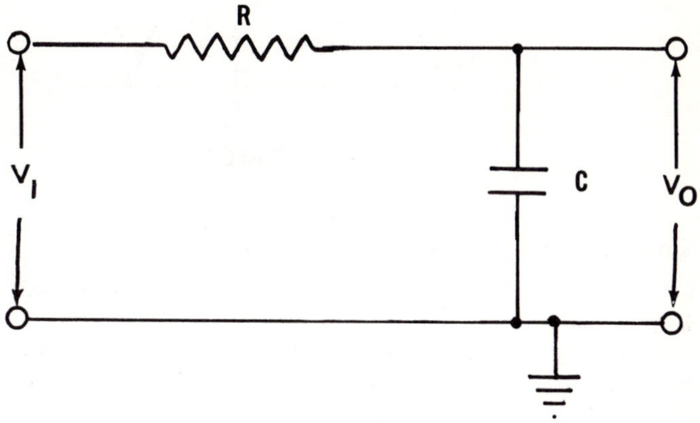

Figure 2-17. The low-pass RC circuit.

The frequency f_2 at which the amplitude of the output will be attenuated to .707 of the amplitude at extremely low frequencies is dependent on the *time constant* of the resistor-capacitor combination, as follows:

$$f_2 = \frac{1}{2\pi RC} \qquad 2.51$$

Therefore, if the resistor in Figure 2-17 is 16,000 ohms (16K Ω) and the capacitor is 1 microfarad (1 μf or 1×10^{-6} farad), then:

$$f_2 = \frac{1}{2\pi \times 1.6 \times 10^4 \text{ ohms} \times 1 \times 10^{-6}} = 10 \text{ cps} \qquad 2.52$$

As the input frequency is increased the reactance of the capacitor becomes smaller and the output signal is more and more attenuated. The *gain* at a given frequency $[\text{output/input}]_f$ of the low-pass network in Figure 2-17 is given by:

$$\left[\frac{\text{output}}{\text{input}}\right]_f = \frac{1}{\sqrt{1 + f/f_2}} \qquad 2.53$$

The output voltage will lead the input voltage by:

$$\phi = -\arctan \frac{f}{f_2} \qquad 2.54$$

Figure 2-18 is a plot of the gain of this low-pass filter as a function of input frequency.

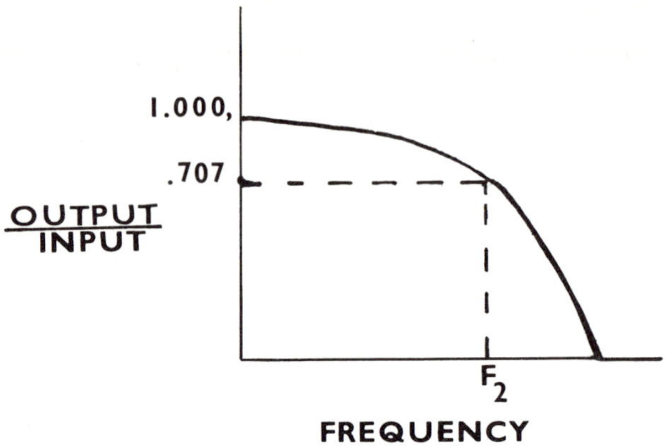

Figure 2-18. Frequency response of a low-pass RC circuit.

The High-Pass RC Filter

Examination of Figure 2-19 reveals that in the high-pass RC filter the position of the resistor and that of the capacitor are reversed. As the frequency of the input is increased in this network the decreasing reactance of the capacitor will allow more and more of the signal to get through to the output terminals.

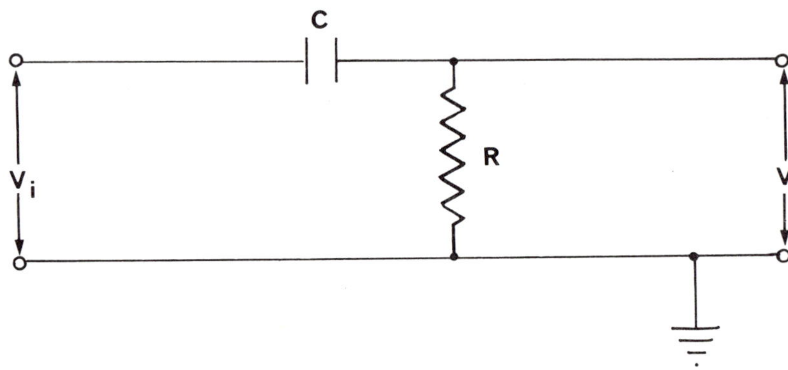

Figure 2-19. The high-pass RC circuit. (From Pulse and Digital Circuits, by J. Millman and H. Taub. Copyright © 1956. McGraw-Hill Book Company. Used by permission.)

If the same values are used for the resistor (16K Ω) and the capacitor (1 μf) as were used for the low-pass filter, the value of the frequency at which the amplitude is .707 times the amplitude at very high frequencies (called the *cut-off* frequency) is still about 10 cps (1/2 RC). In this case, we refer to the cut-off frequency as f_1. The gain of the high-pass filter will approach unity when frequencies higher than the cut-off frequency are used, but will fall off rapidly when lower frequencies are used. The gain at any frequency for the network shown in Figure 2-19 is:

$$\left[\frac{\text{output}}{\text{input}}\right]_f = \frac{1}{\sqrt{1+\{f_1/f\}^2}} \qquad 2.5$$

In this case the output will lead the input by:

$$\phi = \arctan \frac{f_1}{f} \qquad 2.56$$

Figure 2-20 shows the gain of a high-pass filter as a function of frequency.

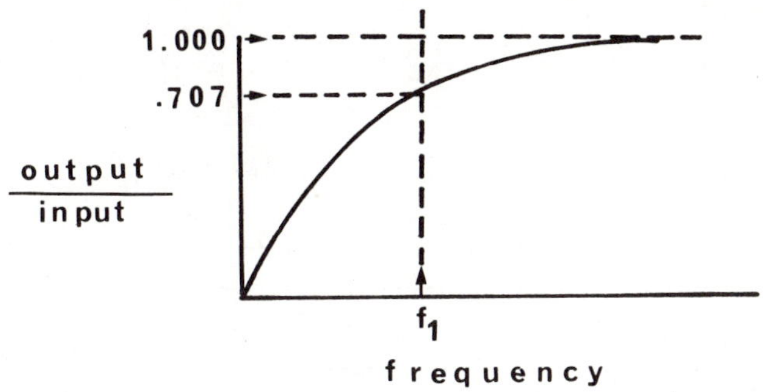

Figure 2-20. Frequency response of a high-pass RC circuit.

The Twin-Tee Network

The twin-tee, or parallel-tee as it is sometimes called, is a convenient network for the elimination of a particular frequency. Sinusoidal signals of most frequencies are attenuated only slightly as they pass through the network, but a relatively small range of frequencies (called a band of frequencies) are diverted away from the output terminals, usually to ground. In some few applications the diverted band of frequencies is utilized for a particular purpose. The twin-tee network is made up of only resistors and capacitors. In general, these components are smaller and lighter in weight than inductors and are more readily available in a variety of values. The twin-tee is most useful for low and audio frequencies, which include most biological applications.

Figure 2-21 is a commonly used twin-tee network.

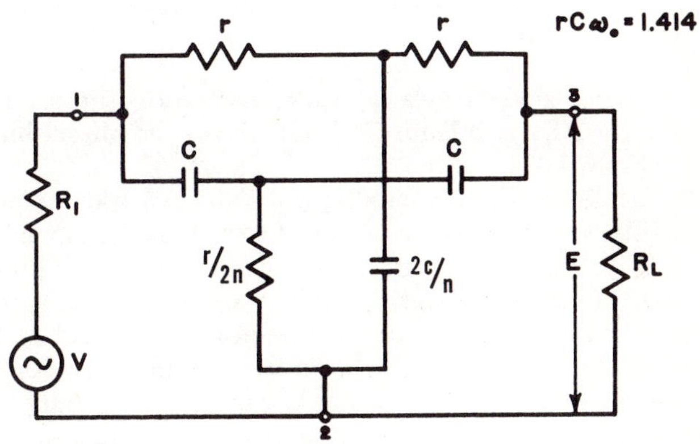

Figure 2-21. The twin-tee network. (Courtesy of White Instrument Laboratories, Austin Tex.)

Several relationships have been worked out for the design of a twin-tee. In Figure 2-21, R_1 is the *effective* resistance of the *source* of the signal and R_L is the *effective* resistance of the *load*. In other words, when the twin-tee is used in a circuit, the resistance which the input terminals apparently see looking backward is the *source* resistance. It is associated with the generator itself, and can be measured. The apparent resistance across the output terminals when the twin-tee is placed in a circuit is called the *load* resistance. The transmission for the schematic shown in Figure 2-21 will be symmetrical when:

$$r = \sqrt{(n+1)\, R_1 R_L} \qquad 2.57$$

where n is determined by the designer's choice of resistor and capacitor values (see equation 2.58 below). The *null* or *rejection* frequency is given by:

$$f_{\text{null}} = \frac{\sqrt{n}}{2\pi r C} \qquad 2.58$$

[Notice that equation 2.58 contains the time constant for a series RC circuit.]

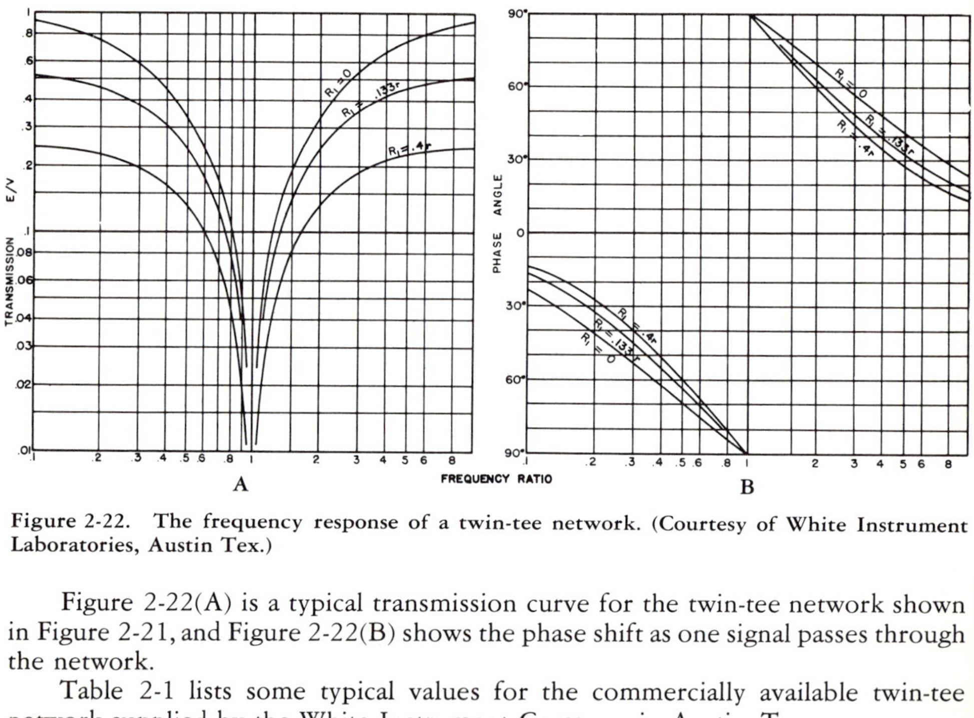

Figure 2-22. The frequency response of a twin-tee network. (Courtesy of White Instrument Laboratories, Austin Tex.)

Figure 2-22(A) is a typical transmission curve for the twin-tee network shown in Figure 2-21, and Figure 2-22(B) shows the phase shift as one signal passes through the network.

Table 2-1 lists some typical values for the commercially available twin-tee network supplied by the White Instrument Company in Austin, Texas.

As a *passive* network, the twin-tee cannot eliminate a *particular* frequency. If, for example, the 60 cps *noise** signal is to be eliminated, it can be seen from Figure 2-22 that in addition to the noise the useful signals have also been attenuated to .707 of peak values at about 30 cps and about 90 cps. Adding a variable resistor in series with $r/2n$ or a variable capacitor in parallel to $2C/n$ (in Figure 2-21) will allow for adjustment of the parameters to yield a somewhat more narrow curve, but not much improvement can be expected unless the twin-tee is built with an active element, *e.g.*, an amplifier.

**Noise* is an unwanted voltage or current signal which is superimposed on the desired signal (see Chapter 12).

TABLE 2-1

Some Values of Resistance and Capacitance Which May Be Used in the Twin-Tee*

Null. f	*r*	*C*
30 cps	75k	.1 mfd
	150	.05
	300	.025
60 cps	37.5k	.1
	75	.05
	150	.025
	300	.0125
120 cps	18.7k	.1
	37.5	.05
	75	.025
	150	.0125
	300	.0062
400 cps	11.2k	.05
	22.5	.025
	45	.0125
	90	.0062
	180	.0031
	360	.0015
800 cps	11.2k	.025
	22.5	.0125
	45	.0062
	90	.0031
	180	.0015
	360	.0008

*These values apply when n in Figure 2-21 is 2.

The twin-tee of any passive filter network does not act as it should theoretically. It is extremely important that the instrumentalist understand this. More will be said about this topic when the over-all frequency response of instruments is discussed later in the book.

There are many other networks with a wide variety of applications. Some examples of these are the T-pad network and L-pad network. The networks can in many cases be used one after the other for special effects. Special combinations of filters can be designed for special purposes, for example, a high-pass filter can be used with a low-pass filter to allow a band of frequencies to pass. This is called a band-pass filter. There are many other pulse-shaping networks employing not only passive components but diodes as well as active elements such as transistors and tubes. Low- or high-pass filters are often used in combination with twin-tee circuits.

REFERENCES

Alternating Current Circuits, Russell M. Kerchner, and George F. Corcoran (John Wiley and Sons, New York, 1951).

College Physics, Ray W. Kenworthy (F. A. Davis Co., Philadelphia 1961).

Electricity and Magnetism, Francis W. Sears (Addison-Wesley Publishing Co., Reading, Mass., 1951).

Transformational Calculus and Electrical Transients, Stanford Goldman (Prentice Hall, New York, 1949).

3

The Measurement of Voltage and Current

principles of measurement
range
sensitivity
accuracy
mechanism errors
scale errors
environmental errors
dynamic errors
reading errors
systematic errors of measurement
random errors
the galvanometer
the pivoted coil galvanometer
the dynamometer
measurement of current
measurement of voltage
the series resistor and galvanometer coil
measurement of resistance
the Wheatstone bridge
the oscilloscope
references

PRINCIPLES OF MEASUREMENT

If one is asked to measure the length of a small polished metal bar, he may do one of several things:

1. If the bar is less than 6 inches long, he may pick up an inexpensive ruler and measure the bar to the nearest 1/16 inch.
2. He may use an engraved ruler and measure the bar to the nearest 1/64 inch.
3. He could use a micrometer and measure the length to the nearest .0001 inch.
4. Or he could use an optical instrument and measure it to .000,001 inch or better.

It would be silly to use a yardstick to measure something about 1/8 inch long, and it would be just as impractical to measure a room with a 6-inch ruler. It is important to choose a measuring instrument with the desired *range* and *degree of accuracy.*

These principles also apply to electronic measurements of every variety. The principles of measurement, discussed in this chapter, apply not only to instruments which measure voltage and current, but to every instrument, and in fact, to every measurement.

RANGE

Every measuring instrument has a scale upon which the reading of the measurement appears. Each scale has a lower and an upper value. The difference between the upper value and the lower value is called the *range.* Many instruments have more than one scale, and each scale has a different range. In this way, the correct range can be chosen for the measurement involved. It is always desirable to choose the scale which will be *mid-range* for the measurement.

The student of instrumentation must decide carefully on the *range* he plans to measure before using an instrument. If the approximate value is not known, an instrument must be employed with a very large range. A smaller range may be used if the lower value is sufficiently high. If for example, a voltmeter is used which has five scales, the instrumentalist should always begin with the highest range.

Thus far, the discussion has been limited to static measurements. Very often, however, it is necessary to measure *dynamic* phenomena, that is, phenomena which vary with time. Here, one must be concerned not only with the range of the parameter being measured, but also with the response of the measuring instrument as a function of time. How fast can the instrument respond to a change of the parameter being measured? This is called *the frequency response* of the instrument. Some instruments, such as the ordinary voltmeter, can measure only DC voltage. Others can measure only AC voltage up to 100 cps, but when used to measure voltages oscillating at 1000 cps respond only to 20 per cent of the true peak-to-peak voltage. The AC voltmeter does not measure DC voltage accurately. Some oscilloscopes, however, are capable of responding accurately to 2000 megacycles (2×10^9 cps). If the instrument responds uniformly, within 5 per cent, to all frequencies from 0 to 30 cps, it is said to be *flat* to 30 cps. Such an instrument could be used to register the blood pressure in the aorta.

The frequency response of an instrument also depends, to a large extent, on mechanical design. A meter consists of an electromagnetic device which moves a pointer. The pointer indicates a position on the scale. The frequency response of such a meter would then depend on the inertia of the moving element, the friction of the pointer mechanism, and the *spring constant* of the spring which restrains the pointer and helps to return it to its zero position.

If the biological investigator has an instrument with which to measure instantaneous blood pressure and that instrument indicates the blood pressure as a voltage output, he must use an instrument which will respond fast enough to include the most rapid *change* in the blood pressure. In terms of *frequency response,* the instrument must be capable of responding to the highest frequency that will occur. In the measurement of blood pressure in animals the instrument must respond not only to the heart rate of the animal but also to the highest Fourier harmonic of the pressure pulse, which can be as great as 30 cps for an aortic blood pressure pulse, 15 cps for femoral blood pressure, or as low as 0.1 cps for pressure developed within the stomach. A great deal more will be said about the *frequency range* of instruments in discussions about *instrument calibration* and in the chapter about *recorders.*

Most instruments have scales which are graduated in a linear fashion, but in some cases, where an extremely large range must be put on a single scale, the scale may have logarithmic graduations like a slide rule. Instruments which measure optical density or the sound intensity often have logarithmic scales.

SENSITIVITY

The *sensitivity* of an instrument may be stated as *the amount of signal required to produce a unit of deflection.* An oscilloscope may have a sensitivity of 1 volt per centimeter of deflection, or an ammeter may have a sensitivity of 0.1 ampere per unit of scale deflection. This is sometimes referred to as *figure of merit.**

The sensitivity may also be stated as the *amount of deflection per unit of signal.* A pressure manometer has a sensitivity which is stated as, for example, 72 μv (microvolts) per 100 mm Hg per volt of excitation. The sensitivity of a thermometer may have units of 1 cm per degree centigrade.

Range and sensitivity are inter-related. The greater the range, the less sensitive an instrument is per unit of deflection. An instrument also may have greater sensitivity in one range than in another.

ACCURACY

The accuracy or precision of a measuring instrument depends, to a great extent, on the errors or lack of errors to which the instrument is susceptible. It is easier to discuss these errors than to discuss the accuracy, *per se.* A more complete discussion of errors associated with recorders will follow in a later chapter. It is important to understand the shortcomings of an instrument in order to choose the proper instrument for a given measurement. Careful calibration of a measuring instrument will usually define its accuracy. In order to improve precision of the instrument, it is important to understand the concepts and terminology in this section. Some errors, especially systematic measurement errors, will not easily show up, and the scientist must be aware that they exist and take special pains to look for them.

Stacy classifies errors into two major groups:

A. *Instrument errors*
   1. Mechanism errors
   2. Environmental errors
   3. Dynamic errors
   4. Reading errors

B. *Measurement errors*
   1. Systematic errors
   2. Random errors

Other authors classify the above errors in a somewhat different manner, but, all in all, the same topics are discussed by each one. It is important to keep in mind that at no time must the measuring instrument be allowed to interfere with or alter the system being measured. This is especially true of biological measurements. For example, if an orifice flowmeter is used to measure blood flow, it is obvious that the flowmeter itself will change the flow that the metering device is attempting to measure, because it is put directly into the blood stream. If a microelectrode is used to measure the potential difference between the inside and the outside of a nerve axon and the microelectrode disturbs the axon membrane, it can give rise to a false measurement.

Biological measurements are highly susceptible to changes caused by measuring instruments. In the case of the orifice flowmeter the disturbance was obvious, but the biological instrumentalist must examine the measuring instrument and the system being measured very carefully to detect the subtle effects which the instrument may have on the system.

---

*The *figure of merit* may be any number of different expressions such as the ratio of *nominal gain* to the *rise time* of an RC coupled amplifier.

Mechanism Errors

These errors are in the instrument itself. A meter, such as a voltmeter, consists of a coil of wire which rotates when a voltage is applied. The coil is commonly mounted on jeweled bearings so that when it rotates it moves the pointer, and a scale, marked in volts, can be read. A small spring is usually attached to the coil and pointer to return them to the zero position.

There are several *mechanism* errors possible in the instrument described above: Friction in the jeweled bearing or a bent pointer may keep the pointer at a lower value; the spring may have non-linear properties and affect the reading by not allowing the pointer to be read correctly. Instruments which use gear assemblies (*e.g.,* the paper drive of a recorder) may have play in the gear chain. Errors in this class need not be mechanical. They can also be electronic, such as a non-linear or faulty amplifier, a leaky capacitor, or an error in design. These errors can usually be detected by proper calibration of the instrument.

Scale Errors

Every instrument must provide some variety of *scale.* The scale may be engraved on a plastic gradicule (such as on an oscilloscope screen); it may be a photograph as in the case of the meter face, or it may be lines printed on a piece of recorder paper. The accuracy of the position of the indicating device (pointer or light beam) is related to the scale accuracy for the over-all precision of the instrument. The scale can only be read to some definite precision which depends on the width of the graduations and the width of the pointer. This is easily seen on an oscilloscope face. If the intensity is increased, the indication dot or line can be made wide enough to cover two engraved divisions on the gradicule. These two divisions could represent a range of several microvolts to several volts, depending on the sensitivity setting of the oscilloscope. Most oscilloscopes also have a focus adjustment. If this is not properly adjusted, one can easily see that the precision of the scale reading is impaired. If an ink pen is writing on a moving chart and too much ink is flowing, the line will be too thick and a scale error will result. If the divisions on a scale are not engraved properly (thin lines with equal spacing), scale errors will obviously result. The advent of the digital meter, which displays the measurement as a series of digits and has no need for a scale or pointer in the usual sense, minimizes this type of error, but it is rather expensive.

Environmental Errors

These errors are likely to occur whenever an instrument is used in an environment different from the one in which it was calibrated. Temperature changes seriously influence the operation of some instruments. Almost every environmental factor imaginable can potentially influence an instrument: humidity, altitude, vibration, increased gravitational acceleration, to name just a few. Many biological instruments must function while immersed in a bag of salt water. This is certainly going to affect many instruments.

Dynamic Errors

The range of frequencies to which an instrument will respond has been discussed in this chapter under the heading of *range.* It is extremely important that any instrument which is to be used to measure a dynamic phenomenon vary with time, *e.g.,* blood pressure and blood flow as a function of the heart rate; measurement of skin temperature, muscle tension. In each case the instrument will require different dynamic properties. Careful thought must be given to dynamic errors, and steps must be taken to eliminate them in the design of the instrument or to avoid them in the use of the instrument.

Dynamic instrument errors are not always obvious and are, in general, not familiar to biologists and physicians with no previous training in instrumentation; a small bubble in the dome of a strain-gauge pressure manometer or in the catheter of such a system can decrease to under 1 cps the frequency response of an instrument which otherwise would be flat to over 40 cps.

The use of a passive filter network, as described in Chapter 2, to eliminate noise, can seriously alter the over-all frequency response of the instrument. Let us say, for example, that a twin-tee is used to eliminate 60 cps noise. The use of this filter will actually affect the frequency response from 30 through 90 cps.

Too good a frequency response is not always advantageous. If an underdamped meter is used to measure a steady voltage the needle may oscillate and make the meter difficult to read. The same may be said for an optical galvanometer. If it is sufficiently underdamped a sudden movement will cause the entire system to oscillate, and this oscillatory reading may of course be in error.

I have seen an instrument designed to measure the movements of the body which result from the recoil from the ejection of blood from the heart (called a ballistocardiograph) which was so sensitive that it also measured the normal vibrations of the building in which it was located.

Reading Errors

Factors which cause reading errors may be the dimensions of the pointer or the graduation, but they may be more subtle. An optical galvanometer has a coil of wire which rotates as a function of the current flowing through it. It has a small mirror mounted in the center of the coil which deflects a beam of light. The beam of light may fall on a graduated scale or on a piece of light-sensitive photographic paper. In both cases, however, the amplitude is read as a linear amplitude, when really it

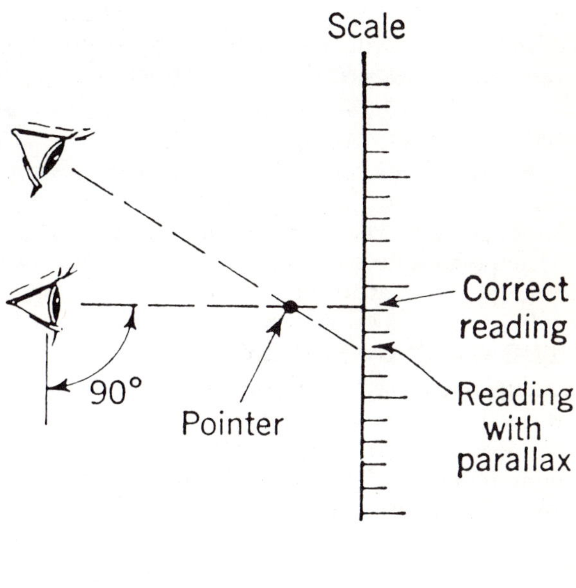

Figure 3-1. Illustrating reading error due to parallax. (From Biological and Medical Instruments, by Ralph Stacy. Copyright © 1960. McGraw-Hill Book Company. Used by permission.)

should be corrected by the term $\theta/\sin\theta$, where θ (measured in radians) is the angle whose sine is given by the measured amplitude divided by the length of the light beam. If the light beam is very much longer than the measured amplitude, then this error is very small, but it does, nevertheless, exist.

Another common type of reading error is due to the fact that, when a scale is read, the eye of the reader may not line up with the pointer and the scale the same way each time. This is shown in Figure 3-1. This reading error is said to be caused by *parallax*. The eye of the instrument reader should sight the scale in such a way as to cause his line of sight to form a 90° angle with the scale. A good meter will often be found to have a mirror just beneath the scale. This mirror is used to sight the pointer. When the line of sight is 90° to the mirror (and hence to the scale) only the pointer and not its image can be seen in the mirror.

Systematic Errors

If, before a series of voltage measurements is made, a voltmeter indicates 1 volt, then the instrument user will be in error by 1 volt on each measurement. This could be caused by a bent pointer incorrectly indicating zero voltage. Such a condition gives rise to a *systematic error*. A *systematic error* would occur if a scientist measuring blood pressure with a strain gauge, for example, did not properly balance the bridge network. The imbalance in the bridge would give rise to a pressure reading which might be too high or too low by a constant 5 mm Hg. In this case, pressure readings *relative* to a reading at one time or place would be correct, but *absolute* reading would be in error.

The accuracy of an instrument may also depend almost entirely upon the method of calibrations. It is often possible to calibrate an instrument by applying an electrical signal to its input and plotting the output as a function of the input signal to determine the repeatability, absolute accuracy, and frequency response. If the instrument is used to measure temperature, blood pressure, or blood flow, it must use a transducer to change the temperature, pressure, or flow into an equivalent input voltage. It should be clear that the ultimate calibration of the instrument can be no better than the accuracy of the known standard used for calibration. If we have no good method for generating a known blood pressure, then we cannot accurately calibrate the instrument used to measure blood pressure. Manufacturers of instruments will often state "Our instrument is accurate to ±2 per cent." In such cases the *method* used for calibration must be seriously questioned.

Random Errors

Many factors may occur which are *random* in nature, such as temperature variation, changes in the supply voltages of an amplifier, or a sudden transient in the house current. These errors will tend to cancel each other out if the measurement occurs over a long enough period. The same principles apply to the cancellation of random errors as to the general application of statistics.

THE GALVANOMETER

The galvanometer is a device which measures the amount of current flowing through a loop of wire. It was first used by Oersted, who placed a coil of wire beneath a compass needle. The needle was held in position because it aligned itself with the poles of the earth. When current was allowed to flow through the loop the compass needle was rotated. Later, Lord Kelvin increased the sensitivity of the

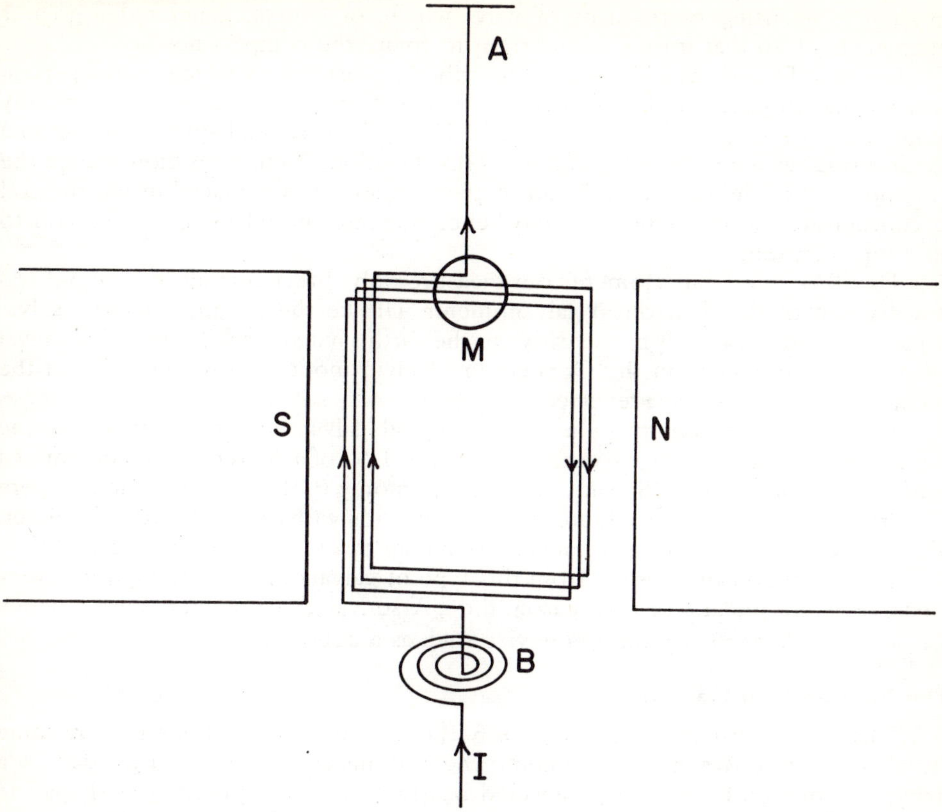

Fig. 3-2A

Fig. 3-2B

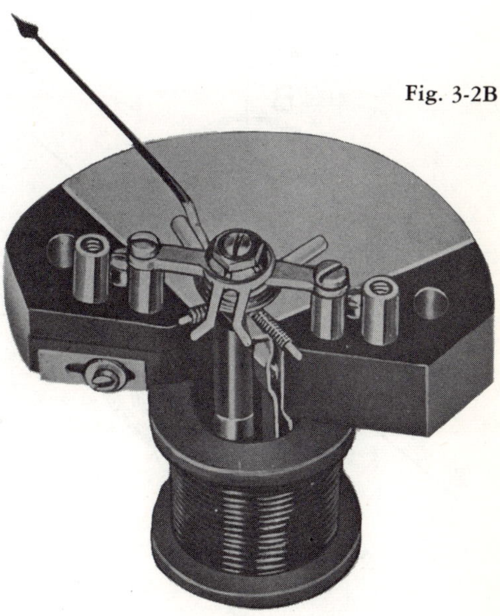

Figure 3-2. A, The d'Arsonval galvanometer. (Courtesy of R. W. Kenworthy, College Physics, F. A. Davis Co.) B, The pivoted-coil galvanometer. (Courtesy of Simpson Electric Co., Chicago.)

instrument by using more turns of wire, which, of course, generated a greater magnetic field, so that it took less current to rotate the compass needle.

In 1883, Deprez and d'Arsonval described an instrument which used a permanent magnet to generate the magnetic field. The current-sensing coil was carefully suspended within the field of the magnet. The coil has a small spring (similar to a watch spring) which returns the coil to its rest position. A mirror is mounted on the coil and used to deflect a light beam. A piece of soft iron is placed inside the coil to concentrate the magnetic field and hence increase the efficiency of the coil to respond to current.

The d'Arsonval galvanometer can measure only direct current. Figure 3-2 (A) is a diagram of the d'Arsonval galvanometer. Unlike the original Oersted galvanometer this instrument is not sensitive to the earth's magnetic field. The d'Arsonval galvanometer differs from the Oersted and Kelvin model in that the roles of the magnetic field and coil are reversed.

The current sensitivity of a suspended-coil galvanometer is defined as the current in the coil necessary to cause a 1-minute (1/60 of a degree) displacement of a light beam 1 meter long. Sensitivities are commonly 0.01 to 0.0001 microampere (10^{-8} to 10^{-10} amperes) per minute of scale deflection with a beam length of 1 meter. Modern galvanometers are not much better than those produced before 1900.

The deflection that results from this type of galvanometer, although it is very sensitive, is not linear. For this reason, the d'Arsonval galvanometer is used mainly for detection of zero current in a device such as a potentiometer.

The Pivoted Coil Galvanometer

This type of galvanometer (Figure 3-2[B]) operates on essentially the same principle as the d'Arsonval movement. The coil, instead of being suspended by a string, is mounted between two jeweled bearings. Soft iron pole-pieces shape the

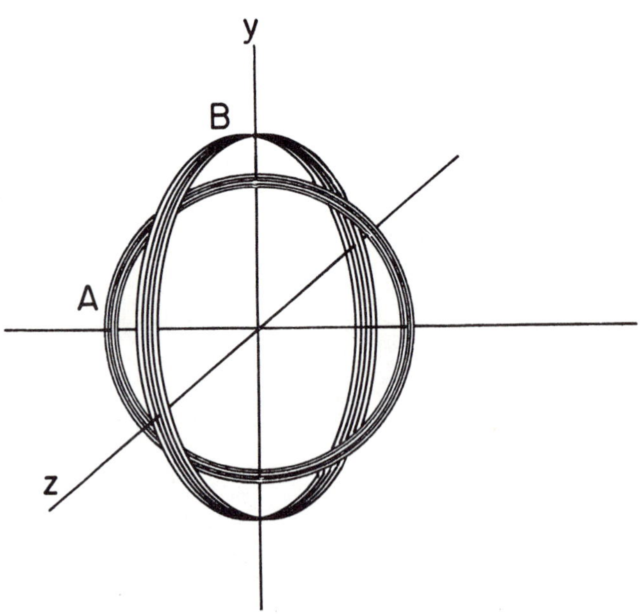

Figure 3-3. The dynamometer. (Courtesy of R. W. Kenworthy, College Physics, F. A. Davis Co.)

magnetic field so that it is always perpendicular to the moving coil. A non-magnetic pointer is attached to the coil to determine the position of the coil. Figure 3-2(B) is a picture of a pivoted coil galvanometer. The movement of the coil with this arrangement is proportional to the current. More precisely, it is proportional to I^2 and hence when used as an ammeter (meter to read amperes) it reads the average of the sum of a number of I^2's. This is called the RMS value and is arithmetically equal to .707 times the maximum current.

THE DYNAMOMETER

If a sinusoidal current were to pass through either of the galvanometers based on the d'Arsonval movement, the oscillations even at 60 cps would occur so rapidly that the pointer would not move. The dynamometer (Fig. 3-3) is a special galvanometer which uses two coils, one fixed to replace the permanent magnet and the other movable to move the pointer. The coils are arranged so that when the polarity changes in the fixed coil, it also changes in the movable coil. The result is that the average deflection is proportional to the RMS alternating current. This type of galvanometer can also be used to measure the product of voltage and current and hence is the basis for the common wattmeter.

MEASUREMENT OF CURRENT

It has already been shown that the galvanometer measures current. A moving-coil galvanometer when placed in *series* with a current-carrying line will measure RMS current. The galvanometer is an extremely sensitive device, and if it is to be used to measure appreciable currents, that is, currents in the milliampere and ampere range, a *shunt* must be used with it. Figure 3-4 illustrates the use of a shunt resistor. The voltage across the coil is equal to the voltage across the *shunt resistor.*

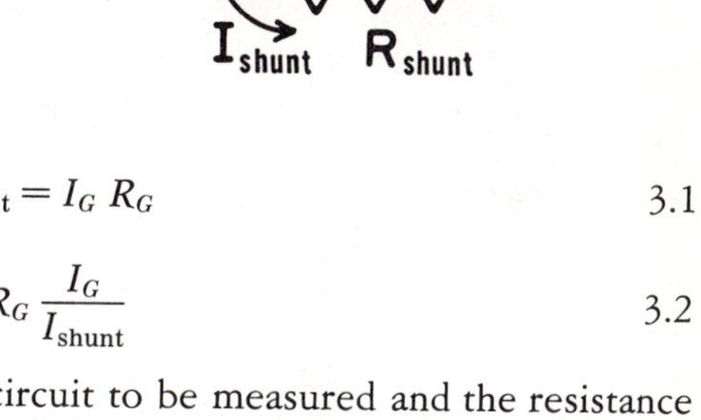

Figure 3-4. The use of a shunt resistor. (Courtesy of R. W. Kenworthy, College Physics, F. A. Davis Co.)

Using Ohm's laws:

$$I_{\text{shunt}}\, R_{\text{shunt}} = I_G\, R_G \qquad 3.1$$

and so on:

$$R_{\text{shunt}} = R_G \frac{I_G}{I_{\text{shunt}}} \qquad 3.2$$

Thus, if 1 ampere is flowing through the circuit to be measured and the resistance of the galvanometer coil R_G is 15 ohms and that particular coil deflects full scale when .01 ampere flows through the coil, then the shunt resistor can be calculated as follows:

$$R_{\text{shunt}} = 15 \times \frac{.01}{.99} = 0.1515 \text{ ohm} \qquad 3.3$$

The current going through the circuit is split; part of it goes through the galvanometer coil, and the balance goes through the shunt resistor. The above ammeter would deflect full scale for 1 ampere, half scale for .5 ampere, and so forth.

MEASUREMENT OF VOLTAGE

The same galvanometer used to measure current can also be used to measure voltage. When used to measure voltage, a very large resistor is placed in series with the galvanometer, and the series resistor and the galvanometer are placed in parallel with the terminals being measured. The galvanometer and series resistor arrangement for measuring voltage are shown schematically in Figure 3-5.

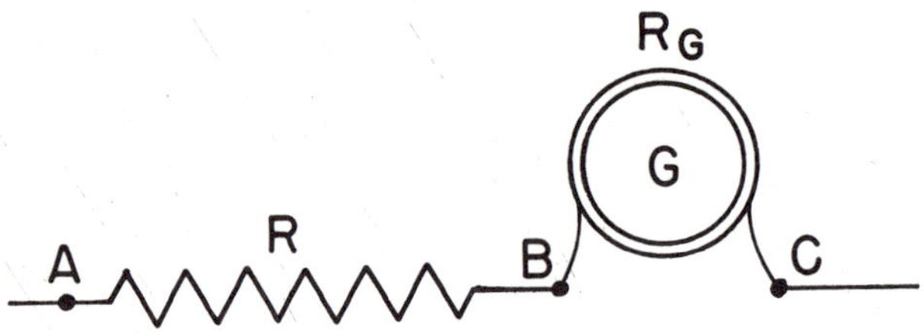

Figure 3-5. The use of a galvanometer and series-resistor arrangement for measuring voltage. (Courtesy of R. W. Kenworthy, College Physics, F. A. Davis Co.)

The galvanometer reads full scale when the current through the coil is 0.01 ampere; if the voltage from A to C is 200 volts then the series resistor must be:

$$R_{\text{series}} = \frac{V_{AC}}{I} = \frac{200 \text{ volts}}{0.01 \text{ ampere}} = 20{,}000 \text{ ohms} \quad 3.4$$

If a 20 K (20,000) Ω resistor is used in series with the galvanometer described above the meter will read full scale when 200 volts are placed across the series resistor and the coil.

THE SERIES RESISTOR AND GALVANOMETER COIL

A voltmeter must have an extremely high series resistance or it will draw current from the circuit in which the voltage is being measured. Voltmeters are rated in this respect by the number of ohms of series resistance per volt of scale reading. Ten thousand ohms per volt is common in an inexpensive voltmeter. Values as high as 6 to 8 megohms/volt of scale reading can be obtained when a vacuum-tube voltmeter (VTVM) is used. When an oscilloscope is used to measure voltage it has an equivalent series resistance of over 1 megohm (10^6 Ω). This is called the input impedance of the oscilloscope. Figure 3-6 is a photograph of a typical commercial volt-ohm-ammeter (VOM). Figure 3-7 is an example of the most recent type of voltmeter—the digital voltmeter.

The ordinary voltmeter is not suitable for measuring DC voltage with great precision. The potentiometer, shown in Figure 3-8, may be used for greater precision, when the voltage of the battery V_s is accurately known. The resistor AD must be uniform and precise. The contact at point C must be movable from one end of AD to the other. Generally the resistor AD is a long coiled bare wire of known resistance and the contact at point C is a metal wiper. The battery E is generally a *standard cell* (calomel half-cell), and its voltage output is stated by the manufacturer to at least *three or four significant figures*.

The unknown battery E (or source of voltage) is connected as shown in Figure 3-8, and a battery known to have a larger voltage than E is placed at B. The slide-

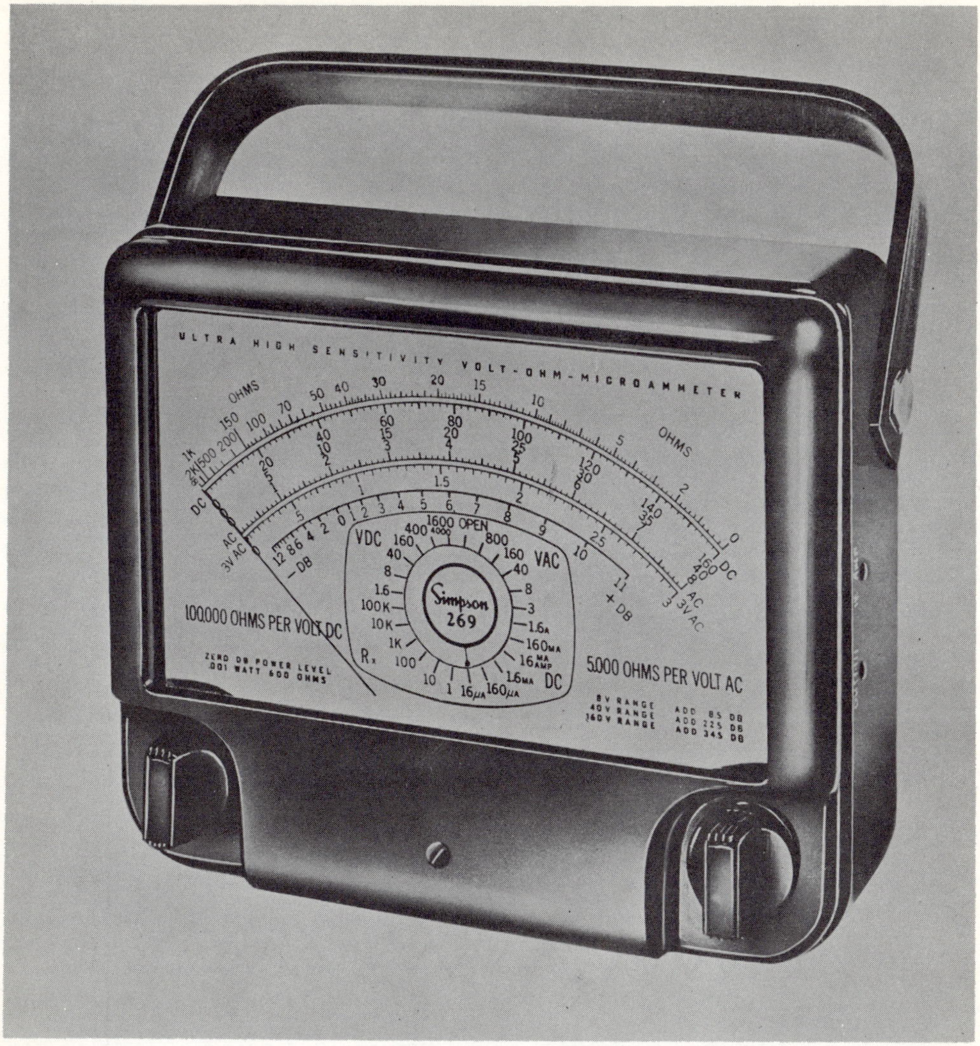

Figure 3-6. A commercial volt-ohmmeter. (Courtesy of Simpson Electric Co., Chicago.)

wire contact C is then moved till *no current* flows through the galvanometer G. This occurs when the current, from battery B, going from C to D is equal to the current flowing from D to C from the unknown battery E. Call this point on the slide wire C_1. A standard cell of known accuracy is then put in place of E (called V_s), and a new point is located on the slide wire at which no current flows through the galvanometer. Call this point on the slide wire C_s. The unknown voltage of battery E may then be found:

$$V_E = V_S \frac{R_{c_1}D}{R_{c_s}D} \qquad 3.5$$

The potentiometer is also a common component in electronic circuitry. When used in circuitry it consists only of a resistance wire with a movable tap. It is commonly used as a voltage divider, but it can be used as a variable resistor by connecting the

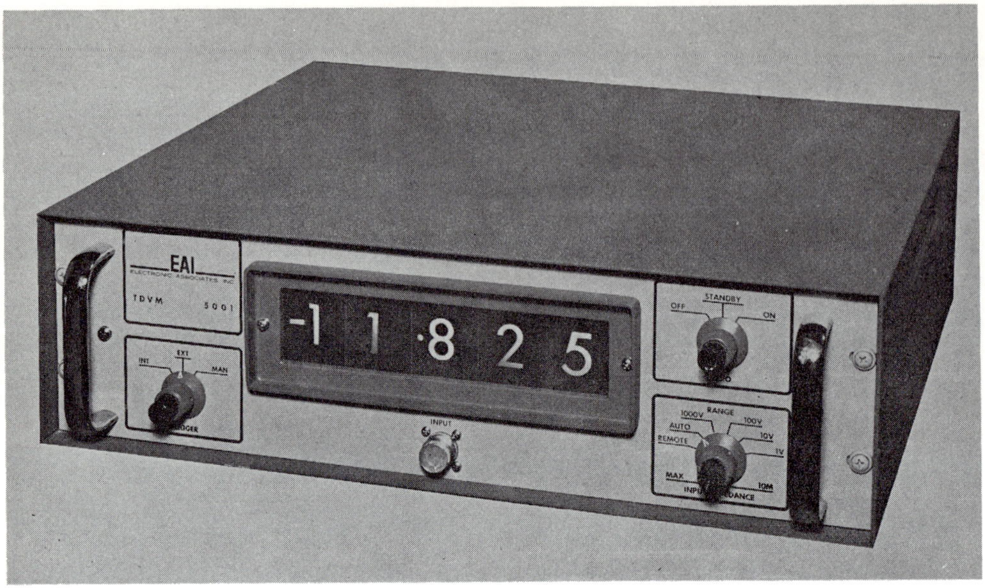

Figure 3-7. The digital voltmeter. (Reproduced with approval by Electronic Associates, Inc., Long Branch, N.J.)

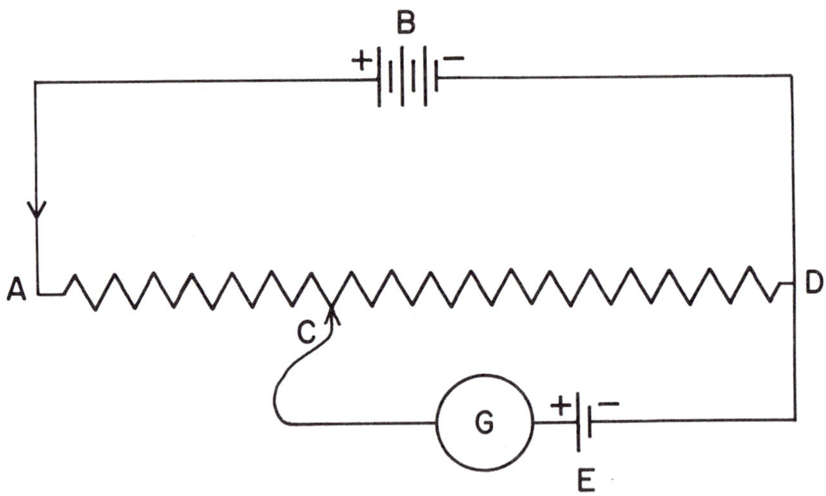

Figure 3-8. A potentiometer. (Courtesy of R. W. Kenworthy, College Physics, F. A. Davis Co.)

movable wiper contact to one end (either *A* or *D*). The symbol for a potentiometer is 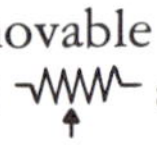 and when used as a variable resistor, it is 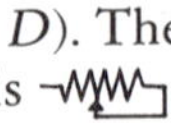 or 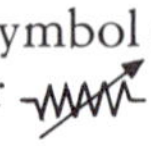 .

MEASUREMENT OF RESISTANCE

Resistance can be measured with a galvanometer and a series resistor as shown in Figure 3-9. When terminals *a* and *b* are *shorted* (connected together), the total resistance in the circuit above consists of the series resistor (R_{Se}) and the resistance of the galvanometer coil (R_G). In the circuit shown, the total resistance would then

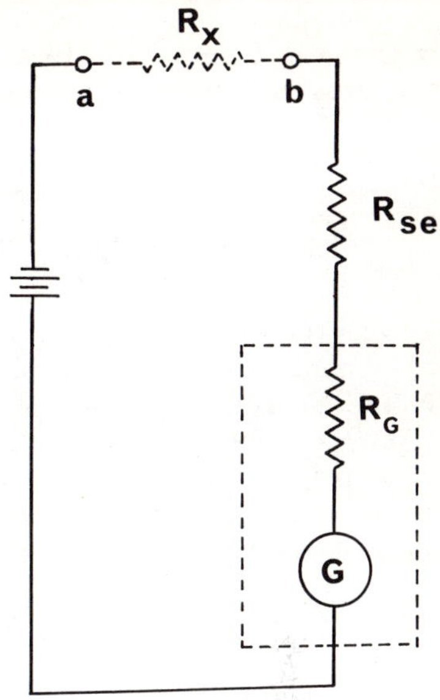

Figure 3-9. The series ohmmeter.

be 8950 ohms. The battery voltage is 9 volts, the current through the coil will be 9/9000 amperes or 1 milliampere (ma), and the meter would read full scale. When a 9K Ω resistor is connected across *a-b,* the total resistance would then be 18K Ω and the current 9/18000 or 0.5 ma, and the meter would read half-scale for a 9K Ω series resistor. The general formula for a series resistor ohmmeter is:

$$R_x = (R_{Se} + R_G)\,\frac{I_1 - I_2}{I_2} \qquad 3.6$$

where: I_1 is the current when *a* and *b* are shorted
I_2 is the current with R_x in the circuit

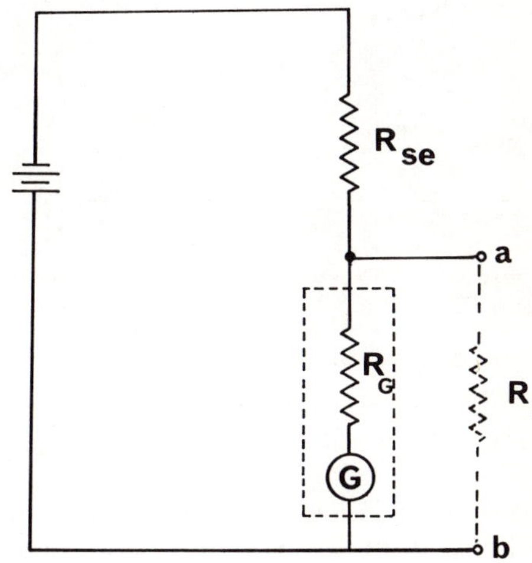

Figure 3-10. The parallel ohmmeter.

Resistance can also be measured, using a shunt resistor with a circuit such as is shown in Figure 3-10. In this case R_x is placed in parallel with a galvanometer coil, and the unknown resistance is calculated as follows:

$$R_x = R_G \frac{I_2}{I_1 - I_2} \quad 3.7$$

where: I_1 is the current through the galvanometer coil without R_x
I_2 is the current through the galvanometer coil with R_x in place

Lower values of resistance are more accurately measured with the shunt type of ohmmeter. Actually, most ohmmeters have series and shunt type of circuits. Figure 3-6 is a photograph of a commercially available volt-ohm-ammeter (VOM). (Never connect a VOM to a circuit to measure voltage or current when the instrument is set to measure ohms as this will destroy the VOM. Furthermore, it is not good practice to leave a VOM set to the ohms position, because if the terminals are shorted, the battery will wear out.)

THE WHEATSTONE BRIDGE

A four-element bridge in which each element is a resistor (three known and one unknown) is called a *Wheatstone bridge* (Fig. 3-11).

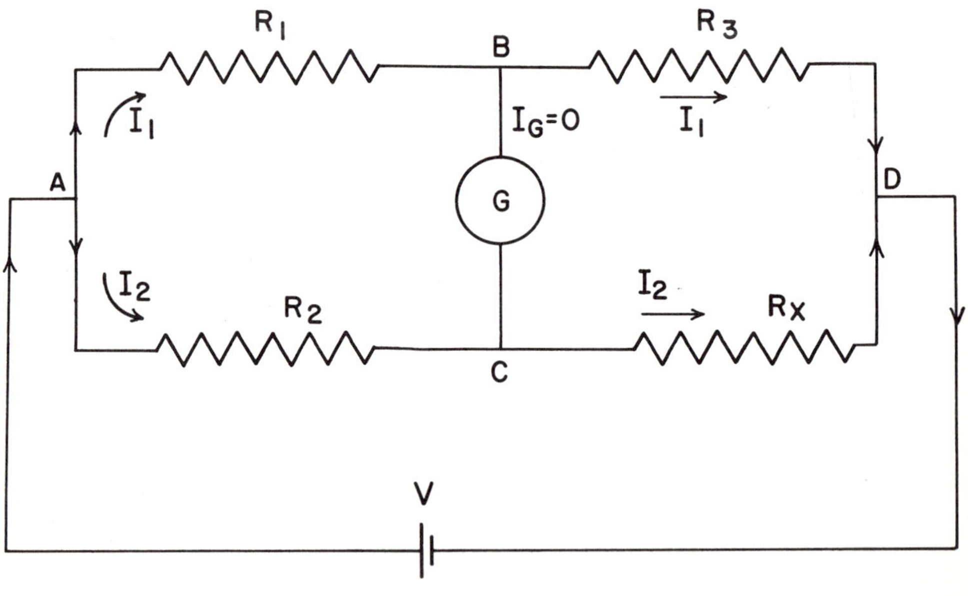

Figure 3-11. The Wheatstone bridge. (Courtesy of R. W. Kenworthy, College Physics, F. A. Davis Co.)

The current entering the bridge at A is split, as in any parallel circuit. I_1 goes through R_1 and R_3, and I_2 goes through R_2 and the unknown resistor R_x. The values of R_1 and R_3 are adjusted as in a potentiometer so that the voltage drop across R_1 is equal to the voltage drop across R_2. When this adjustment is completed no current will flow from B to C or *vice-versa* as both B and C are at the same potential. It may

Figure 3-12. A commercial impedance bridge. (Courtesy of General Radio Co., West Concord, Mass.)

be necessary to choose a new R_2 so that $R_2 + R_x$ is almost equal to $R_1 + R_3$. The unknown resistance R_x may be calculated from:

$$\frac{I_1R_1}{I_1R_3} = \frac{I_2R_2}{I_2R_x} \qquad 3.8$$

from which:

$$R_x = \frac{R_2R_3}{R_1} \qquad 3.9$$

Unknown capacitance, inductance, or impedance can be determined for any given frequency by using modifications of the Wheatstone bridge, such as the Hay bridge, the Shering bridge, and the Wein bridge. Descriptions of these bridges can be found in most reference books. Figure 3-12 is a photograph of the impedance bridge manufactured by the General Radio Company.

Bridges have many other applications in medical electronics. The most commonly used instrument for measuring blood pressure consists of a Wheatstone bridge in which the unknown resistor is replaced by a resistance wire called a *strain gauge* which, when stretched, changes its resistance. The bridge is balanced with the strain gauge in a set position (exposed to a set pressure such as atmospheric pressure). Any change in pressure changes the resistance of the bridge and generates a potential difference between *B* and *C* in Figure 3-10. This change in potential difference causes a current to flow from *B* to *C* which can be detected with a suitable galvanometer and linearly related to pressure change. The strain gauge manometer and other uses of the strain gauge as well as many other specific applications of the *bridge* will be discussed in more detail in a later chapter.

THE OSCILLOSCOPE

It is important to introduce the student of electronics to his most versatile single tool—*the cathode-ray oscilloscope (CRO)*—at the earliest possible time in his study of medical electronics. The oscilloscope provides a visual image of variation of voltage as a function of time without disturbing the circuit being studied or designed. There are many different oscilloscopes on the market. Some of these are for general laboratory use, and others may be used only for specific applications. In each case, however, the basic principle of the oscilloscope is the same. Every oscilloscope contains, as its basic component, *a cathode-ray vacuum tube (CRT)* such as the one shown in Figure 3-13.

A piece of metal (C) called the *cathode* emits electrons when heated by the filament (FIL). These electrons form a cloud near the cathode. A large voltage, positive with respect to the cathode, causes the negatively charged electrons to accelerate toward the screen at the right (FS). The electrons are then collimated to form a narrow beam, which is focused by the application of current through specially shaped electrodes to form a narrow beam of electrons which is accelerated again to attain an even higher velocity.

A vacuum is maintained in the tube to prevent the electrons from hitting air molecules which would tend to scatter them and keep them from hitting the screen. In order to impart enough kinetic energy (velocity) to the electrons, it is necessary to use extremely high voltages. For this reason, the beginner should not attempt to take apart an oscilloscope.

The collimated, focused beam then passes two pairs of deflection plates. The first of these are the *horizontal deflection plates* (HP). These plates move the focused spot from left to right across the face of the tube screen. The electronic circuitry is designed so that the beam moves slowly from left to right then rapidly returns to the left, and again moves toward the right. This is called the *sweep circuit.* Most oscilloscopes provide a control to allow variation in the time that it takes the beam to move from left to right. The control is usually calibrated in units of time per centimeter of distance moved horizontally across the face. Modern oscilloscopes are available which will allow a sine wave of 5000 megacycles/sec (5×10^9 cps) to be seen on the oscilloscope. The type of sweep circuit described above is called *internal*, because the sweep circuit is *triggered* by a circuit on the inside of the oscilloscope.

The sweep of the oscilloscope may be set to be triggered by an external signal if desired. If, for example, an electrocardiogram (ECG) is being displayed on the oscilloscope face, the sweep may be triggered by its QRS complex. In this case,

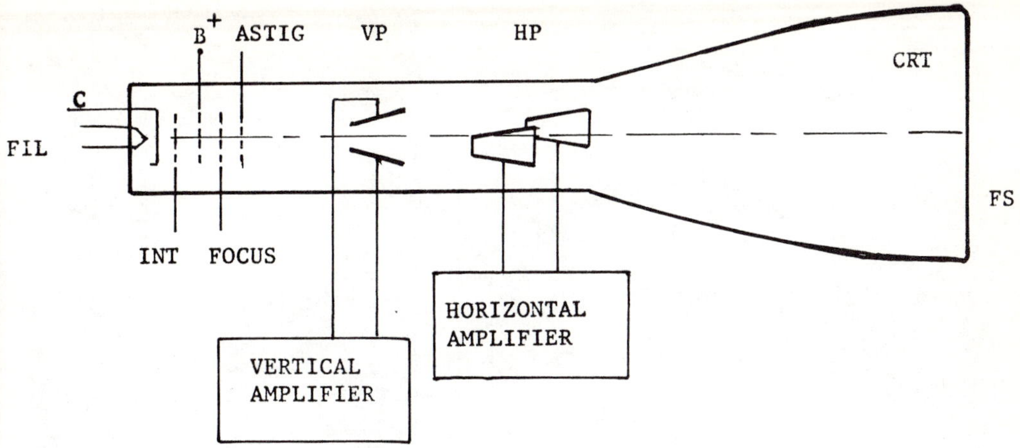

Figure 3-13. The cathode-ray tube.

when the sweep is completed, the beam is again rapidly moved back to the left and the next QRS may be seen. In this way, the wave shape of the electrocardiogram may be studied in detail and the time for one complete cycle may be determined. The sweep may also be set to be triggered by the P wave or the T wave. The control for setting the positive or negative voltage at which the sweep will trigger is called the *trigger level.* In this paragraph, the sweep as described would be triggered externally. There are a few cases where the sweep is not used, and the horizontal plates act in the same way as the vertical plates. An example of this would be if one were to use the oscilloscope to display the voltage, in a given circuit, vertically while using the current in that circuit to drive the horizontal amplifier, resulting in a Lissajous figure.

The *vertical deflection plates* (VP) cause the electron beam to move vertically as a linear function of the voltage applied to the input terminals of the oscilloscope. Switches are available on the control panel to allow selection of an appropriate sensitivity in volts per centimeter of distance moved *vertically* on the screen face.

The electron beam which hits the face of the screen is visible because the interior of the tube is painted with a fluorescent chemical which emits visible light when its molecules are struck by electrons. Some fluorescent chemicals will emit light for a longer time, following electron bombardment, than will others. These are said to have long *persistence.* It is important to have a long-persistence screen to allow time to study the waveform, but it is also important to have short persistence if the oscilloscope screen is to be photographed. It is possible to get more than one persistence on the same screen. Since the visible light emitted from a short-persistence fluorescent paint is a different color from that emitted from a longer persistence paint, one or the other may be filtered out, as desired.

It is also possible to examine more than one part of a circuit at the same time. This is done in one of two ways. The first and most obvious is to build two or more sets of all the parts shown in Figure 3-13 into a single vacuum tube. Such a tube is referred to as having two electron guns. An oscilloscope with two electron guns is called a *dual-beam* oscilloscope. Figure 3-14 is a photograph of a commonly used dual-beam oscilloscope which is particularly well suited for use in the field of medical electronics.

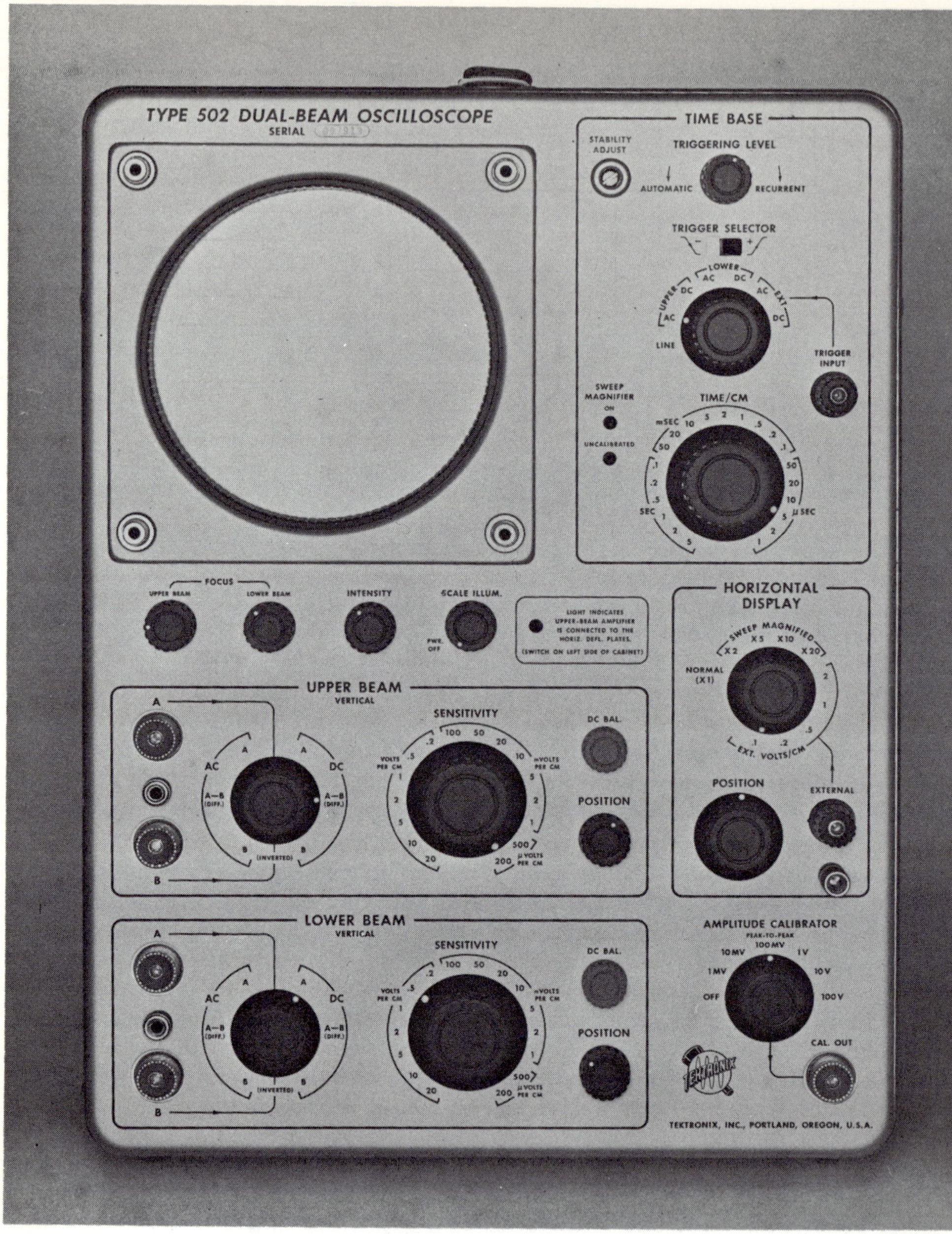

Figure 3-14. A commercial oscilloscope particularly well suited for use in medical electronics. (Courtesy of Tektronix, Inc., Beaverton, Ore.)

A special circuit, called an *electronic switch*, may be employed to sample any number of signals and present them at different vertical positions on a single-beam oscilloscope. The electronic switch, however, must move from one signal to the next very rapidly, and if the signal itself has a high enough frequency, the electronic switch may not be fast enough to present all the details of the waveform of the signals. An oscilloscope equipped with such an electronic switch may appear, at first glance, to be a dual-beam oscilloscope.

The description above will probably not familiarize the student with the oscilloscope. The only way he can do this effectively is to sit down with an oscilloscope and its instruction manual and experiment with the controls. It is very difficult to break an oscilloscope by using it improperly. One word of caution is in order, however. Do not leave a small spot focused on the screen without sweep and with high intensity as the fluorescent chemical can be burned off the face at that point. *It is virtually impossible for the serious student to gain an appreciation of medical electronics without using an oscilloscope at every opportunity.*

REFERENCES:

Biological and Medical Electronics, Ralph W. Stacy (McGraw-Hill Book Co., New York, 1960).
Dynamical Analogies, Harry F. Olson (D. Van Nostrand Co., Princeton, N.J., 1958).
Technology of Instrumentation, Eve B. Pearson (D. Van Nostrand Co., Princeton, N.J., 1957).

4

The Rectifier and the Diode

the vacuum-tube rectifier

the source of electrons—thermionic emission

the operation of the rectifier

some applications of the solid state diode

the semiconductor diode

zener diode

tunnel diode

references

The discovery which was to initiate the science of electronics was made by Thomas Edison in 1883. Edison discovered that if he put a small plate inside one of his new electric lamps and then connected the plate (through a galvanometer) to the positive side of the 120-volt DC filament, current would flow. No current would flow, however, if the plate were connected to the negative side of the filament circuit. Edison's lamp is shown in Figure 4-1. The electron had not as yet been described, and Edison had no knowledge of electron emission or the charge on the electron. This phenomenon was known as the *Edison effect,* which is, in fact, the principle of the vacuum-tube rectifier.

THE VACUUM-TUBE RECTIFIER

An evacuated glass tube with two elements or electrodes is called a *diode.* One of these elements *emits electrons*, and the other element *collects* them. In the next chapter the effects of adding one or more elements will be discussed. The conversion of alternating current into direct current is called *rectification.* A vacuum-tube diode was referred to as a rectifier for many years before the invention of the semiconductor device. In this book the term *diode* shall be reserved for the semiconductor device, and *rectifier* shall refer to the vacuum-tube version unless otherwise indicated. It should be clearly understood that both the *rectifier* and the *diode* rectify alternating current into direct current.

The construction of a rectifier is shown in Figure 4-2. Basically it consists of a *filament* which when heated emits electrons and a metal which collects the electrons when it is positively charged. These parts are mounted in a glass stem, which acts

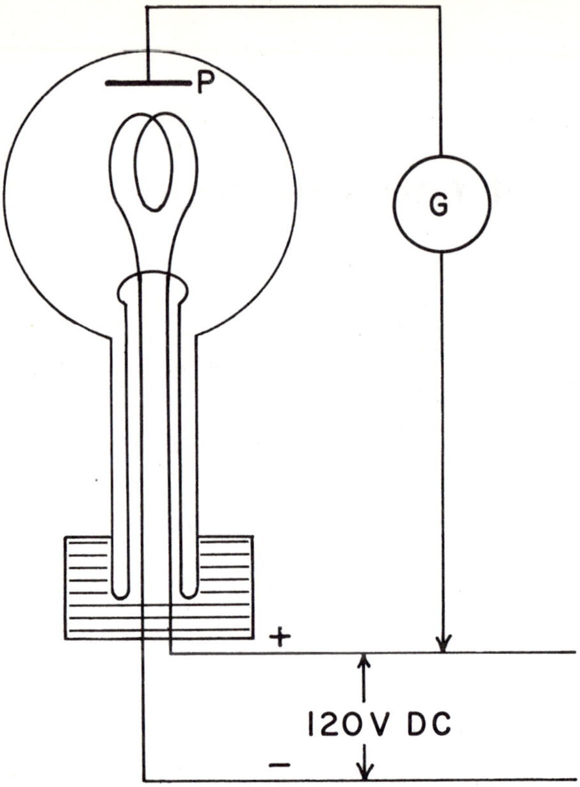

Figure 4-1. Edison's lamp. (Courtesy of R. W. Kenworthy, College Physics, F. A. Davis Co.)

as a support, and the whole unit is enclosed in a glass envelope. A small piece of metal, called a getter, is placed inside the envelope. The glass envelope is then evacuated and sealed. When the getter is heated, it absorbs most of the few remaining molecules, thus improving the vacuum. As the getter is heated, it evaporates, leaving a silvered mirror on the inside of the tube. This can easily be seen on every vacuum tube.

THE SOURCE OF ELECTRONS—THERMIONIC EMISSION

When a wire is heated, more kinetic energy is imparted to the electrons which are free to travel from atom to atom. In other words, these free electrons have greater energy when they collide with electrons in the outer shells of atoms. When these electrons reach the surface of the metal, they may have enough energy to leave the metal. The higher the temperature of the metal the more likely this is to occur. The amount of work that it takes to release an electron from the surface is related to the metal itself, and is called the *work function* of the metal. Some metals such as tungsten have a lower work function than others and hence are used for this purpose. These metals are often coated with an oxide of thorium or an oxide of alkaline earth (barium oxide) for even better electron emissivity.

The electrons which have left the surface of the metal form a *cloud* near the surface. This cloud, of course, has a negative charge which tends to repel the electrons leaving the metal. Some of these electrons re-enter the metal. This *electron cloud* is free to be influenced by any existing electrostatic field.

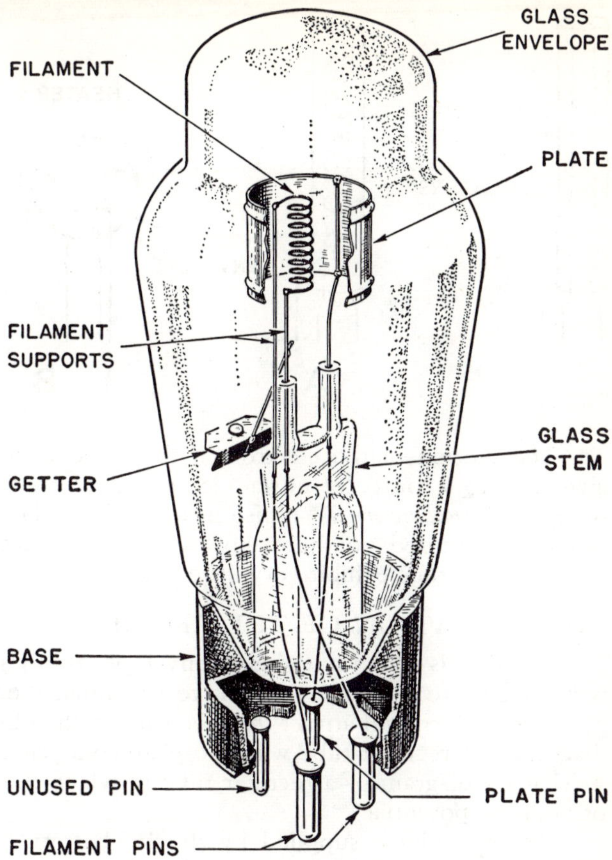

Figure 4-2. A vacuum-tube rectifier. (Reproduced by permission of the Department of the Army.)

Figure 4-2 shows one type of filament, the coiled filament. The filament also may be in the shape of an *inverted* V or an *inverted* W. The current supplied to the *high-resistance* filament wire heats the wire in the same way as a lamp or a toaster, that is, the heat loss is proportional to the product of the square of the current and the resistance (I^2R). When the filament is supplied by alternating current, current in the filament fluctuates enough to vary the emission as a function of the frequency of the current through the filament. In ordinary use, it is convenient to heat the filament with 60-cps house current as it is inexpensive and readily available. If a 1-cps alternating current signal is to be rectified into pulsating direct current, using a rectifier whose filament is heated with 60-cps house current, then there will be a tendency to find 60-cps *ripple* on the rectified wave. To avoid this the filament is surrounded by a metal shield which has a low work function and is properly coated to emit electrons easily. The energy radiated from the filament is then used to heat the metal sleeve, called the *cathode*. This arrangement is shown diagrammatically in Figure 4-3. The *cathode*, once it is heated, does not cool in concert with the filament, and hence it can maintain a relatively constant *electron cloud* and still be heated by 60-cps alternating current. In addition, the cathode is insulated from the heater and its entire surface can be maintained at a potential which may differ from that of the filament.

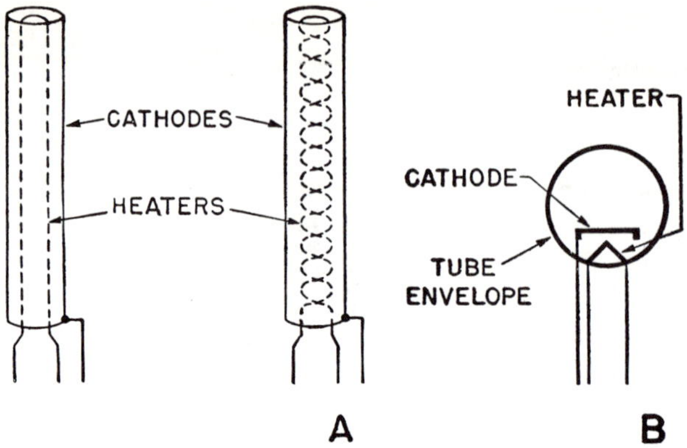

Figure 4-3. Indirectly heated emitter and schematic symbol. (Reproduced by permission of the Department of the Army.)

The motion of the electron in the cathode, the filament, and the tube itself gives rise to a source of unwanted signal, referred to as *noise.* This type of noise is called *thermionic noise.** Noise in general will be discussed in a later chapter. It should be pointed out here, however, that semiconductor devices do not need filaments and thus exhibit very little if any thermionic noise.

THE OPERATION OF THE RECTIFIER

If a plate is added to the glass envelope, as in Figures 4-1 and 4-2, and the plate is connected to a source of positive potential, then an electrostatic field will be generated which will attract the electrons in the electron cloud. Figure 4-4(A) is a diagram of a rectifier tube with the plate connected to a positive potential; Figure 4-4(B) is a diagram of a rectifier tube in which the plate is connected to a source of negative potential.

As the voltage supplied by the B^+ battery (plate circuit) is increased, the current from the cathode to the plate increases. The relationship between *plate current* and *plate voltage* is characteristic of the particular type of rectifier tube. Such a curve is called the *static characteristic curve.* The curve will also vary with the temperature of the cathode. Figure 4-5 shows a *family* of curves for a common rectifier.

*See page 309.

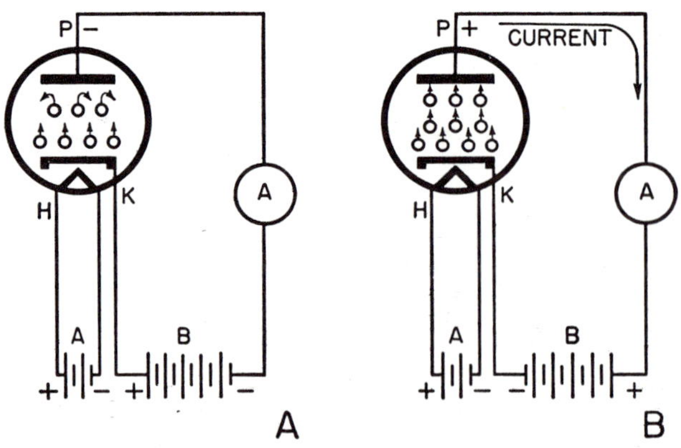

Figure 4-4. Different conditions of current flow through a vacuum-tube rectifier. (Reproduced by permission of the Department of the Army.)

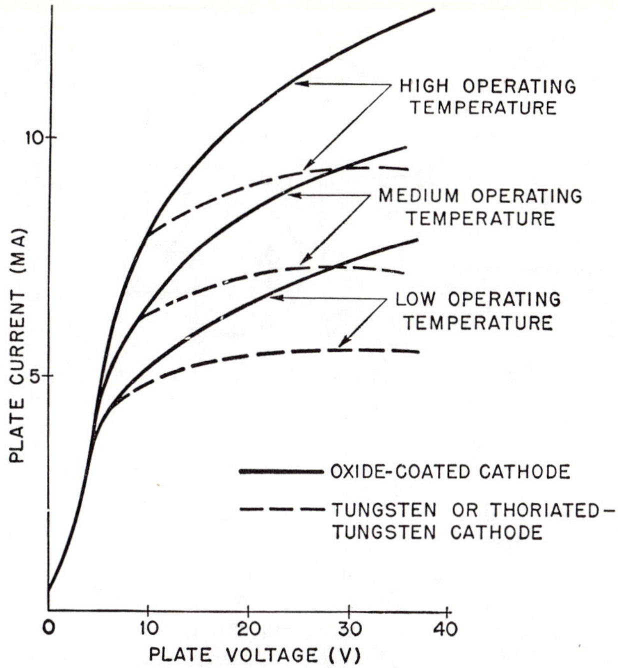

Figure 4-5. Characteristic curves for a typical vacuum-tube rectifier. (Reproduced by permission of the Department of the Army.)

A rectifier tube can conduct current in only one direction. Therefore, if a rectifier tube is used as an element in a circuit, such as the one shown in Figure 4-6, any portion of the input voltage which makes the plate negative with respect to the cathode will stop the conduction of current through the tube. If the input current is sinusoidal, as in Figure 4-6, then the current through the resistor (called the *load resistor*) will flow during only half of each cycle. This is called *half-wave rectification.*

Let us examine the circuit shown in Figure 4-6 another way. The rectifier tube, when used as a circuit element, has *two* resistance values, one, when the plate is positive with respect to the cathode and, the other, when the plate is negative. When the plate is positive, the resistance is called *forward resistance.* When the plate is negative, it is called *back resistance.* The *back resistance* is very much higher than the forward resistance. It will be remembered that the sum of the voltage drop across two resistors in series equals the source voltage. Thus, if the back resistance is very high, almost all of the voltage drop will appear across the tube and almost none

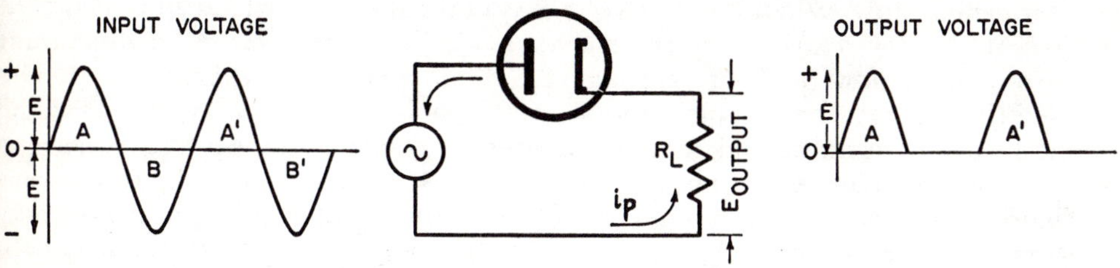

Figure 4-6. Diagram illustrating unidirectional current flow in a vacuum-tube rectifier. (Reproduced by permission of the Department of the Army.)

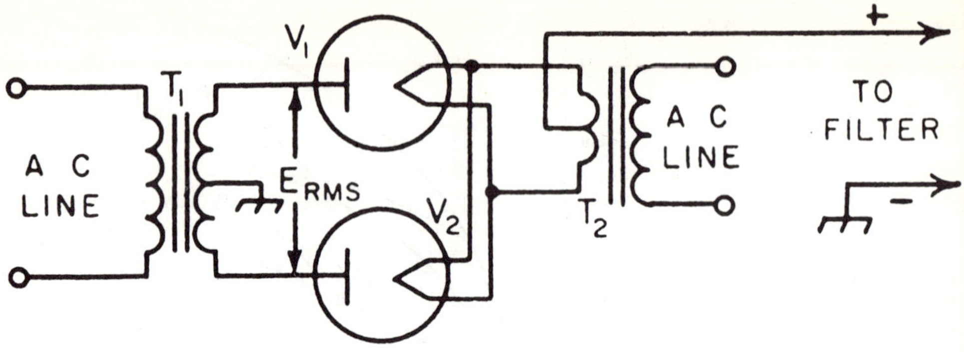

Figure 4-7. A full-wave rectifier. (Courtesy of Radio Amateur's Handbook.)

across the resistor. The source voltage is almost all across the load resistance during conduction of the tube. A vacuum-tube rectifier has a much higher *back resistance* than does a semiconductor diode.

Figure 4-7 shows a circuit in which two rectifiers are used as a *full-wave rectifier.* In this case the alternating current is more efficiently converted to pulsating direct current. Although the output voltage is the same, the frequency of the output from the full-wave rectifier is twice that from the half-wave rectifier, and it is therefore easier to *filter* (convert to smooth direct current with little or no ripple). It is necessary to use a *center-tapped* transformer to supply the AC input wave to the two rectifier tubes in Figure 4-7. It is also necessary to use a center-tapped transformer to supply the filaments. Only half of each transformer and one rectifier tube operate at any given instant. When the input signal is such that the plate of V_1 is positive, then the plate of V_2 is negative. At this instant V_1 will conduct and V_2 will not. When V_2 is positive with respect to its filament, it will conduct. The result is that half of each cycle is contributed by each tube. Two rectifier elements may be combined into one glass envelope, which is called a *dual diode.* The voltage output may be doubled by using four rectifier elements instead of two. This is called a *full-wave bridge rectifier.*

THE SEMICONDUCTOR DIODE

The semiconductor diode consists of a layer of P-type material and a layer of N-type material. The region between the P-type and N-type materials is called the *junction.* Figure 4-8(A) shows the diode *biased* with a battery in such a way as to allow the flow of current, whereas in Figure 4-8(B), for all practical purposes, current cannot flow. When the bias voltage is in the direction which allows current to flow, it is called *forward bias;* the converse is called *reverse bias.* A small amount of current does flow backward through the diode when it is reverse biased. The characteristic curve for a typical diode is shown in Figure 4-9. Notice that the forward current is measured in milliamperes (ma), while the reverse current is measured in microamperes (μa). If the reverse-bias voltage supplied by the battery in Figure 4-8(B) is sufficiently high, the free electrons will acquire enough energy to irreversibly break down the junction, that is, unless some arrangement is provided to limit the flow of current in the circuit. The voltage at which breakdown occurs is called the *avalanche voltage* of the diode (see page 114).

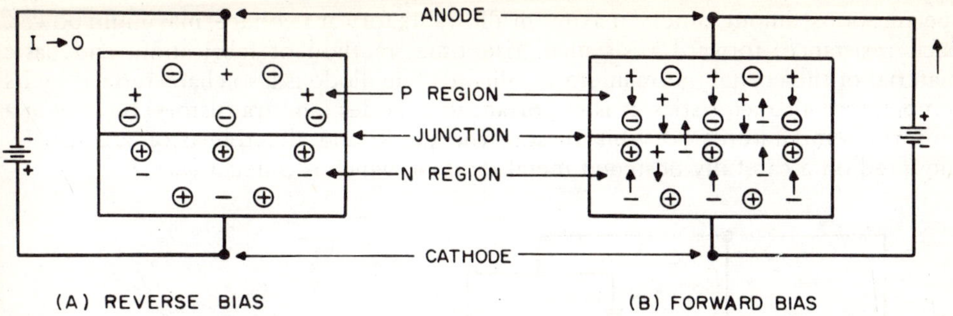

Figure 4-8. Conduction in a PN-junction diode. (Courtesy of General Electric Co., Syracuse, N.Y.)

The solid state diode does not need a heater or a filament, and hence it does not need a separate power supply to heat the cathode. One of the most frequent failures of the vacuum-tube rectifier is the breaking of the filament. This has a tendency to occur because as the filament is turned on and off, the metal is heated and cooled rapidly; eventually the metal becomes fatigued and fails. The filament of the vacuum-tube rectifier must be turned on long enough to get hot before it will function. The semiconductor device functions immediately as it has no filament.

The diode has been given the symbol —⊳|—. The arrow is the anode and the line is the cathode. Current, in the conventional sense, flows in the direction of the arrow with less resistance than in the reverse direction. (Electrons, however, flow in the other direction.) Since a diode has one junction, diodes are classified with the prefix 1N followed by a number which is assigned, in order, to diodes as they are designed (e.g., 1N34A or 1N626). Diodes are characterized by the following

Figure 4-9. Chart of current through and voltage across a PN junction. (Reproduced by permission of the Department of the Army.)

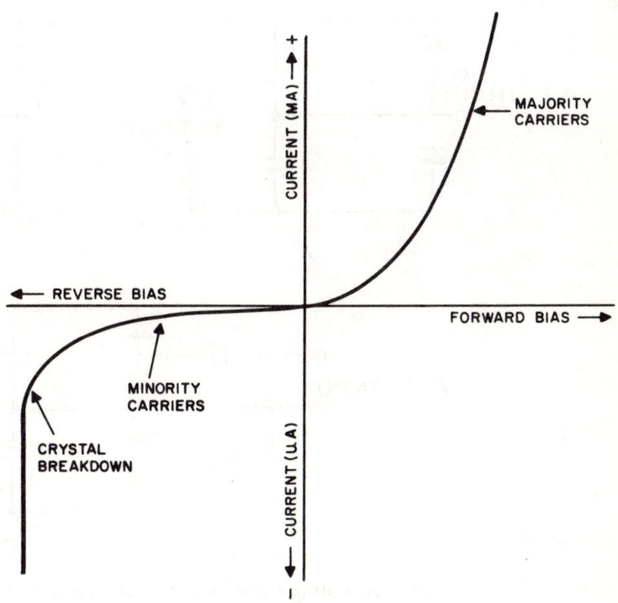

specifications, among others: maximum allowable forward voltage, maximum power, back resistance, forward resistance, rise time, method of fabrication, and basic material of the crystal (germanium or silicon). The diode is also characterized by its temperature characteristics. It is important that diodes (and transistors) be kept at a uniform temperature. For this reason, the diode and the transistor are usually mounted on a carefully designed metal radiator, which is called a *heat sink.*

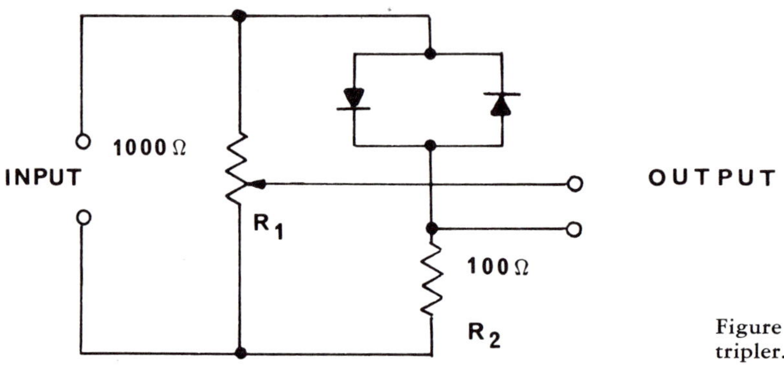

Figure 4-10. Frequency tripler.

Some Applications of the Solid State Diode

The diode can be used in place of a vacuum-tube rectifier in the circuits shown in Figures 4-6 and 4-7. There are many other combinations of diodes which are extremely useful and which the electronicist should keep in mind. Full-wave rectification yields a signal which, although it is not a sine wave, has twice the frequency of the output of the half-wave rectifier (called a frequency doubler).

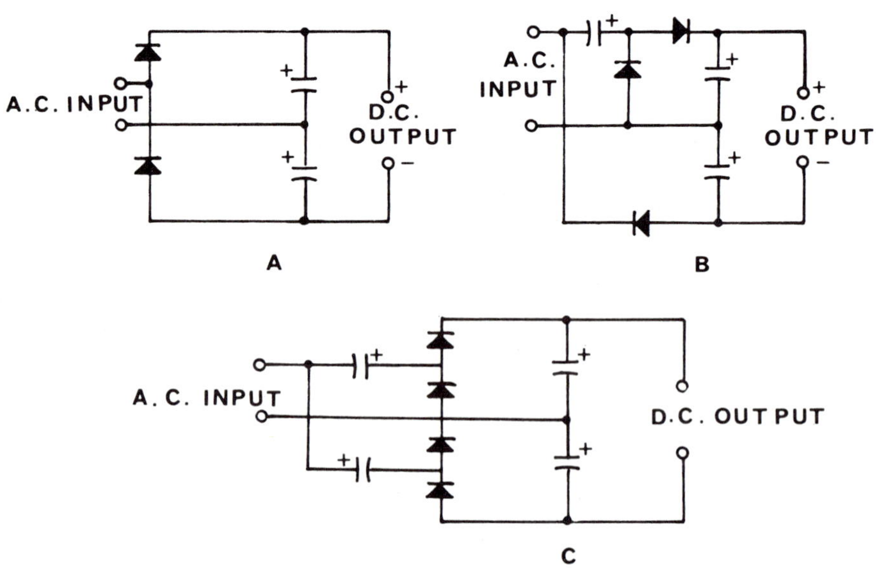

Figure 4-11. A, voltage doubler; B, voltage tripler; C, voltage quadrupler.

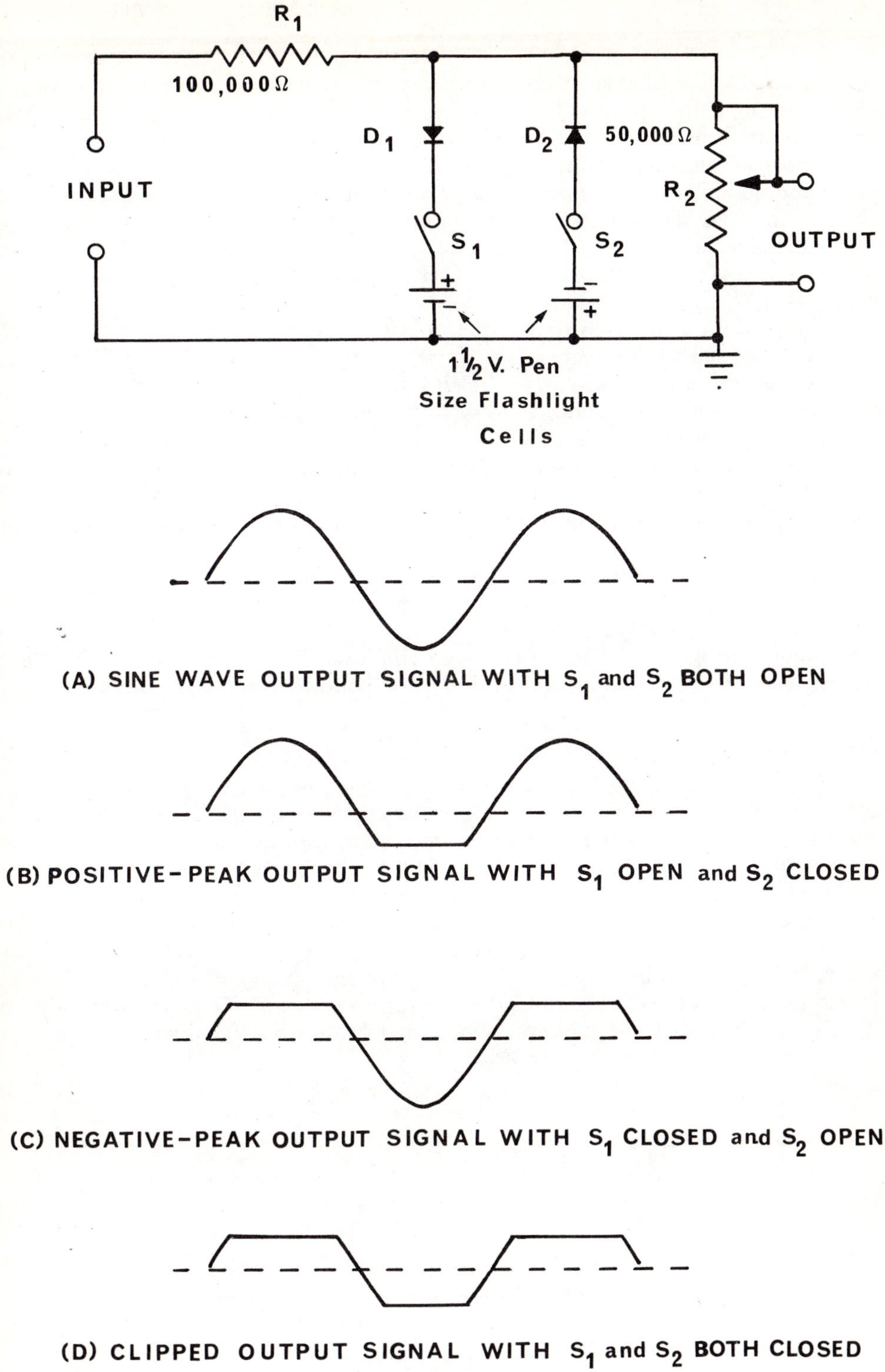

Figure 4-12. A clipping circuit.

A sine-wave signal is half-wave rectified so that only the positive peaks appear at the output. The same sine-wave signal may be treated so that only the negative peaks appear at the output. The two resulting sets of half-wave rectified signals can be treated in two ways: (1) they could be added with no change in phase to yield a normal full-wave rectified wave, or (2) one set of peaks could be shifted 90° in phase and added to the original signal. The latter case would result, after suitable filtering, in a DC signal with twice the amplitude as that from the filtered full-wave rectifier. Such a circuit is called a *voltage doubler* and is shown in Figure 4-11(A). A *voltage tripler* is shown in Figure 4-11(B), and a *voltage quadrupler* is shown in Figure 4-11(C).

The diode may be used to *clip* or to *limit* the signal on the top or the bottom or both, as is shown in Figure 4-12. If the battery in the S_1 side is 9.0 volts, the diode is *forward biased* and current readily flows to ground. When the sine wave is less than 9 volts, the diode is *reverse biased* and appears as a high resistance to ground, thus allowing the signal to pass directly to the output. In this way, the sine wave is limited to a maximum amplitude of 9 volts. The bottom of the input can be clipped in the same way by reversing the diode and the battery. The 100K Ω resistor must be in the circuit to prevent a short circuit to ground when the diodes are forward biased and are passing the signal to ground.

There are many other uses for diodes in electronic circuitry. The student who knows the properties of electronic components and elements can manipulate voltages, currents, and wave shapes to suit his needs. Several kinds of diodes with special applications have evolved and are being used in modern equipment. These are the *zener* diode, the *tunnel* diode, the silicon-controlled rectifier, and others.

Zener Diode

When the diode was first discovered, some of the first diodes showed a tendency to have zero resistance at a particular reverse voltage. These diodes were discarded, until a use was found for them. Let us assume that a particular diode breaks down (has zero reverse resistance) when there is a 12-volt drop across the diode. This diode, called a *zener* diode, is used, for example, to regulate the voltage in a circuit such as the one shown in Figure 4-13. The basic voltage supplied to the circuit may be 20 volts, but the voltage across the zener diode and the load resistor will remain at 12 volts, even if the supply voltage should increase or decrease. This is possible because as soon as the 12-volt breakdown voltage is exceeded, the current going through the junction is increased, but the voltage across the junction

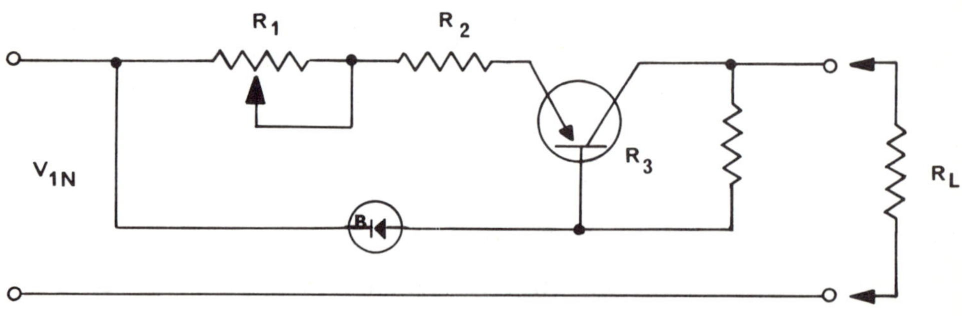

Figure 4-13. Simple zener regulating circuit.

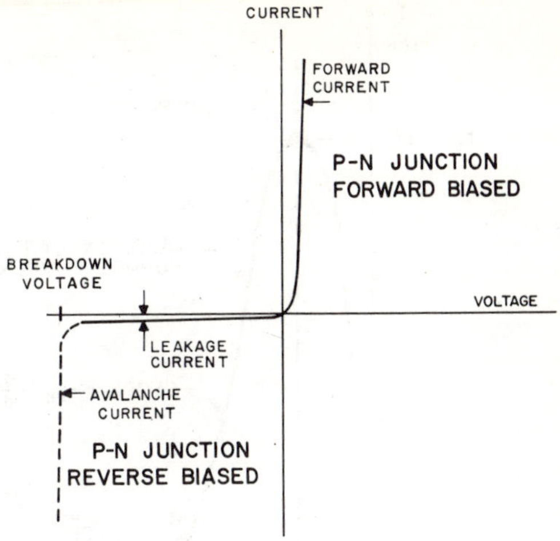

Figure 4-14. General characteristic curve. (Courtesy of General Electric Co., Syracuse, N.Y.)

remains essentially constant at 12 volts. Figure 4-14 shows the characteristic curve for a typical zener diode.

Many uses have been developed for the zener diode. It can be used to regulate and stabilize a voltage. It can be used to protect a meter or to provide a fixed-bias voltage for a transistor or tube element. The zener diode may be used in a circuit to replace a battery. The zener diode also is extremely useful as a wave-shaping device. Two zener diodes connected back to back (anode to anode) from a source of voltage to ground will *clip* the top and bottom (as does the circuit shown in Figure 4-12, but the batteries will not be necessary).

Tunnel Diode

If the amount of impurity is increased in an otherwise conventional PN diode and the junction is made very thin, the resulting semiconductor device is known as a *tunnel* diode. Its name is derived from the fact that because of the high conductivity of the material and the thin junction, electrons are able to cross the junction without the amount of energy that would be expected from classical physical theory. The electron appears to *tunnel* through the energy barrier.

Current will begin to flow in the tunnel diode at a much lower forward voltage than it will in a conventional diode. But as the energy of the electrons is increased, by raising the forward-bias voltage, the current across the junction first reaches a *peak* and then decreases to a *valley point* before beginning again to increase. The explanation for this phenomenon is found in quantum mechanics. In order for an electron in the N region to cross the junction and enter a *hole* in the *P* region, it must have an energy which falls within a certain specified band of energies. Electrons with other energy levels are, in the language of quantum mechanics, *forbidden* to enter these holes. Hence, as the forward bias is increased, the energy of the electrons in the N region increases, first into a region which allows the electrons to cross the barrier and enter holes and so the current increases, then into a forbidden region and the current decreases, then the current again begins to climb till the energy has exceeded the barrier. Here the tunneling effect is no longer necessary, and current increases with increasing forward bias until the avalanche voltage is exceeded. A typical characteristic curve for a tunnel diode is shown in Figure 4-15.

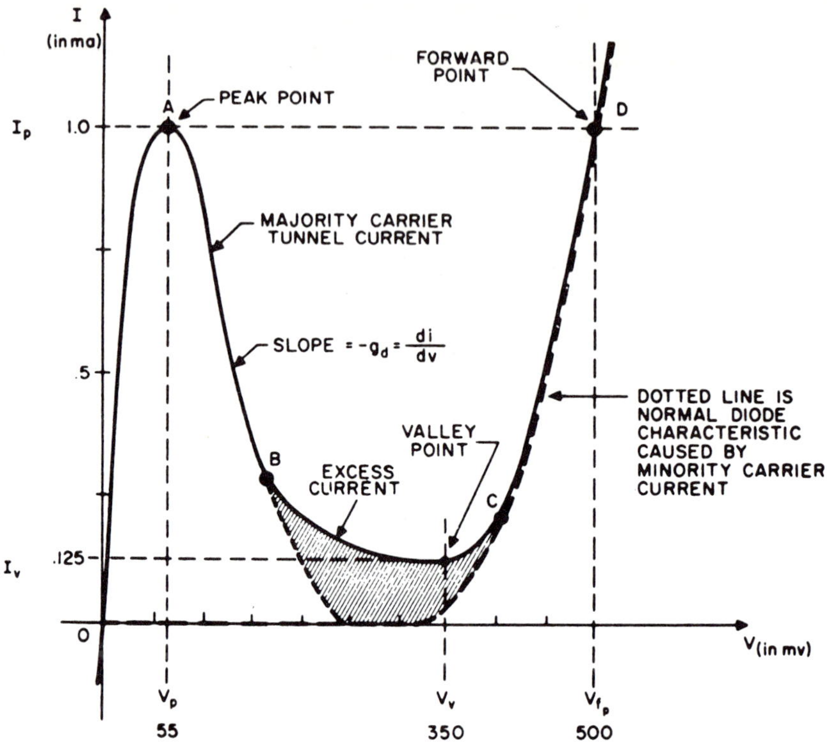

Figure 4-15. Tunnel-diode characteristic curve. (Courtesy of General Electric Co., Syracuse, N.Y.)

According to Ohm's law, when the voltage across a resistive material, such as the N and P materials in a diode, is increased the current should increase. Examination of Figure 4-15 reveals that this does not occur between the *peak* and the *valley.* Since the current is decreasing with increasing voltage, the resistance of the tunnel diode is said to be *negative.*

It is this property of the tunnel diode that is the basis for its application in electronics. Amplifiers and oscillators can be designed with tunnel diodes. In such designs many components may be eliminated, thus simplifying the circuit and making its size much smaller and possibly even less expensive.

The student interested in reading more about the tunnel diode and other new semiconductor devices is encouraged to consult the references.*

REFERENCES:

Basic Theory and Application of Electron Tubes (Department of the Army, TM 11-662 to 16-1-255, 1952).

Basic Theory and Application of Transistors (Department of the Army, TM 11-690, 1959).

*The student of medical instrumentation should be familiar with the gallium arsenide infrared-emitting diode and its companion the PIN photodiode (see page 201).

G. E. Rectifier Guide, Gutzwiller, F. W., et al. (General Electric Co., 1961).

G. E. Transistor Manual, 6th Ed. (General Electric Co., 1962).

G. E. Tunnel Diode Manual (General Electric Co., 1961).

International Rectifier Corporation Engineering Handbook, J. T. Cataldo, Ed. (International Rectifier Corporation, 1959).

Radio Amateur's Handbook, revised annually (American Radio Relay League, Newington, Conn.).

5

Amplification

the vacuum-tube triode
the phase inverter
push-pull operation
the cathode follower
multiple stages and examples of circuits
feedback
operational amplifier
the transistor
the transistor as an amplifier
multiple stages
biasing a transistor circuit
coupling transistor stages
examples of transistor amplifier circuits
the power amplifier

THE VACUUM-TUBE TRIODE

In 1907, Lee De Forest described the use of a third element in the vacuum tube. This third element, called a control grid, entirely changed the character of the infant science of electronic engineering and began a new era.

The control grid is placed between the cathode and the plate. When the plate is wired to the cathode, through appropriate circuitry, to form the plate circuit, a current will flow from the plate to the cathode through a transformer or a resistor (often called a *load*). When the *control grid* is made negative with respect to the cathode, the electrons passing near the control grid will experience a decelerating force. The result is that fewer electrons will reach the plate, and hence the plate current will be decreased.

As the potential difference between the control grid and the cathode, called the *grid bias*, is made more negative, fewer electrons reach the plate (Fig. 5-1). As the potential of the control grid becomes more positive with respect to the cathode (or less negative), more electrons reach the plate and the plate current is increased. In this case, however, some current may flow from the grid to the cathode. This is called *grid current* and is to be avoided in most applications of the vacuum-tube amplifier as it increases the consumption of power. Figure 5-2 is an illustration which shows the effective physical arrangements of the cathode, the control grid, and the plate in the triode. Figure 5-3 shows the vacuum-tube triode and its parts as generally shown in circuit diagrams.

When an alternating voltage such as that shown in Figure 5-4(A) is applied to the vacuum-tube amplifier across the input resistor shown in Figure 5-3, and the grid is biased (battery C) with −6 volts relative to the cathode (Fig. 5-4[B]), the

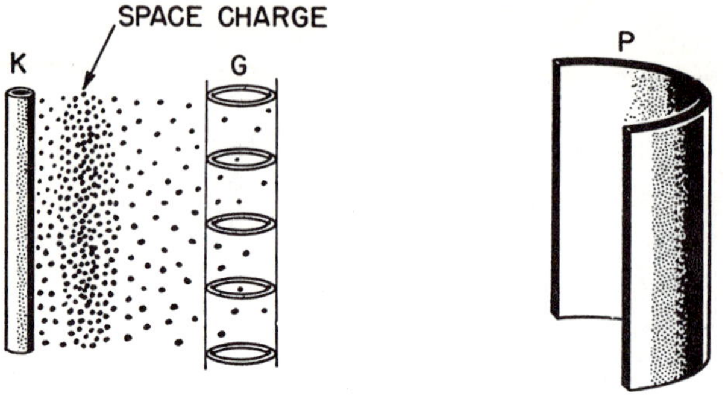

Figure 5-1. Imaginary view of electrodes in a triode. (Reproduced by permission of the Department of the Army.)

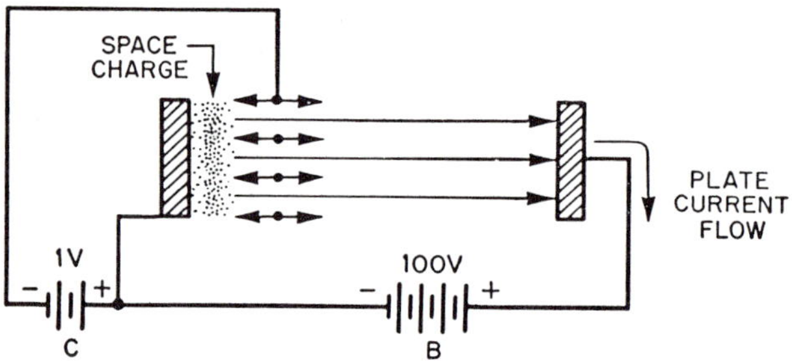

Figure 5-2. Electron flow in a triode with the control grid negative to the cathode. (Reproduced by permission of the Department of the Army.)

grid-cathode voltage and the plate current output are as shown in Figures 5-4(C) and 5-4(D) respectively. Notice the phase relations (that is, the relative position of a curve with respect to another along the abscissa) among the input signal, the plate current, and the grid voltage.

It is clear that when the grid voltage changes, the current in the plate circuit must also change. A triode, like other components of a circuit, must obey Ohm's law. If the plate voltage is constant and its current is changed, it is clear that the apparent resistance of the tube element must have changed as a function of the grid voltage. This is referred to as the *plate resistance* (r_p). Do not, however, confuse the plate resistance with the *load* resistance identified as *output* resistance in Figure 5-3.

The reciprocal of resistance is called *conductance* and is measured in terms of the unit, the mho (ohm spelled backward). The conductance of a vacuum tube is often called the *transconductance* and is an important parameter in determining the properties of a given vacuum tube. The transconductance of a triode is the quotient of a small change in plate current divided by the grid voltage change responsible for the plate voltage change, assuming that all other tube parameters are held constant.

The graph which appears in Figure 5-5 illustrates the change in plate current as a function of the grid voltage for one section of a typical triode when the plate voltage and the size of the load resistor are constant. This graph is one of the *transfer-characteristic curve* of the vacuum-tube triode and is an important tool in circuit design. Figure 5-6 is the *plate-characteristic curve* formed by plotting the plate current as a function of the plate voltage. In this set of curves, the grid voltage appears as a parameter of each curve. When the grid voltage reaches a certain negative value, depending on the tube type, the plate will receive only a minimum

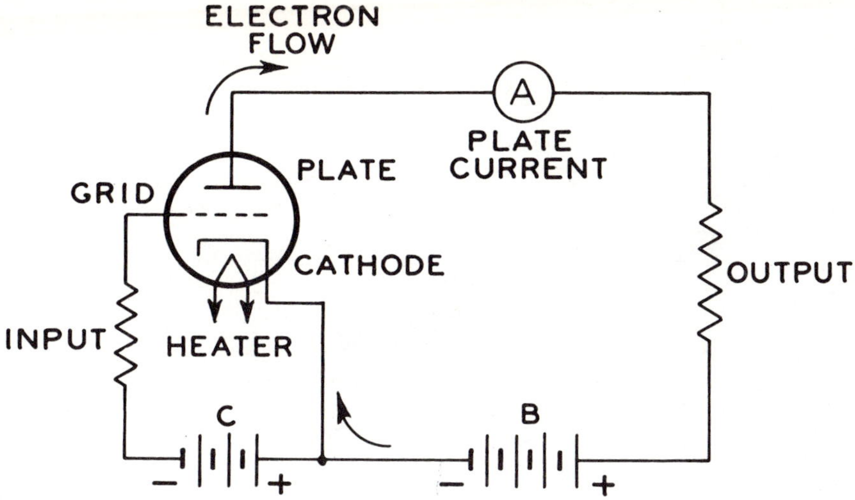

Figure 5-3. A vacuum-tube triode amplifier. (Courtesy of Radio Corporation of America, Harrison, N.J.)

of electrons. The grid voltage at which this occurs is called the *plate cut-off voltage.* Characteristic curves for almost every variety of vacuum tube may be found in one or more of the tube manuals listed at the end of this chapter. Characteristic curves of some of the most popularly used tubes can be found in the appendix. A special oscilloscope has been developed by the Tektronix Company which will present on its face any of the characteristic curves for any tube using almost any desired circuit parameters. Thus, the designer may *breadboard,* that is, put the circuit together on the top of a chassis in a temporary fashion, and see a family of characteristic curves on the oscilloscope face.

When sinusoidal voltage is applied to the grid, the current through the load resistor or load transformer will be the same as the plate current and will vary as illustrated in Figures 5-4 through 5-6. The change in current in the load resistor will generate a change in voltage across the load. The amplification or *voltage gain* (μ) is thus:

$$\mu = \frac{\text{change in voltage across load}}{\text{change in grid voltage}} \quad 5.1$$

Figure 5-7 illustrates the change in amplification (μ) as well as the plate resistance (r_p) and the transconductance (g_m) as a function of the grid volts (I_b). Thus, if a given voltage is required for a particular application, the necessary plate current and grid bias can be easily determined from the dynamic characteristic curves published by the manufacturers.

The plate current passing through the output or load resistor (Fig. 5-3) develops a voltage drop across it. This voltage drop varies directly with current through the resistor and is in phase with it. The voltage measured between the plate and the cathode is, however, 180° out of phase with the voltage measured across the output or load resistor. The current through the resistor is equal to the plate current. The voltage amplification or gain is the change in the voltage drop across the load resistor

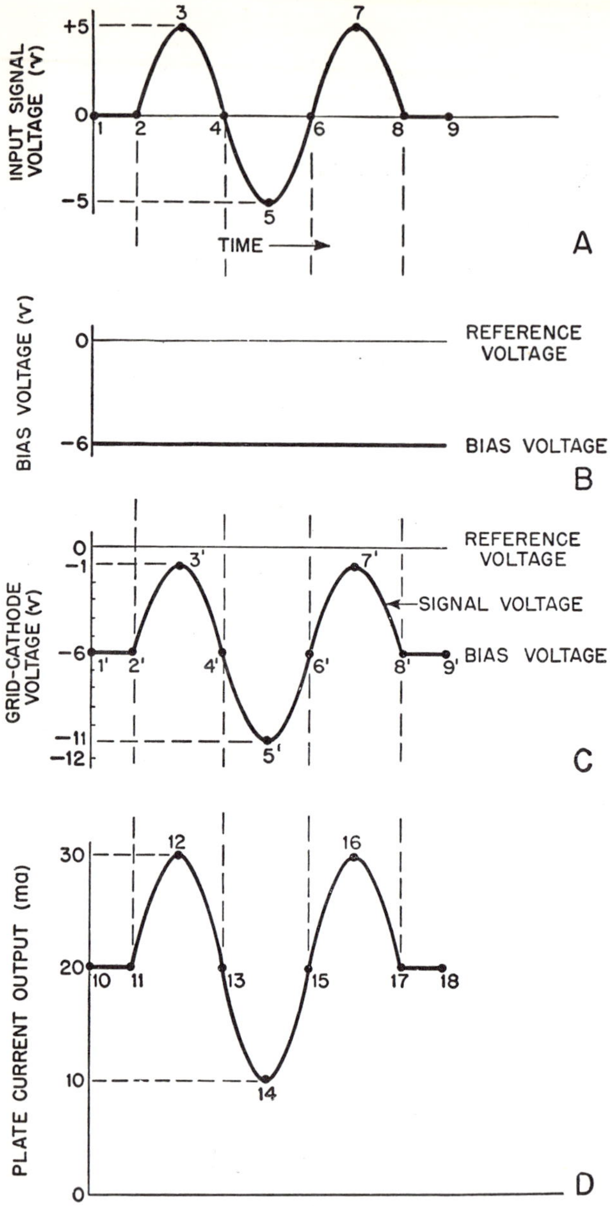

Figure 5-4. Wave shapes illustrating bias, signal voltage, and plate current. (Reproduced by permission of the Department of the Army.)

per unit of input signal voltage. The equation for the gain may be expressed in two ways:

$$\text{voltage gain} = \frac{\mu \times R_L}{R_L + r_p} \tag{5.2}$$

or:

$$\text{voltage gain} = \frac{g_m \times r_p \times R_L}{1{,}000{,}000 \times (r_p + R_L)} \quad 5.3$$

where: μ is the amplification factor, r_p is the resistance of the plate, R_L is the load resistance, and g_m is the transconductance in micromhos.

The useful gain of the tube is actually somewhat less than the amplification factor because of the presence of the load resistor. This is especially true when the load resistance is low compared to the plate resistance. The voltage gain approaches the amplification factor as the load resistance gets very high (Fig. 5-8).

Figure 5-5. Exploded view of a typical transfer-characteristic curve. (Courtesy of Radio Corporation of America, Harrison, N.J.)

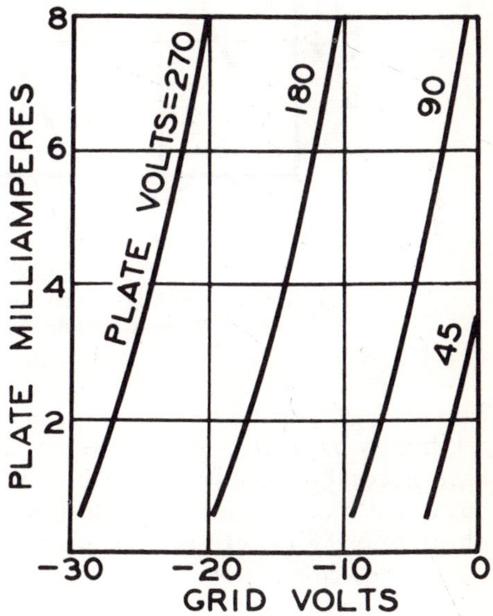

It is sometimes useful to calculate the *plate efficiency* of a particular vacuum-tube triode. The plate efficiency is the ratio of the power output (AC voltage across the load times the AC current through the load resistor) to the product of the average DC plate voltage and the DC plate current, when operating at its highest input voltage.

$$\text{Plate efficiency } (\%) = \frac{P_o}{E_p \times I_p} \quad 5.4$$

where: P_o is the AC output power, E_p is the DC plate voltage, and I_p is the plate current.

If, in addition to the sinusoidal grid signal shown in Figure 5-8, a source of DC voltage such as a battery is included in the grid circuit, the DC voltage of the battery can be varied to cause the input signal to oscillate around any grid voltage. This DC voltage is referred to as the *bias voltage.*

The grid bias may be supplied by a battery or a zener diode, in which case it is called a *fixed bias*. Figure 5-8 is an example of fixed bias. The grid bias may, however, be supplied by a voltage developed across a resistance as a result of the plate current.

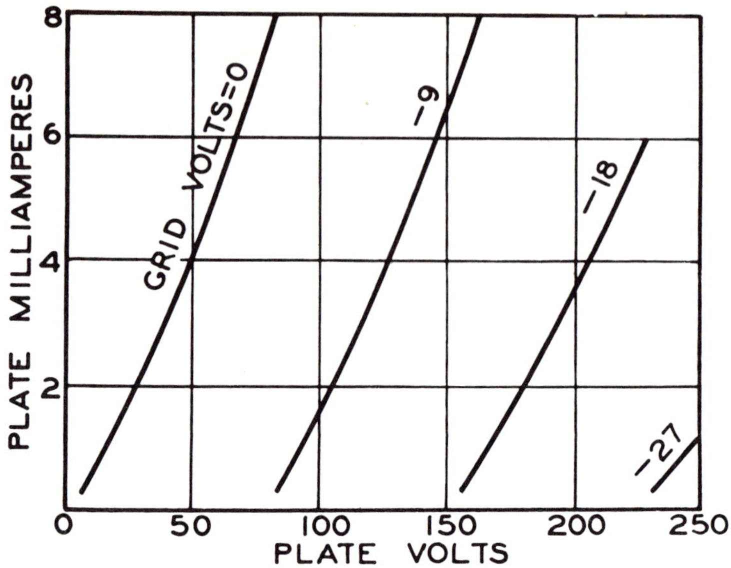

Figure 5-6. Exploded view of a typical plate-characteristic curve. (Courtesy of Radio Corporation of America, Harrison, N.J.)

This latter type of bias is called *self-bias.* Commonly a combination of fixed and self-bias is used. It is true, however, that in the last few years, since zener diodes have become available and relatively inexpensive, the zener diode is being used more and more. Figure 5-14 illustrates several ways of employing self-bias. In this illustration the load resistor is missing because it is a cathode-follower stage, which will be described later in this chapter.

When the bias voltage is adjusted so that the input grid signal falls on the center portion of the characteristic curve of a particular vacuum tube, that tube is being used in Class A operation. Notice that the center part of the curve is relatively linear while the upper and lower portions are non-linear. Figure 5-9(A) and 5-9(B) illustrate the tube when used as Class A and Class B amplifiers respectively. Notice that in each case the projected input signal falls on a different portion of the grid-voltage-versus-plate-current curve. It should also be noted that, in Class B operation, half of the grid signal is cut off and does not appear in the output. Class B amplifiers are operated so that the grid bias almost equals the plate cut-off voltage. They are used because the plate efficiency is only 20 to 30 per cent, and, as a result, smaller power supplies may be used. Classes AB_1, AB_2, and Class C operations are illustrated in Figure 5-10. Vacuum tubes in all classes except Class A are rarely used as amplifiers in medical or biological instruments because of their inherent high distortion.

THE PHASE INVERTER

A triode circuit constructed as in Figure 5-11 with a resistor in the plate circuit and a resistor of equal magnitude (in ohms) in the grid-cathode circuit is called a phase-inverter stage. A sinusoidal voltage signal applied to the grid may be shifted 180° in phase when the output voltage is measured between the plate and ground.

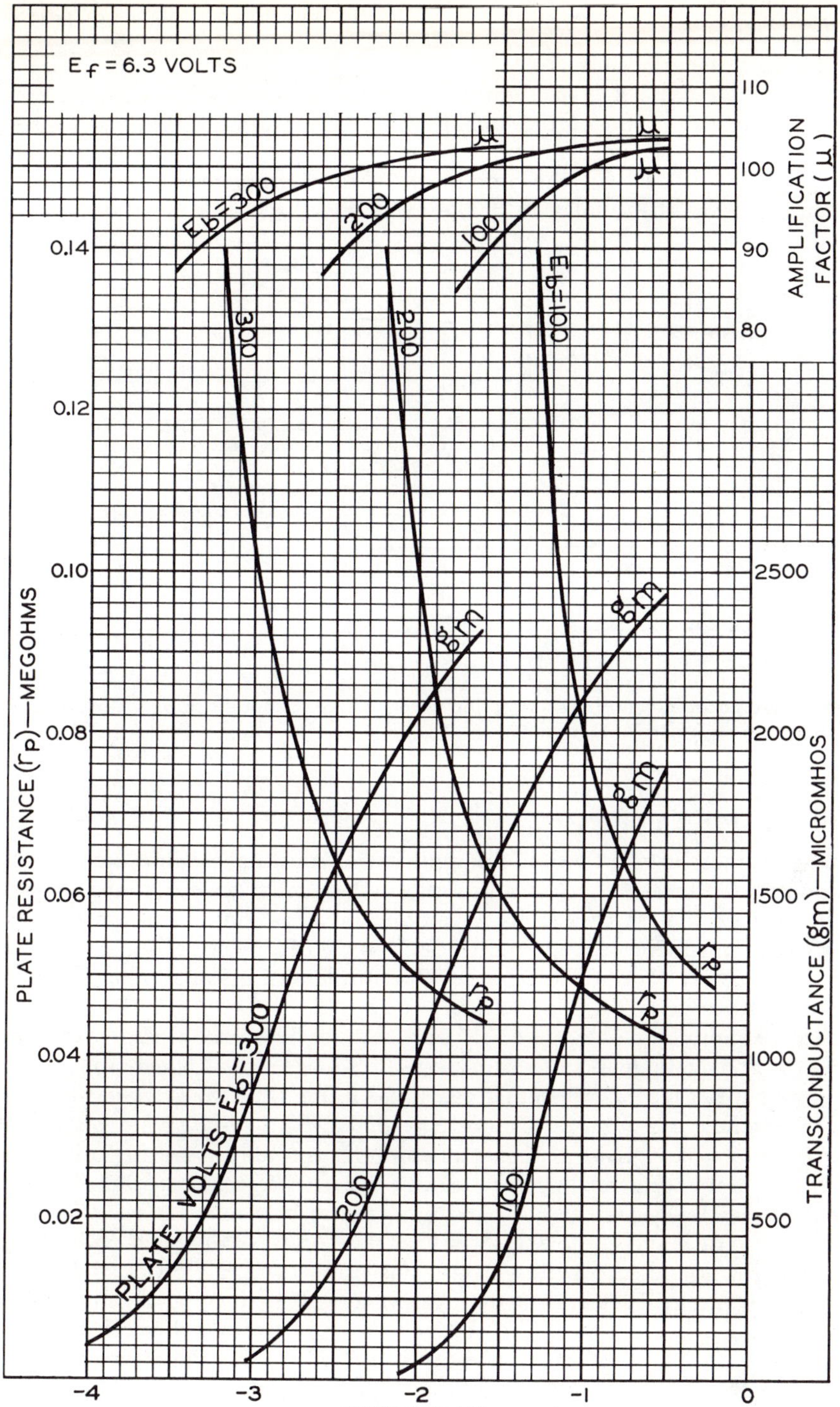

Figure 5-7. Plate resistance, transconductance, and amplification factor as a function of grid volts for the 12AX7 tube. (Courtesy of Radio Corporation of America, Harrison, N.J.)

Such a shift in phase amounts to an inversion of the input signal. It is sometimes necessary to *isolate* one stage in a multistage amplifier circuit from another stage without changing the phase of the signal. When the output voltage is measured across the resistor in the grid-cathode circuit, it will be in phase with the input voltage signal. In either case, the amplitude of the output will be equal, for all practical purposes, to the amplitude of the input signal.

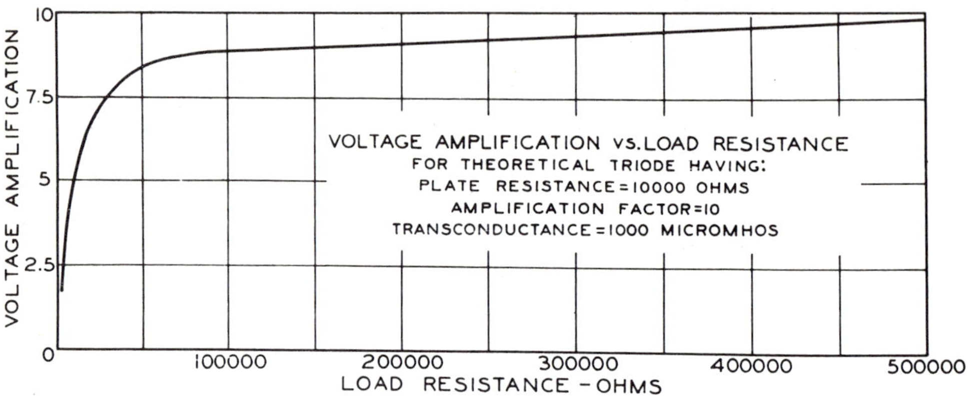

Figure 5-8. Voltage amplification as a function of the load or output resistance. (Courtesy of the Radio Corporation of America, Harrison, N.J.)

PUSH-PULL OPERATION

Many amplifiers used in biological applications must have minimum distortion and maximum amplification. In addition, the output must have a minimum of extraneous signal such as is caused by variations in thermionic emission or signals which originate from fluorescent lights or from a nearby x-ray machine. These extraneous signals are referred to as *noise.* The problem of noise elimination will be considered in more detail in Chapter 12.

Two triode amplifiers, having equal gain, may be arranged as shown in Figure 5-12. Such an arrangement is called a *push-pull* circuit. Signals which appear at the input *in phase* are rejected, whereas signals which are 180° *out of phase* are added. Since noise signals are random, they are rejected. The *in-phase* signals are also known as *common mode* signals. A push-pull amplifier is rated by the amount of common mode signal it can reject. A push-pull amplifier has a greater power output and a higher plate efficiency than two tubes connected in parallel.

Examination of Figure 5-12 will show that the output of the push-pull amplifier is the voltage developed across the output transformer, which is proportional to the current through the primary coil of the transformer. When the grid of the tube V_1 swings positive, the current from the plate of V_1 to its cathode through half of the primary of the output transformer increases, but when the signal applied to the grid of V_2 is *in phase* with the signal applied to the grid of V_1, then the grid of V_2 will swing negative and the current from the plate of V_2 to its cathode, through the other half of the primary coil of the output transformer, will decrease. In other words, when the signals are applied to the grids of V_1 and V_2, the increase in one is balanced by the decrease in the other. When, however, the signals are 180° *out of phase,* the

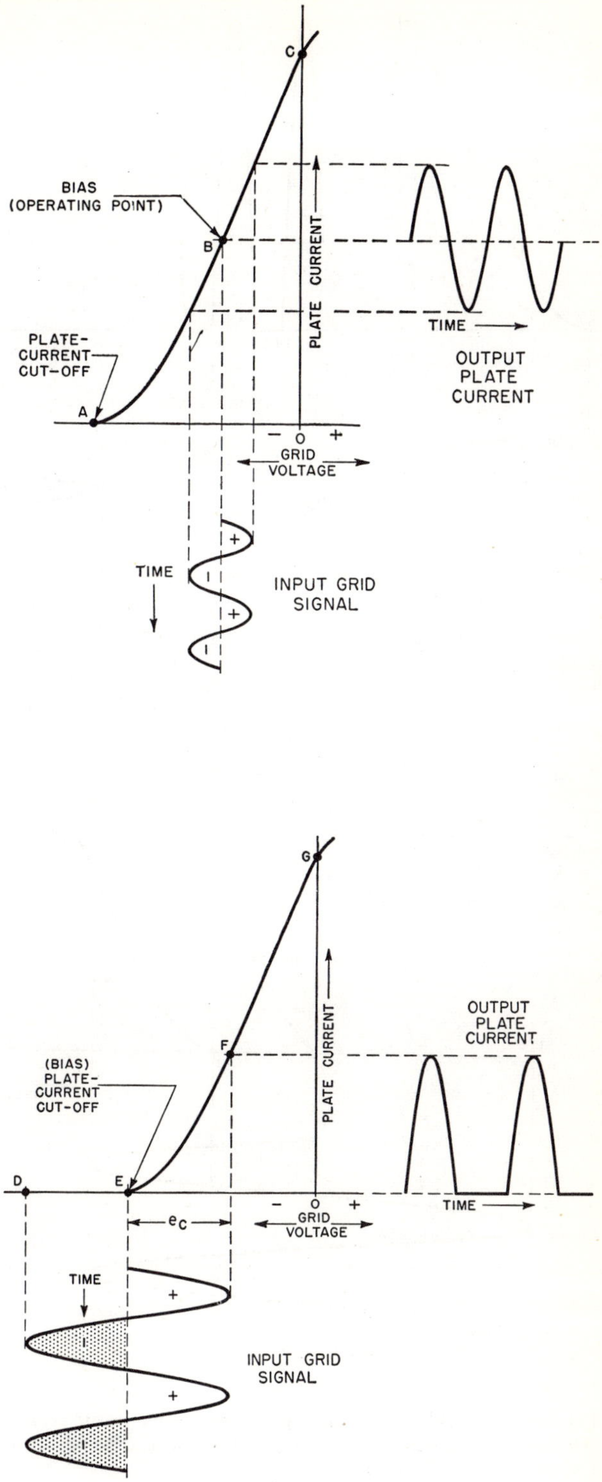

Figure 5-9. The relationship between the input signal and the output signal during Class A and Class B operation. (Reproduced by permission of the Department of the Army.)

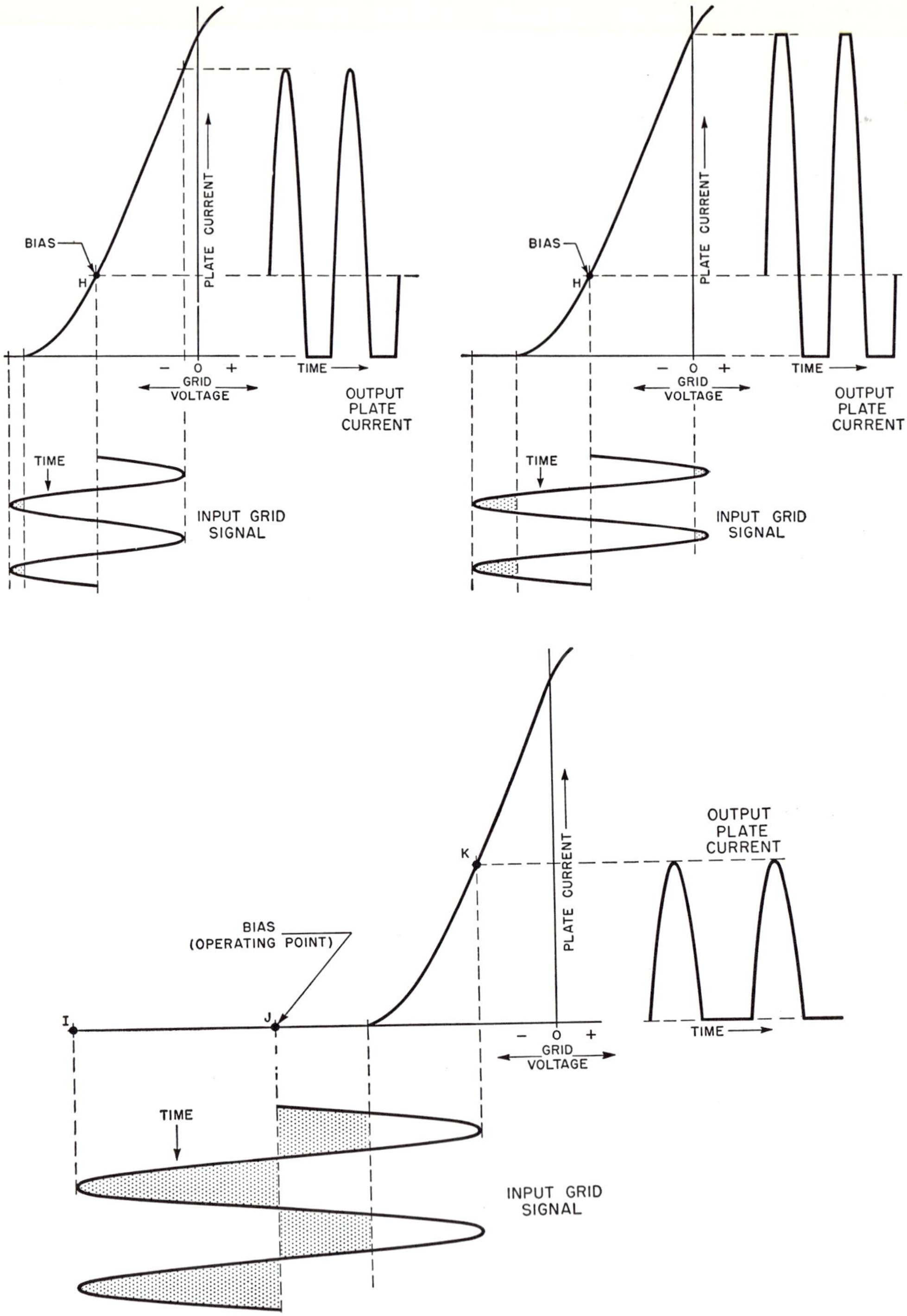

Figure 5-10. The relationship between the input signal and the output signal during Class AB_1 (top left), AB_2 (top right), and C (bottom) operation. (Reproduced by permission of the Department of the Army.)

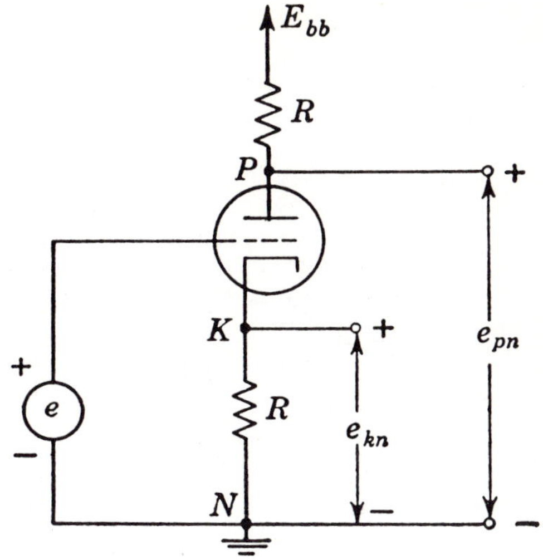

Figure 5-11. The phase inverter. (From Pulse and Digital Circuits, by J. Millman and H. Taub. Copyright © 1956. McGraw-Hill Book Company. Used by permission.)

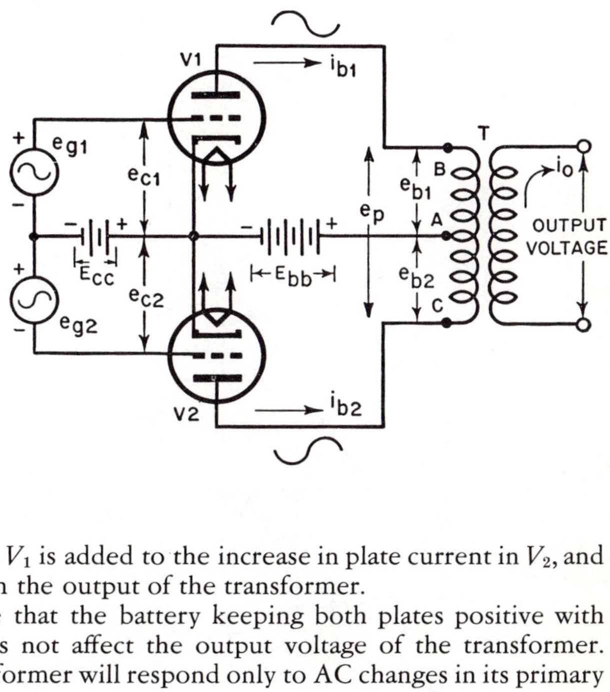

Figure 5-12. Push-pull triode amplifier circuit. (Reproduced by permission of the Department of the Army.)

NOTE: *E* and *e* in the above figures stand for voltage.

increase in the plate current in V_1 is added to the increase in plate current in V_2, and the result is a positive swing in the output of the transformer.

The reader should notice that the battery keeping both plates positive with respect to their cathodes does not affect the output voltage of the transformer. The output of the output transformer will respond only to AC changes in its primary coil. Therefore, the battery supplying the plate voltage does not affect the output. If the output of the push-pull amplifier in Figure 5-12 is used as the input to another amplifier, it would be said that the two amplifiers or stages of amplification were transformer coupled. The dynamic characteristic curve of the push-pull amplifier in Class A operation is shown in Figure 5-13 (above). The dynamic characteristic curve of the same amplifier is shown in Class B operation and when the grid bias is set at plate current cut-off (Fig. 5-13 [below]). Notice the distortion of the output waveform. This distortion is easily corrected by adjusting the bias voltage. This

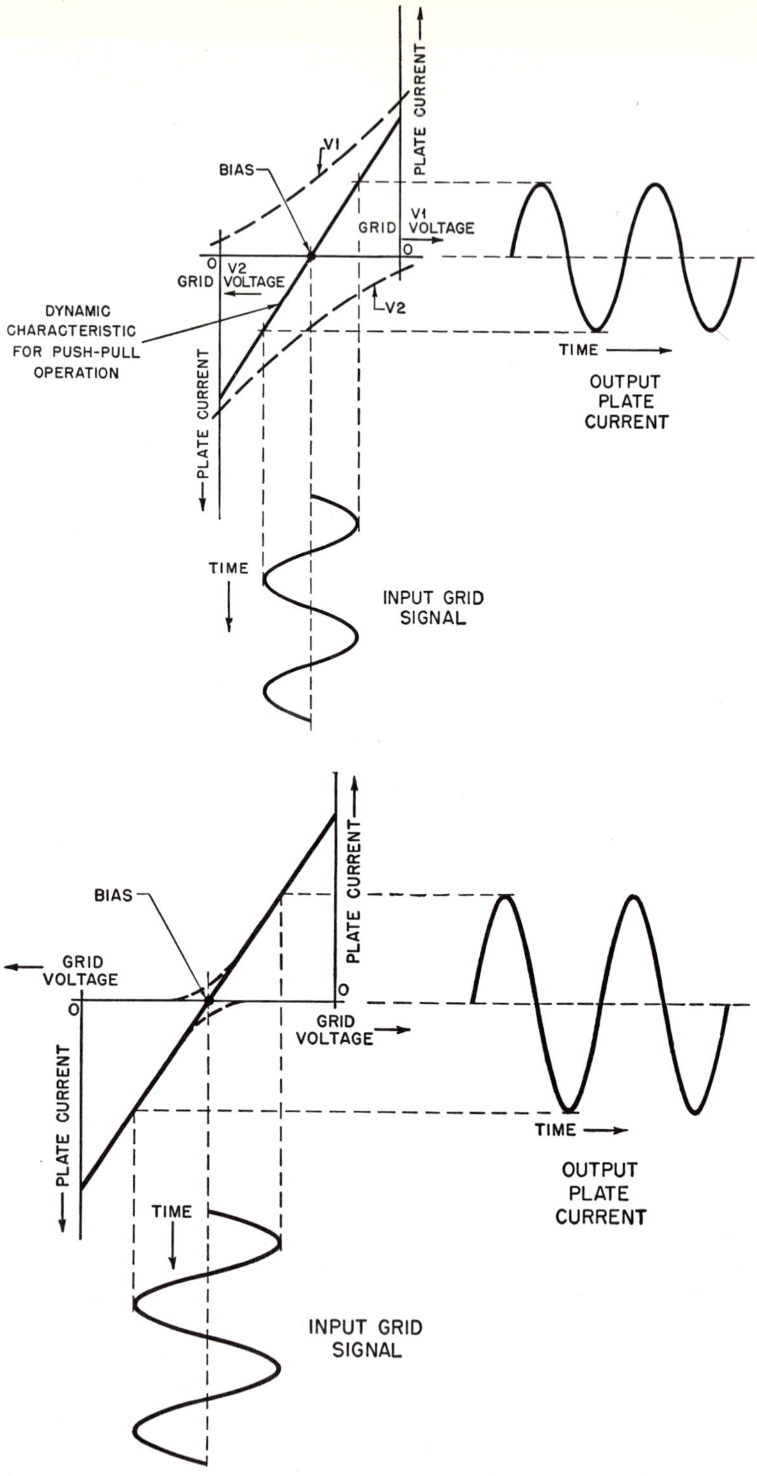

Figure 5-13. Dynamic characteristics of a push-pull triode amplifier in Class A (above) and Class B (below) operation. (Reproduced by permission of the Department of the Army.)

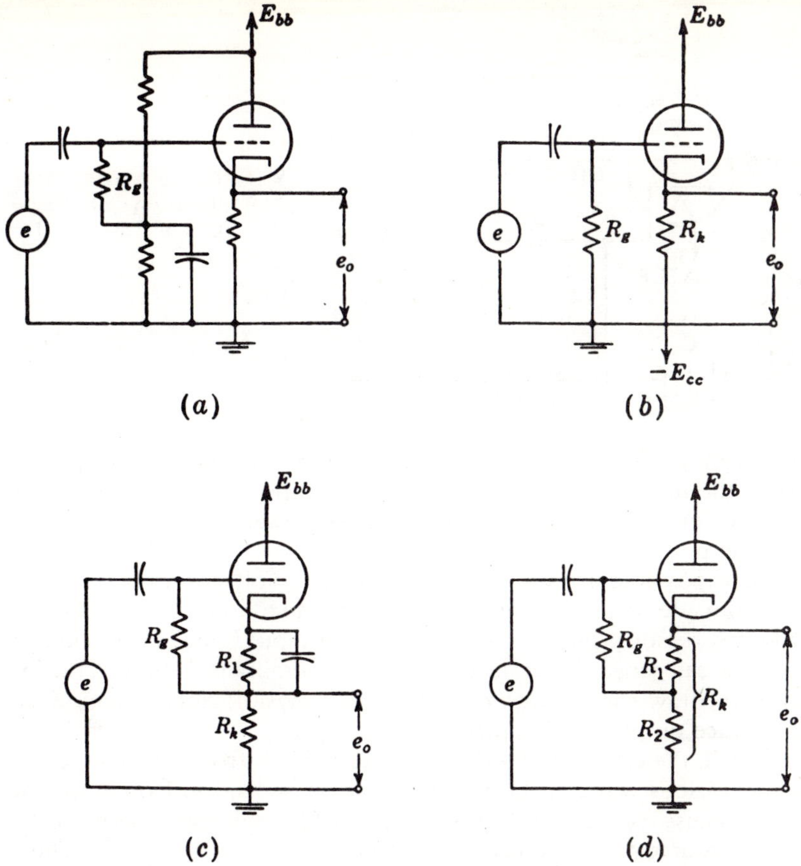

Figure 5-14. Four biasing arrangements for a cathode follower. (From Pulse and Digital Circuits, by J. Millman and H. Taub. Copyright ©1956. McGraw-Hill Book Company. Used by permission.) (*E* and *e* stand for voltage.)

may mean changing the value of a resistor or adjusting the plate voltage among other things. If the push-pull amplifier must amplify DC signals, then, of course, the output transformer will have to be replaced by an output resistor.

THE CATHODE FOLLOWER

The cathode follower is an amplifier which has its load resistance in the cathode circuit. Several varieties of cathode-follower circuits are shown in Figure 5-14. This figure also serves to demonstrate several different ways to bias the grid of a vacuum-tube triode. An *ideal* cathode follower has a gain of one, and the input signal is reproduced with no inversion of the signal in the output. It is useful as a stage in an amplifier because it has a high-input impedance and a low-output impedance. Even though the supply voltages may vary somewhat, the output of the cathode follower still faithfully follows the input signal even with large swings in the input signal. Thus the cathode follower adds stability to a multistage amplifier.

When it is necessary to carry the output of an amplifier over long cables, with a great deal of capacitance, the low-output impedance prevents or minimizes the

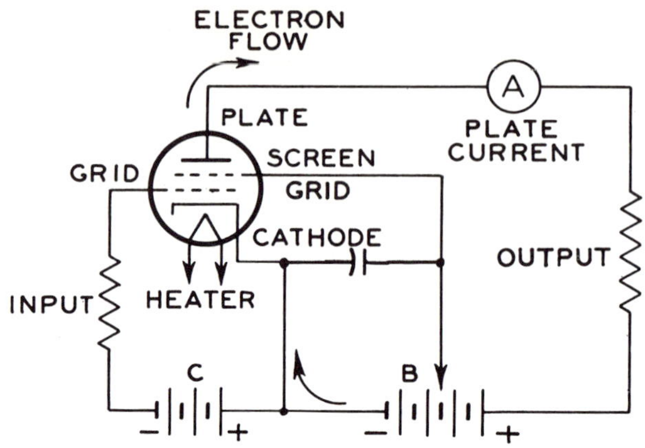

Figure 5-15. The tetrode. (Courtesy of Radio Corporation of America, Harrison, N.J.)

noise which may develop in the cable. When the output of one amplifier circuit is to become the input to another amplifier circuit the cathode follower often is used as a buffer to prevent signals generated within the second amplifier from having an effect on the first.

When a physiologist wants to measure the voltage developed between the inside and outside of a nerve membrane, a cathode follower may be used as an input stage for two reasons: First, an extremely high input impedance is needed so that the amplifier can sense the voltage across the membrane without drawing current. If it were to draw current, it would alter the physiological system. Secondly, the low-impedance output of the cathode-follower stage is important so as not to cause *noise pick up* in the cable. Stray electromagnetic fields from lights, especially fluorescent lights, various wires in the room and in the walls, and from a myriad of other sources will cause extraneous current to flow which will appear as noise. This occurs because the sources of noise cause a voltage drop across the cables. The higher the inherent impedance of a cable the greater the noise current for a given noise voltage (Ohm's law).

MULTIGRID TUBES—THE TETRODE AND THE PENTODE

Although widely used in medical electronics, the triode does not satisfy all needs in the design of electronic circuitry, especially in the area of communications, which includes transmitters and receivers.

The triode has a fairly large inherent *interelectrode capacitance* which is much less of a problem at low frequencies than at high frequencies. To avoid or minimize this problem an electrostatic shield can be added to the plate of a triode. This is called a *screen grid.* Such a vacuum tube having a cathode, a control grid, a screen grid, and a plate is called a *tetrode.* The voltage of the screen grid is kept somewhat less than the plate voltage with respect to the cathode. This may be done in many ways, one of which is shown in Figure 5-15. Since the plate is shielded from the control grid, the plate current is less dependent on the plate voltage, and the result is increased voltage amplification. A capacitor from the screen grid to the cathode is necessary to avoid degeneration or loss of gain at high frequencies.

The *pentode,* in addition to the screen grid, also contains a *suppressor grid.* This latter grid is kept much closer to the plate than the screen grid and is maintained at cathode potential. When electrons are accelerated to high velocities, as they are in a vacuum tube, they cause electrons in the surface layers of the plate to become

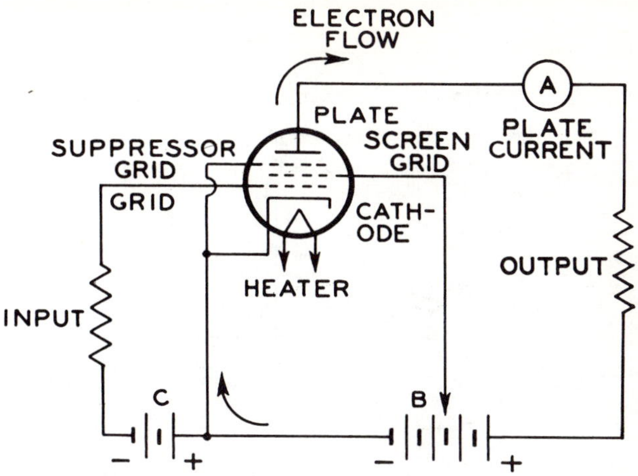

Figure 5-16. The pentode. (Courtesy of Radio Corporation of America, Harrison, N.J.)

displaced and leave the plate. This tends to decrease the gain of the tube. The *suppressor grid,* being at cathode potential, forces the electrons back to the plate. The result is that much greater amplification can be attained in a pentode than in a triode. Figure 5-16 illustrates a pentode, and Figure 5-17 shows the characteristic curves for the pentode section of the 6AW8A triode-pentode.

MULTIPLE STAGES AND EXAMPLES OF AMPLIFIER CIRCUITS

Thus far, the vacuum tube has been considered as a single element in a circuit. In general use, however, it is necessary to combine two or more of the units already described in order to accomplish some purpose. In some cases, the combination of units results in a general purpose amplifier while in others the application is quite limited.

The voltage developed as a result of the current passing through the plate circuit may be developed across a resistor as shown in Figure 5-8, or it may generate a voltage in a transformer as shown in Figure 5-12.

One of the most common types of coupling is the resistor-capacitor (RC) coupling shown in Figure 5-18. With this arrangement, the voltage from the first triode V_1 (called the *first stage*) is developed across the load resistor, labeled R_L. The coupling capacitor prevents the grid of V_2 from being biased at the same voltage as the plate of V_1. It is often called a *blocking capacitor.* The coupling capacitor usually has a rather low reactance (.001 to .1 microfarad). The voltage passing through the coupling capacitor C_c acts as a signal generator for the grid of the second stage V_2. The resistor labeled R_g is chosen to set the bias of the grid of V_2 with respect to its cathode at a normal operating level. Typical values for this resistor R_g range from 500K Ω to several megohms. Notice that if Figure 5-18 were redrawn the coupling capacitor C_c and the grid-bias resistor R_g would form a voltage divider network. The dashed capacitor C_T illustrates the total shunt capacitance of the circuit. It represents some loss of signal. This shunt capacitance is often called *stray capacitance* and results from the physical placement of wires and components as well as interelectrode capacitance.

It is sometimes necessary to amplify DC signals. When this is the case it must be pointed out that the DC signal cannot pass through a coupling capacitor as this would block the DC signal. In such a case, a coupling resistor must be used in place of a coupling capacitor. It is then necessary to set the grid bias of the following stage,

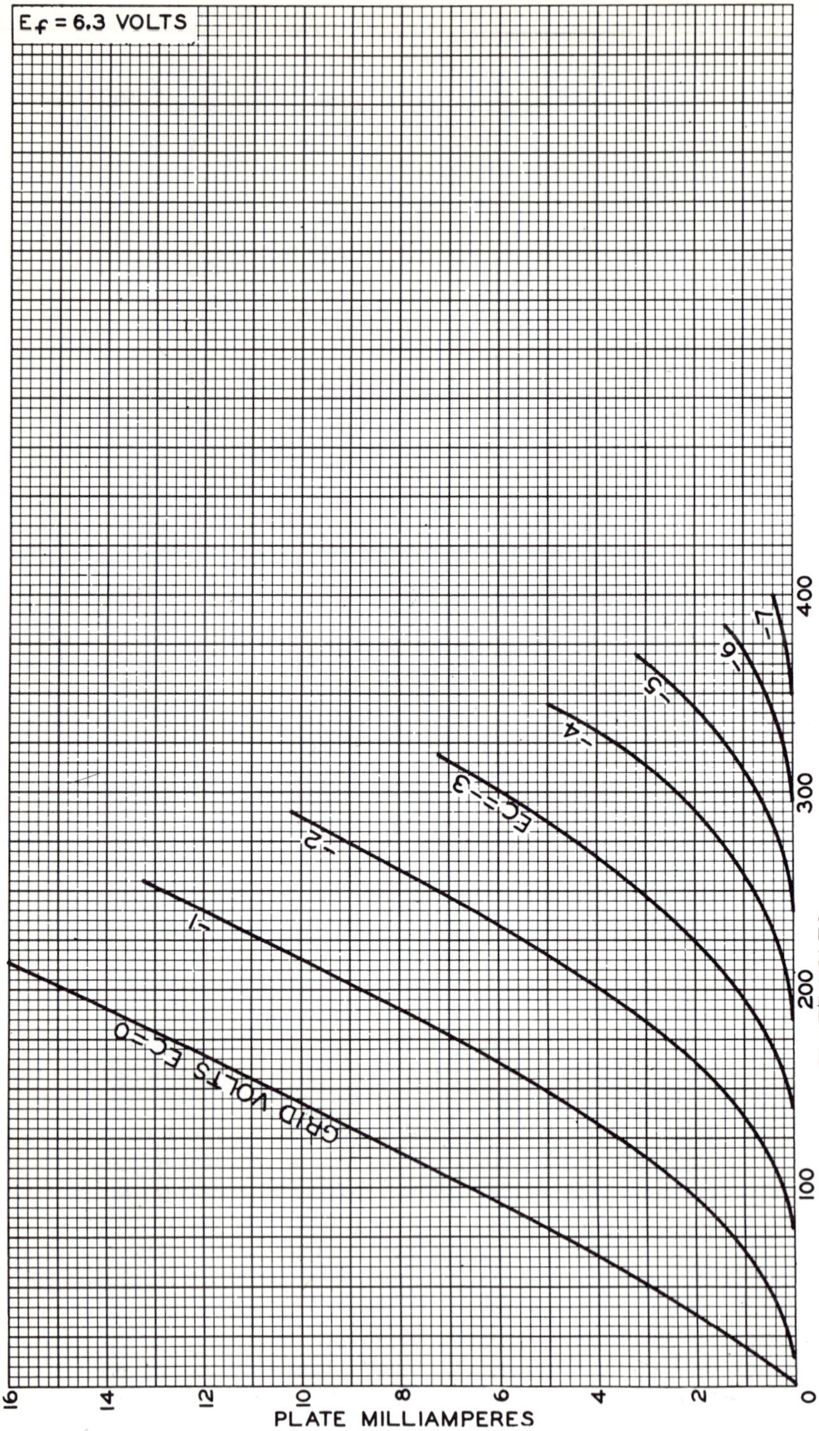

Ef = 6.3 VOLTS
GRID VOLTS EC=0
-1
-2
EC=-3
-4
-5
-6
-7
PLATE VOLTS
0
100
200
300
400
PLATE MILLIAMPERES
16
14
12
10
8
6
4
2
0

Fig. 5-17A

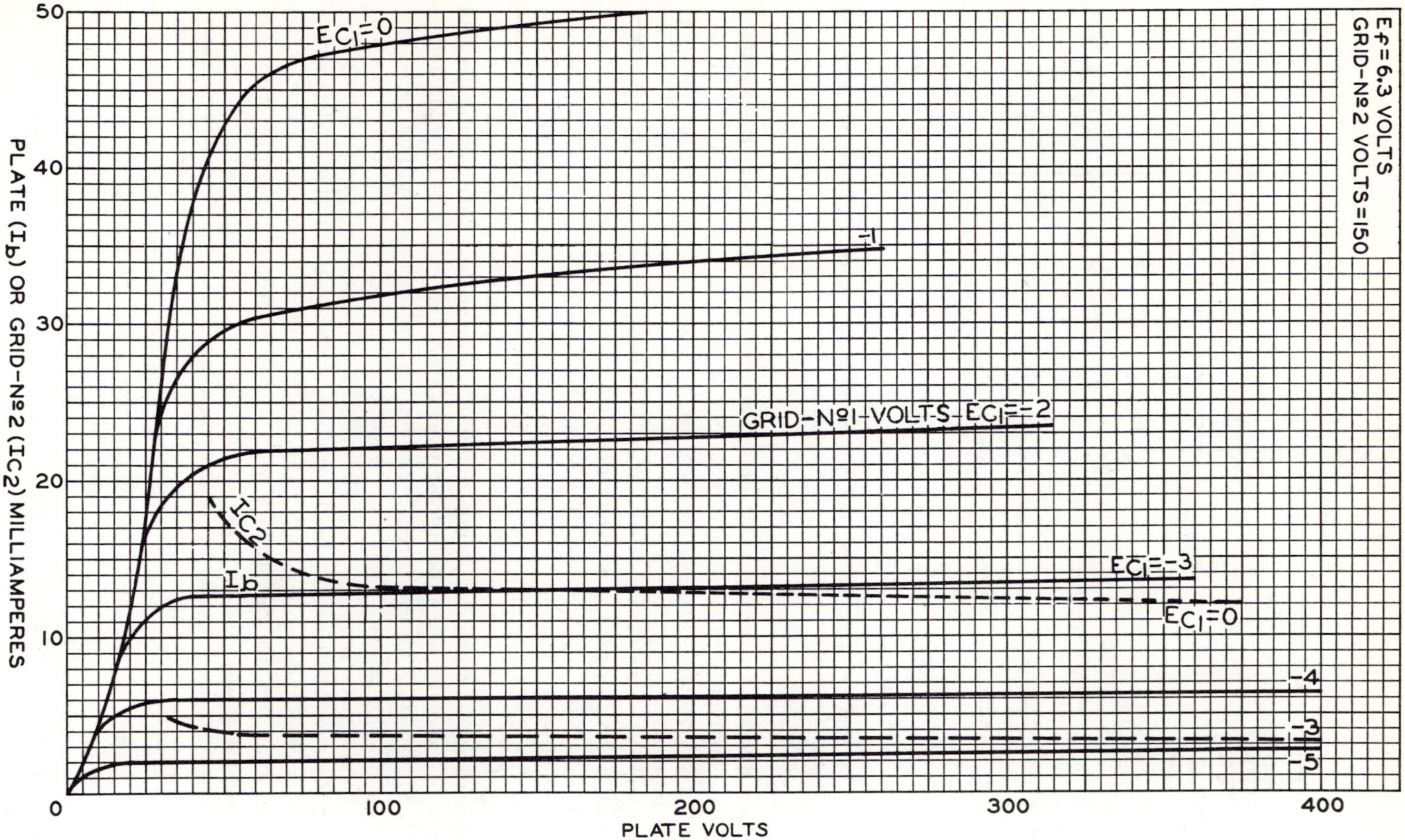

Fig. 5-17B

Figure 5-17. A and B, The characteristic curves for the pentode section of the 6AW8A. (Courtesy of Radio Corporation of America, Harrison, N.J.)

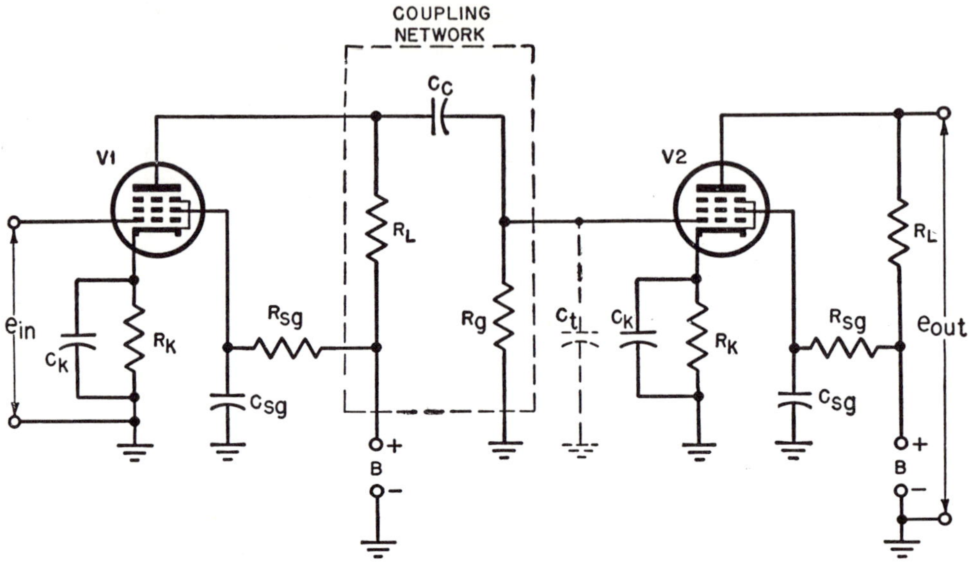

Figure 5-18. RC-coupling network between two triode amplifiers. (Reproduced by permission of the Department of the Army.)

using a special form of voltage divider. Here a zener diode, which has a fixed voltage drop across it and very low resistance, or a neon light bulb (e.g., NE-51), which also has a large fixed voltage drop across it, is used to keep the grid of the following stages at the correct potential.

The coupling network may also consist of a transformer, as shown in Figure 5-19(A), or a resistor-capacitor-inductance (RCL) network, as shown in Figure 5-19(B). Any network may be used to couple two stages of an amplifier. It must be remembered, however, that not all signals may be transmitted through any network. As was discussed in Chapter 2, the coupling network is also a filtering network. The over-all frequency response of the amplifier depends to a great extent on the coupling network or networks chosen. Figure 5-20 shows the relative voltage gain which may be expected when the R-C networks shown in Figure 5-21 (A, B, and C) are used. Figure 5-22 shows three examples of transformer-coupling networks. (Note that only the primary coils are shown [L_p] in Figures 5-21 and 5-22.) The vacuum tube is represented by $-\mu e_g$ which is the amplified and inverted output of the previous stage. The student should examine some diagrams of complete amplifier circuits, such as those in Figure 5-23 and 5-24, to identify the coupling networks used. With some practice the student will be able to predict the range of the frequency response of the amplifier. Notice that the first stage of Figure 5-24 consists of a high-gain twin triode connected push-pull. Since this stage eliminates in-phase signals such as noise it is important that the gain of each half of the triode be equal. The differential balance is provided for this purpose. The first stage is coupled to the second with a 0.5-μf capacitor. Notice the adjustable capacitor in the plate circuit just before the coupling capacitor. This capacitor allows some signal to go to ground. It therefore adjusts the high end of the frequency response. The capacitor will allow more high-frequency signal to go to ground than signals of low frequency. This occurs because the reactance decreases with an increase in frequency. A similar

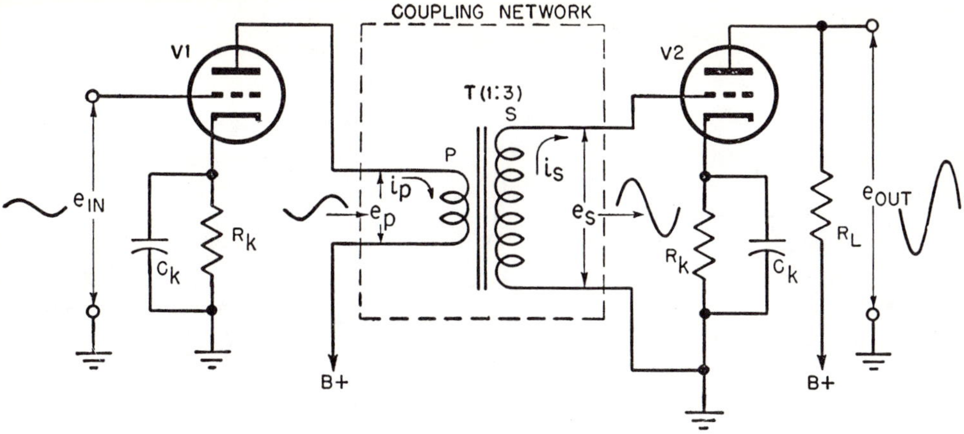

Fig. 5-19A

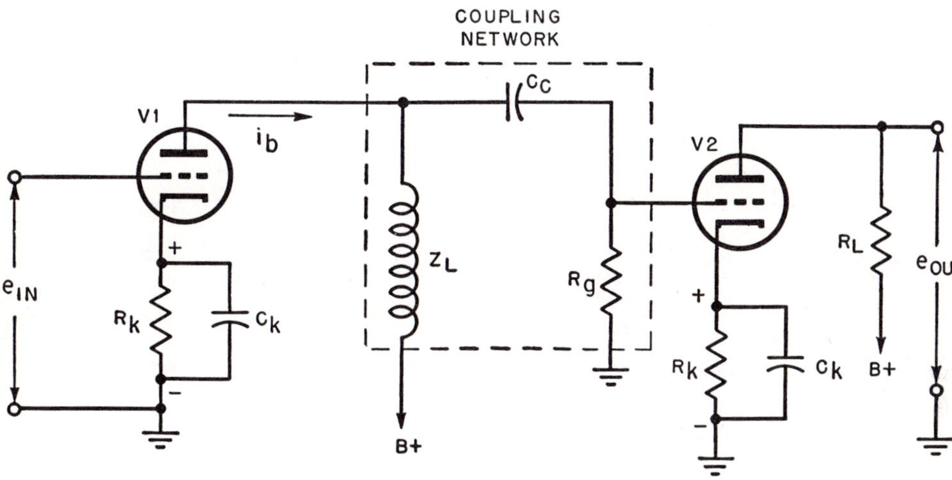

Fig. 5-19B

Figure 5-19. A, transformer-coupled amplifier using triodes. B, RCL-coupled amplifier. (Reproduced by permission of the Department of the Army.)

arrangement is seen with capacitor C10 in stage V2B and again in V3A. The control over the low frequency is regulated by the coupling capacitor between stage V2A and stage V3A. Stage V3B is a cathode follower.

FEEDBACK

It is often necessary to correct an amplifier to compensate for various inherent defects, some of which are due to the vacuum tubes and their characteristics and some to other components used to fabricate the amplifier. Sometimes the properties which the designer wants a particular amplifier to have are possible only with such correction. The specific correction referred to above is called *feedback.* As may be seen in Figure 5-25, some of the resultant output signal (sometimes voltage and sometimes current) is fed, using a suitable coupling network, back into the input of the amplifier.

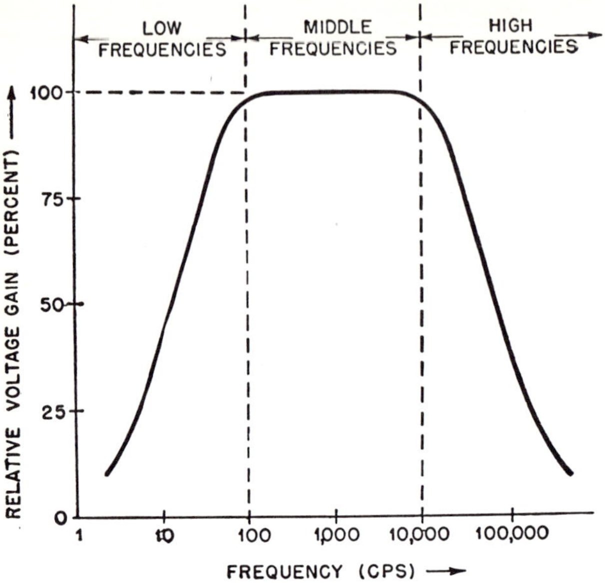

Figure 5-20. Voltage gain characteristic of typical RC-coupled audiofrequency amplifier. (Reproduced by permission of the Department of the Army.)

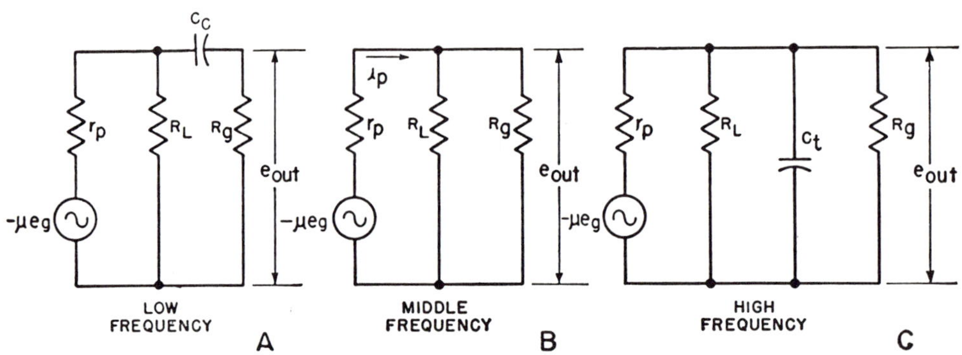

Figure 5-21. Equivalent circuit of RC-coupled amplifier. (Reproduced by permission of the Department of the Army.)

If the signal is fed into the input in such a way as to add to the input (Fig. 5-26[A]) the signal is called *positive feedback, regeneration, or bootstrapping.* The term to *bootstrap* means to pull the output of the stage up by its bootstraps and drive the tube to the cut-off portion of its characteristic curve. This will be described in the next chapter under the heading of *bistable or flip-flop circuits.* The positive feedback signal is *in phase* with the input signal.

If, however, the signal is fed back into the input in such a way as to cause a decrease in the input signal, the feedback is termed *negative feedback, degeneration, or inverse feedback* (Fig. 5-26[B]). Such a signal is *out of phase* with the input. Negative feedback is used to stabilize an amplifier. It also decreases the distortion and improves the linearity of the output with respect to the input signal.

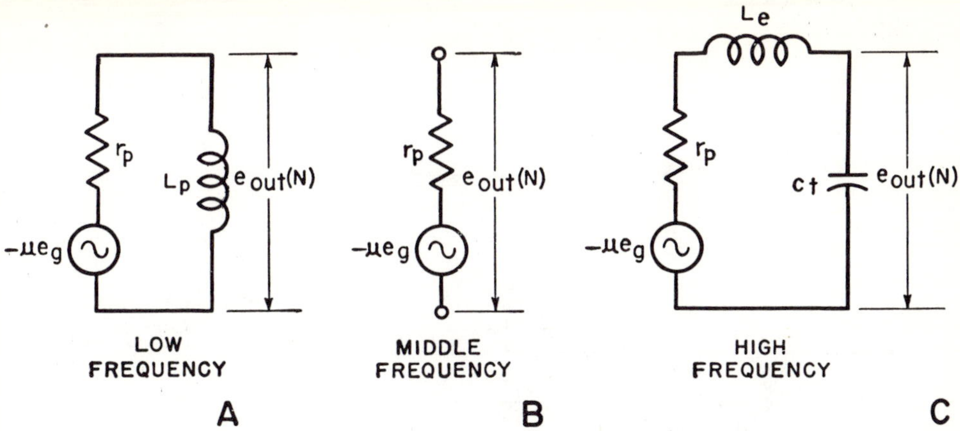

Figure 5-22. Equivalent circuits of transformer-coupled amplifier. (Reproduced by permission of the Department of the Army.)

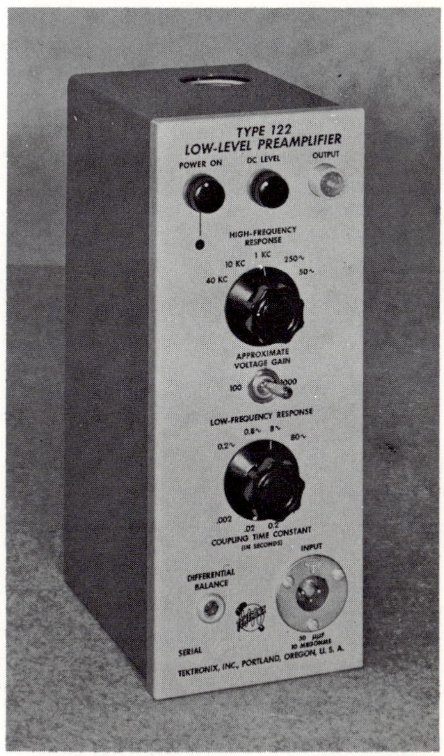

Figure 5-23. Tektronix type 122 low-level preamplifier. (Courtesy of Tektronix, Inc., Beaverton, Ore.)

Figure 5-27 illustrates one specific application of negative feedback. Here, one triode section of a 12AT7 is set up as a cathode follower. In this circuit the input signal is capacitively coupled to a twin-tee filter,* which has the unique property of zero transmission at a particular frequency. The output of the twin-tee supplies

*See the twin-tee network, pages 45 to 47.

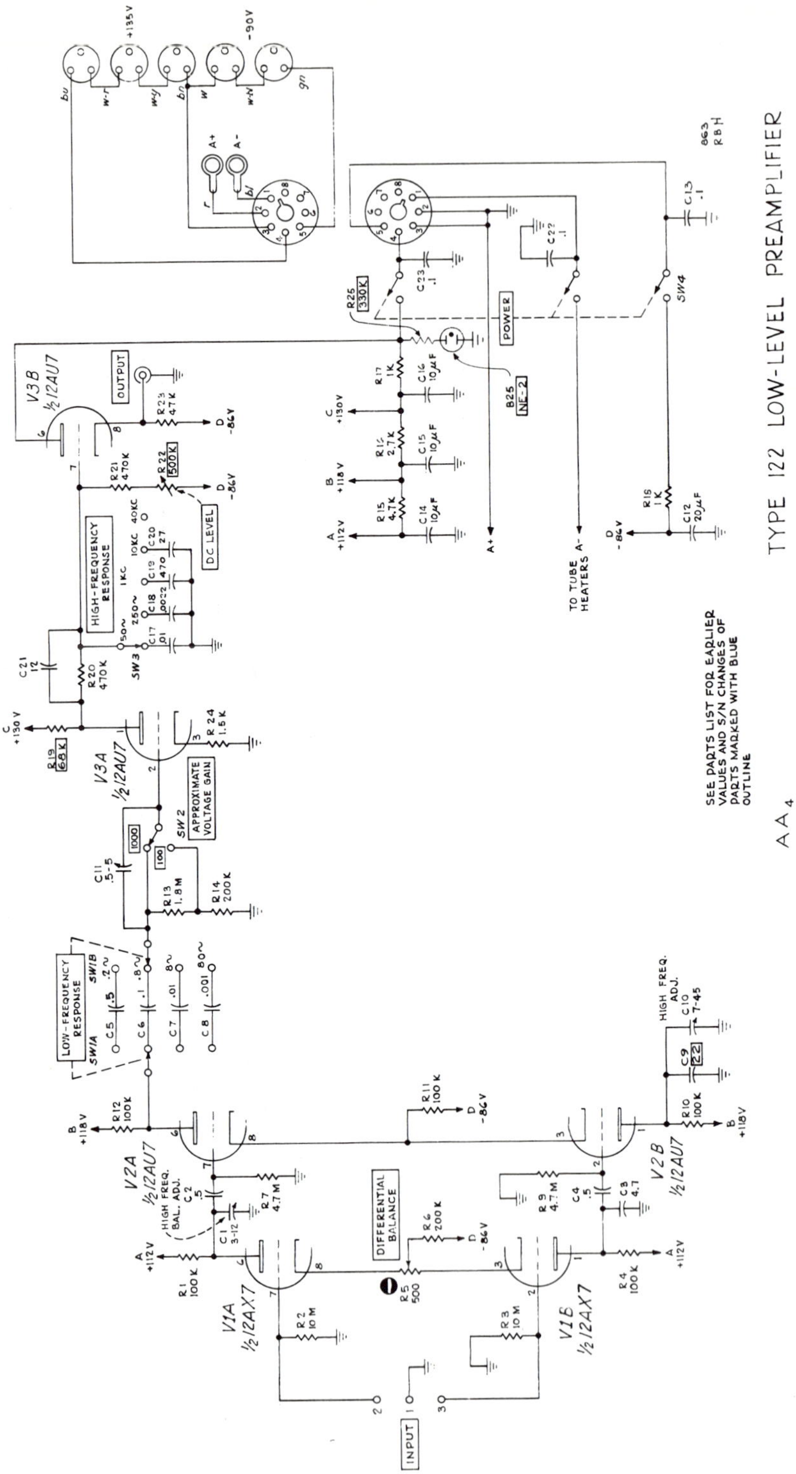

Figure 5-24. Schematic for Tektronix low-level preamplifier. (Courtesy of Tektronix, Inc., Beaverton, Ore.)

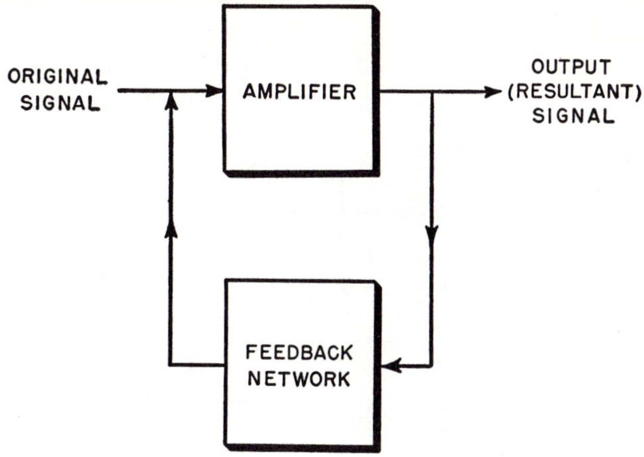

Figure 5-25. Block diagram of feedback amplifier. (Reproduced by permission of the Department of the Army.)

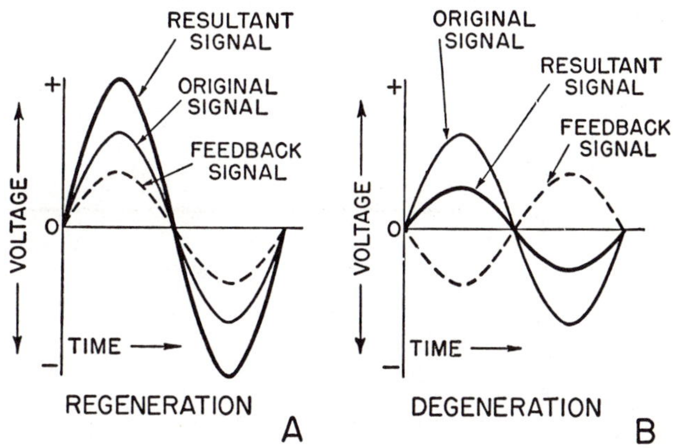

Figure 5-26. Phase relationships of signals showing regeneration (positive feedback) and degeneration (negative feedback). (Reproduced by permission of the Department of the Army.)

the signal to the grid of the cathode follower. Notice that the output of the cathode follower is fed back into the input through a 1-megohm resistor. The unwanted frequency, after being eliminated from the input, is fed directly into the output. This has the effect of canceling whatever signal of that frequency that should happen to appear at the grid of the tube. Such a circuit is called a *notch amplifier.* Figure 5-28 illustrates the improvement that the notch amplifier has in the elimination of a particular frequency. This is very important to the designer of biological instruments. As has been discussed previously, a frequency response of 30 cps, without loss of signal, is entirely adequate for most biomedical applications. If the designer must design an amplifier and eliminate the unwanted 60-cps noise which is picked up as a result of transmission from the house current wires, transformers, or fluorescent lighting he could pass the output through a twin-tee network which would eliminate 60 cps. It may be seen in Figure 5-28, however, that the signal would be reduced about 15 per cent at 6 cps (0.1 × 60 cps), and about 55 per cent at 30 cps (0.5 × 60 cps). The notch amplifier output would be almost flat (almost no loss) at 30 cps.

Other examples of feedback will be pointed out in the circuits that appear later in this book and will be discussed at that time. Electronic feedback is analogous to the physiological feedback which one sees in almost every body system. The liberation of circulating thyroid hormone, which is activated by thyroid-stimulating

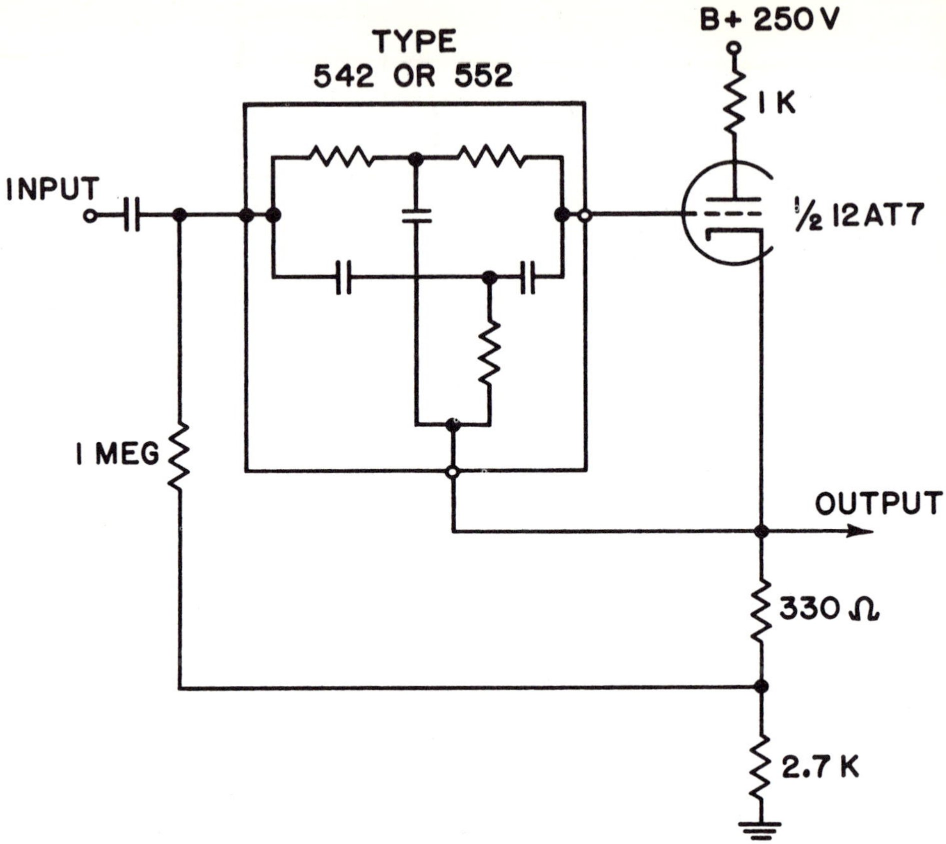

Figure 5-27. The notch amplifier. (Courtesy of the White Instrument Laboratories, Austin, Tex.)

hormone from the pituitary, causes a decrease in the secretion of thyroid-stimulating hormone, which in turn causes less circulating thyroid hormone to be released. This feedback loop tends to stabilize the concentration of circulating thyroid hormone in the blood. The above is an example of a chemical feedback system. The effect on the iris of an increase in the amount of light entering the eye is an example of a more biophysical feedback system. It is obviously impossible to describe all the examples of feedback systems in the body at this time.

THE OPERATIONAL AMPLIFIER*

A very useful form of the feedback amplifier is the operational amplifier shown in Figure 5-29(A). In this illustration, the operational amplifier is represented by the triangle and consists of a very high-gain amplifier ($\mu = 10^6$), usually three or more stages. The resistor R_2 in Figure 5-29(A) refers some of the output voltage back to the input. This feedback arrangement is somewhat different from the usual feedback circuit because the signal has a path through R_1 and R_2 to the output even

*Miniature operational amplifiers (transistorized and short-circuit proof) are now available. These amplifiers may be used as a basic module for almost any electronic application (*Handbook of Operational Amplifier Applications,* Burr-Brown Research Corporation, Tucson, Ariz.; see Appendix VI). These multipurpose operational amplifiers are also available from other manufacturers, such as Data Device Corporation, Hicksville, N.Y., for example.

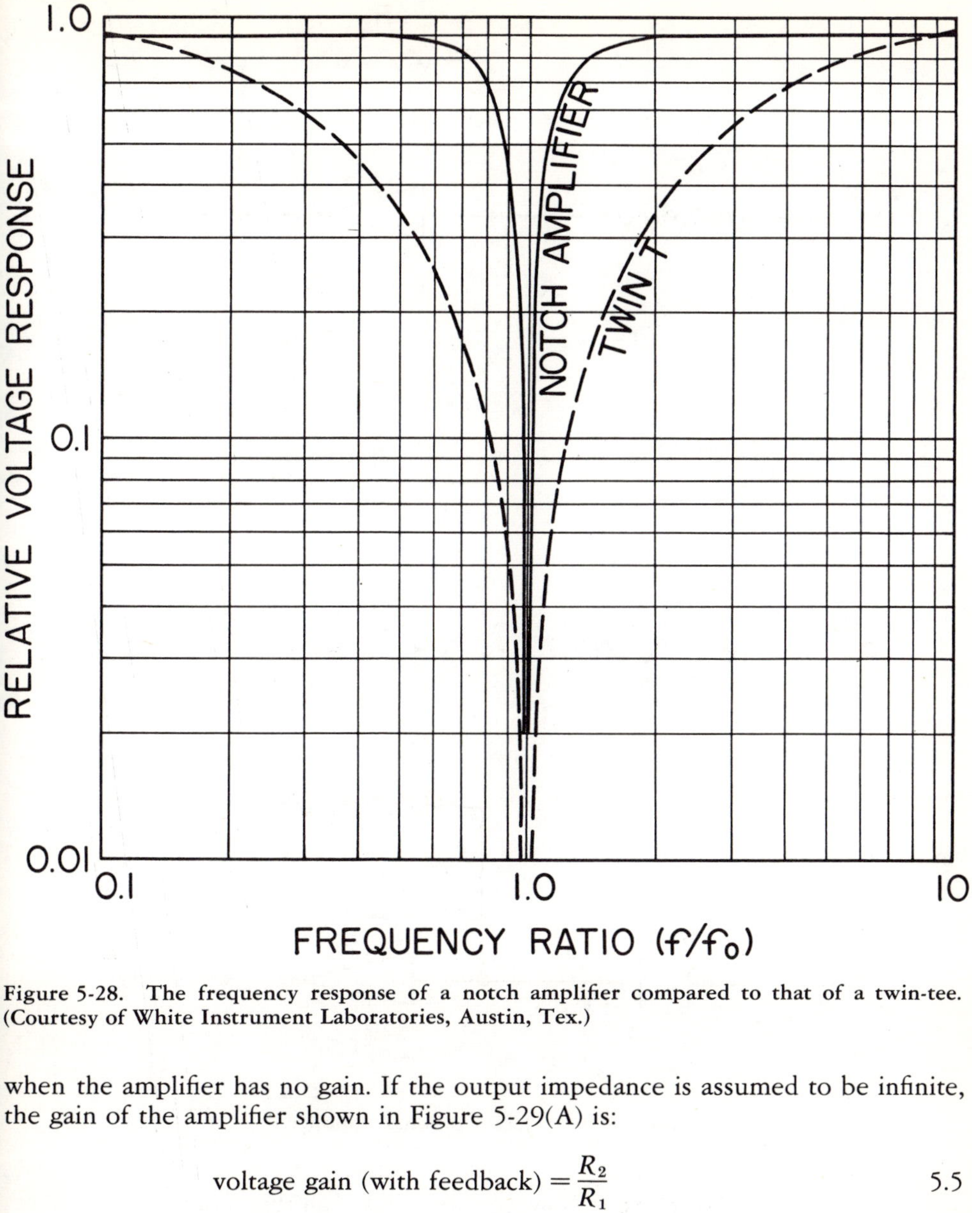

Figure 5-28. The frequency response of a notch amplifier compared to that of a twin-tee. (Courtesy of White Instrument Laboratories, Austin, Tex.)

when the amplifier has no gain. If the output impedance is assumed to be infinite, the gain of the amplifier shown in Figure 5-29(A) is:

$$\text{voltage gain (with feedback)} = \frac{R_2}{R_1} \qquad 5.5$$

From equation 5.5, it may be seen that the output signal will be inverted with respect to the input. Notice also that by properly closing R_2 and R_1, the operational amplifier could be used to *multiply* the input by a constant, namely, R_2/R_1. Multiplication of two variables, however, is more complicated and cannot be done with operational amplifiers alone. The most common analog multiplier utilizes a magnetic circuit and a Hall effect voltage generator. (See reference section at the end of this chapter.)

Since the output of the amplifier is fed back to the input, it can be shown that, when the gain of the amplifier is very high, there is effectively a short between the

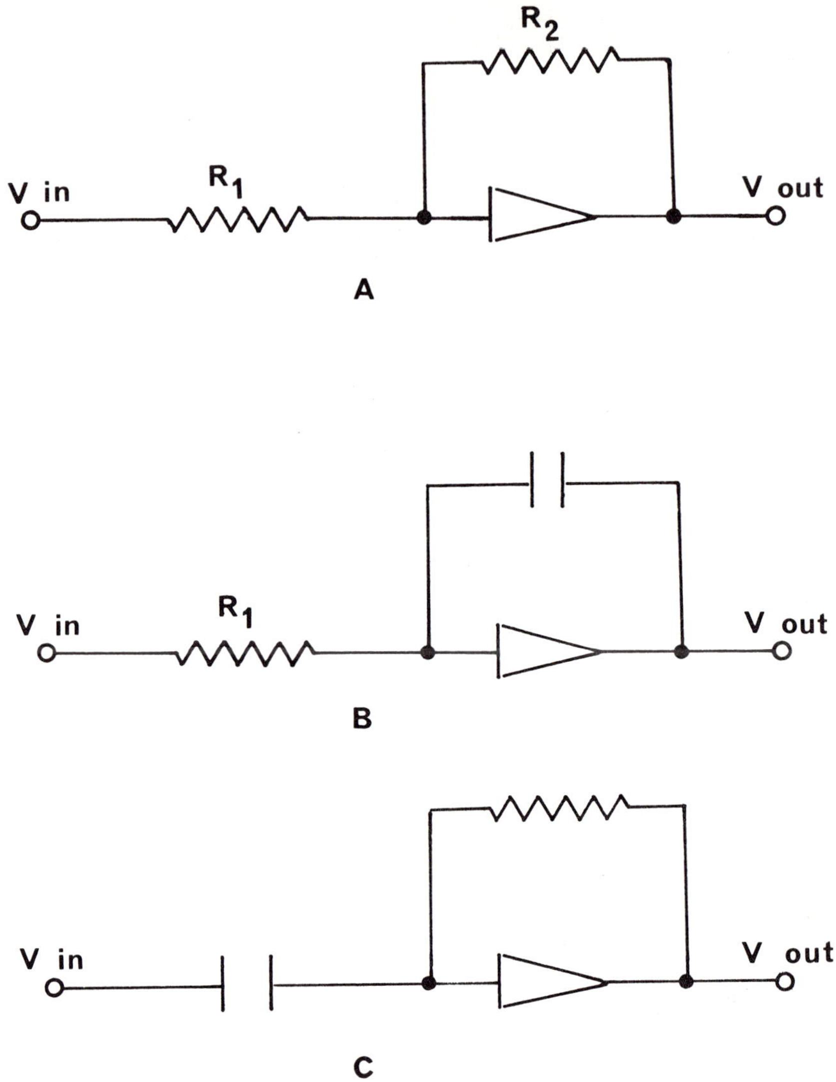

Figure 5-29. The operational amplifier.

two input terminals. No current, however, flows through this short, and it is therefore referred to as a *virtual short* (to ground) or simply as *virtual ground*. Any current passing through R_1 continues past the virtual short and goes through R_2.

The operational amplifier may be used to perform a variety of mathematical operations. If, however, $R_1 = R_2$ in the Figure 5-29(A), the unit acts as a *sign changer*. The use of the operational amplifier as a multiplier (by a constant) has already been discussed. By substitution of appropriate impedance networks for R_1 and R_2, the amplifier becomes a *phase shifter*.

If a capacitor is substituted for R_2 in Figure 5-29(A), as shown in Figure 5-29(B), then:

$$V_{out} = -\frac{1}{C}\int idt = -\frac{1}{RC}\int Vdt \qquad 5.6$$

Thus, the output voltage V_{out} is proportional to the integral of the input voltage signal, and the unit becomes an *integrator.*

If, on the other hand, the resistor and the capacitor are interchanged, as shown in Figure 5-29(C), then the output voltage (V_{out}) is:

$$V_{out} = -RI = -RC\frac{dV}{dt} \qquad 5.7$$

The output voltage is proportional to the derivative of the input signal and is called a *differentiator.*

The operational amplifier can also be used to add or subtract (add with a sign change). An analog computer is a number of such operational amplifiers connected so as to solve a particular problem.

The student should bear in mind that, when a circuit is being designed, it often becomes necessary to alter the signal waveform in one of the ways described above. For example, the designer may wish to convert a square wave into a triangular waveform. This is easily done by incorporating an integrator into the circuit because the integral of a square waveform is a triangular wave.

THE TRANSISTOR

Since the advent of the point-contact transistor in 1948 and the junction-transistor in 1949, and more recently the drift, the mesa, the planar, and the field-effect transistor, the use of transistors and other semiconductor devices has exploded.

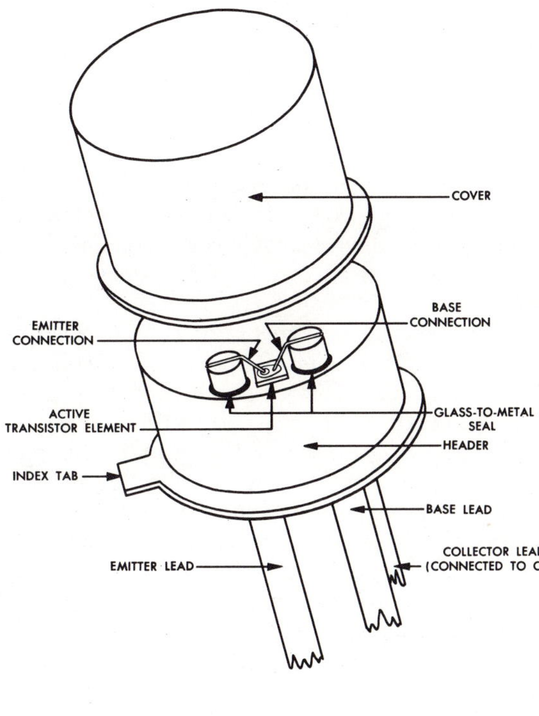

Figure 5-30. Internal structure of the transistor. (Courtesy of Radio Corporation of America, Harrison, N.J.)

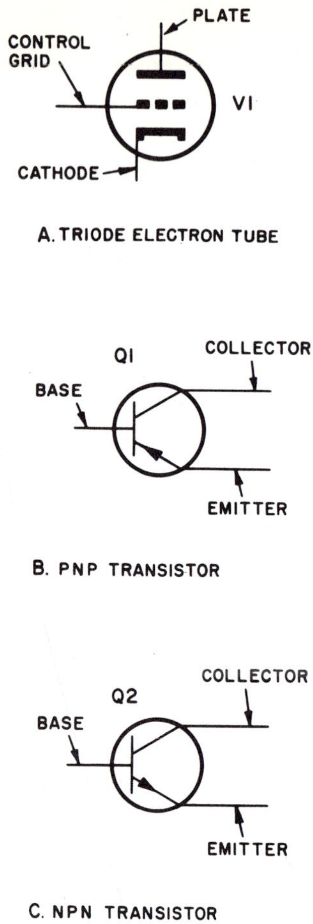

Figure 5-31. Transistor and triode electron tube reference designations and graphical symbols. (Reproduced by permission of the Department of the Army.)

Transistors may be used as components in all kinds of amplifiers, oscillators, transmitters, receivers, and electronic switches, to mention only a few applications. New semiconductor devices, such as the silicon-controlled rectifier, the tunnel diode, and the integrated microcircuit module, have expanded the use of these devices even further.

Transistors, because of their small size, enable instruments to be fabricated which are small or even tiny when compared to the unit in which vacuum tubes are used. Since transistors have no filaments, they require no appreciable "warm-up time." In general, the transistor is more efficient than a vacuum tube. The transistor requires lower operating voltage (equivalent to the plate voltage in a tube), and thus power supplies can be much smaller. Semiconductors, when properly constructed, can be immersed in water and can withstand greater shocks than vacuum-tube units. In general, transmitter circuits have a longer life than their tube equivalents.

The inherent noise generated within a vacuum tube is, however, substantially less than for a similiar transistor device. This is especially true at lower frequencies, which are generally used in biological or medical instruments. Amplifiers which must increase signals of less than 1 microvolt must, at the present time, be made

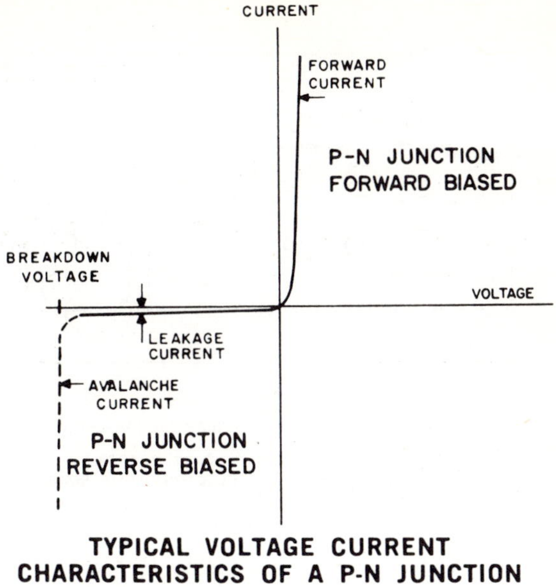

Figure 5-32. Conduction in a PN-junction diode. (Courtesy of General Electric Co., Syracuse, N.Y.)

either of vacuum tubes or a combination of vacuum tubes and transistors. (This is called a *hybrid* circuit.) It is fairly certain, however, that in the very near future advanced transistors will be developed which will be equal to or better than their vacuum-tube equivalents. It will be some time, however, till the cost of the very low-noise transistor will approach the present cost of the low-noise tube amplifier. Many applications, however, do not require low-noise circuitry and, thus for many circuits, the problem of noise may never arise.*

The present transistor has another serious drawback. It is relatively sensitive to environmental temperature. This will be seen later in this chapter. Increased temperature causes increased current to flow through the transistor. This increased current raises the junction temperature of the transistor, which increases the current still further. Unless the current is limited by means of a resistor, the transistor can develop *thermal runaway* and destroy itself. Other temperature effects can be essentially eliminated with special feedback and bias arrangements, using thermistors (temperature-sensitive resistors) and diodes.

THE TRANSISTOR AS AN AMPLIFIER

The transistor behaves essentially like a valve, and may be compared to the vacuum-tube triode. Figure 5-30 illustrates the physical arrangement of the basic parts of the transistor. The symbols for the triode and the two basic types of transistors, namely, the PNP and NPN transistors, are shown in Figure 5-31 as they appear in circuit schematics.

It will be recalled from Chapter 1 that there are two types of semiconductor materials, P type and N type. Figure 5-32 illustrates the flow of current when P-type and N-type materials are joined together to form a PN-junction diode.

*By the time this manuscript was set in type, low-noise transistors and other solid-state devices have become available and are reasonably priced.

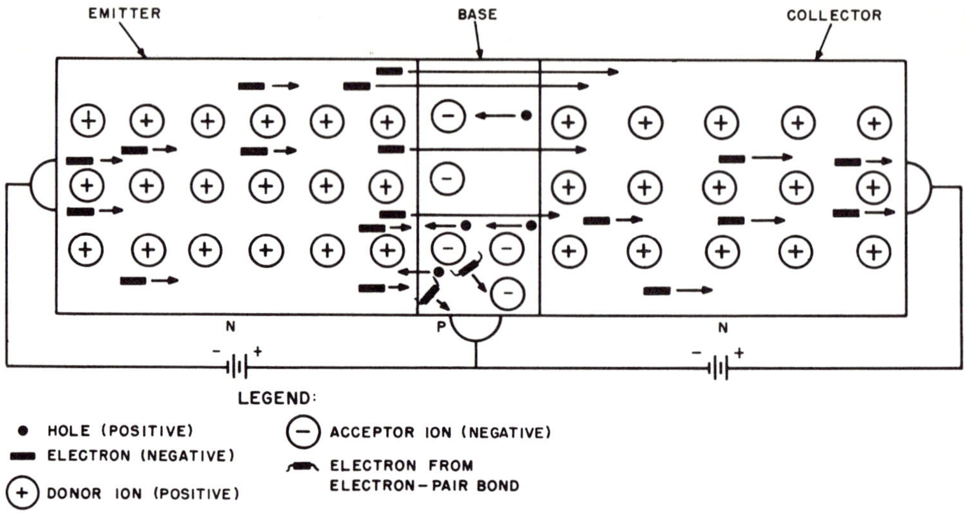

Fig. 5-33A

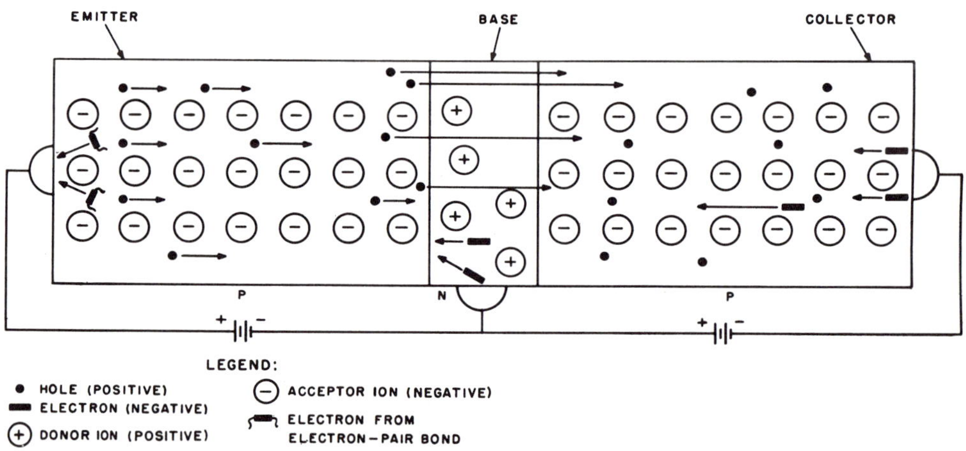

Fig. 5-33B

Figure 5-33. A, Simultaneous application of forward bias between emitter and base and reverse bias between base and collector of an NPN transistor. B, Simultaneous application of forward bias between emitter and base and reverse bias between base and collector of a PNP transistor. (Reproduced by permission of the Department of the Army.)

Notice that a battery may be connected so that the polarity does not enhance current flow. This is called *reverse bias,* and only a small leakage current will flow. The opposite polarity is called *forward bias* and allows much more current to flow (Fig. 5-32).

The NPN transistor, shown in Figure 5-33(A), illustrates the flow of current when forward bias is applied between the emitter and the base, and reverse bias is applied between the collector and the base. Notice that in the NPN transistor the current flow is mostly electrons. Figure 5-33(B) shows a PNP transistor with the

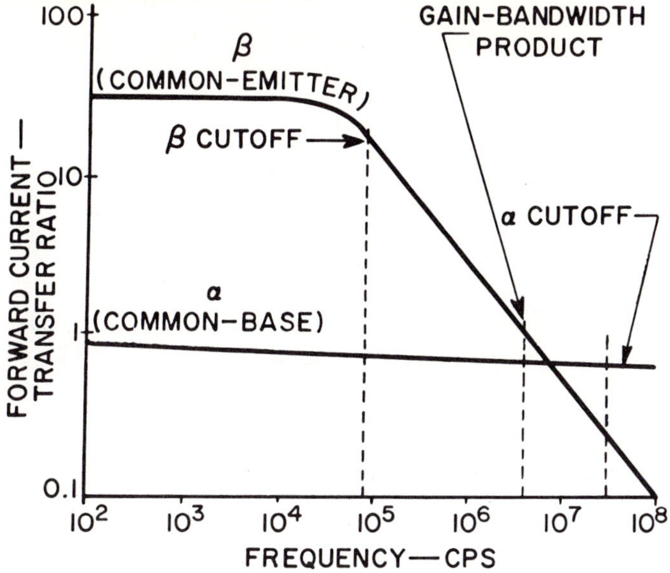

Figure 5-34. Forward current transfer ratio. (Courtesy of Radio Corporation of America, Harrison, N.J.)

same bias arrangement. Notice that the emitter-to-collector carrier in the PNP transistor is the hole rather than the electron.

The transistor may be compared to the triode. Electrons in the triode flow from the cathode through the grid to the plate. In the transistor the main current flow is from the emitter through the base to the collector. In the vacuum-tube triode, the plate voltage is always positive with respect to the cathode or no current will flow. The transistor equivalent to the plate is the collector, and it may be positive or negative with respect to the emitter, depending on whether it is an NPN or PNP transistor. The DC electron current flows against the direction of the arrow on the emitter. The base-emitter junction of a transistor is always forward biased while the collector-base junction is always reverse biased. Current flow in the transistor is thus by *injection* of the carrier into emitter, *diffusion* through the base region (usually less than 0.001 inch thick), and *collection* at the collector before returning to the external circuit.

As the voltage between the base and the emitter is increased in an NPN transistor, the positive charge in the base region will increase. This will increase the number of electrons flowing across the base, but some of the holes will also get across the barrier. Therefore, in order to keep the amount of positive charge in the base constant, some current will have to flow in the emitter-base circuit. The ratio of the current which flows in the collector (I_c) to the base current (I_B) is called the forward current transfer ratio or the current gain of the transistor (h_{FE}):

$$h_{FE} = \frac{I_C}{I_B} \qquad 5.8$$

For AC signals, the forward current transfer ratio for the common-emitter configuration is referred to as *beta* and is defined as:

$$\beta = \frac{i_C}{i_B} = 49 \qquad 5.9$$

The ratio of the collector current to the emitter current for the common-base configuration is called *alpha* and is:

$$\alpha = \frac{i_C}{i_E} = 0.98 \qquad 5.10$$

As the frequency of the signal is increased the forward current transfer ratio will decrease as shown in Figure 5-34. Notice that there is a frequency for the common-emitter circuit where the gain (forward current transfer ratio) begins to fall off. This is called *beta cut-off*. There is also a frequency at which the gain of the common-base configuration begins to decrease. This is called *alpha cut-off*. Both of these cut-off points are characteristic of a given transistor.

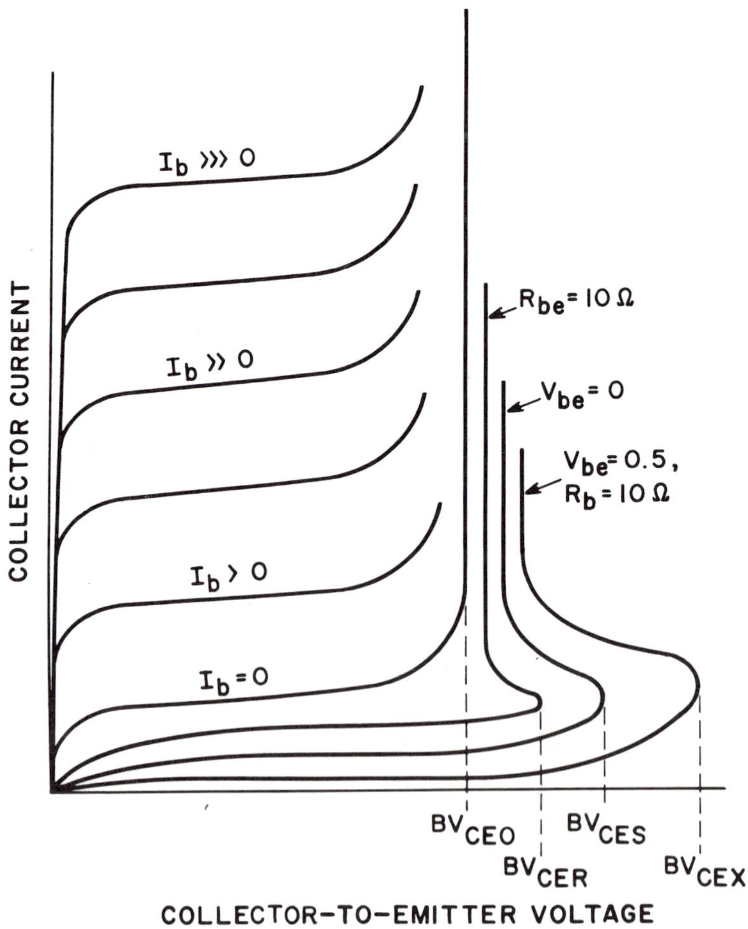

Figure 5-35. Breakdown voltages. (Courtesy of Radio Corporation of America, Harrison, N.J.)

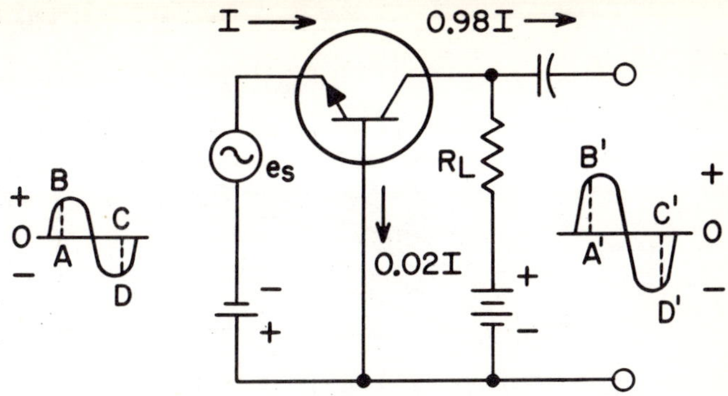

Fig. 5-36A

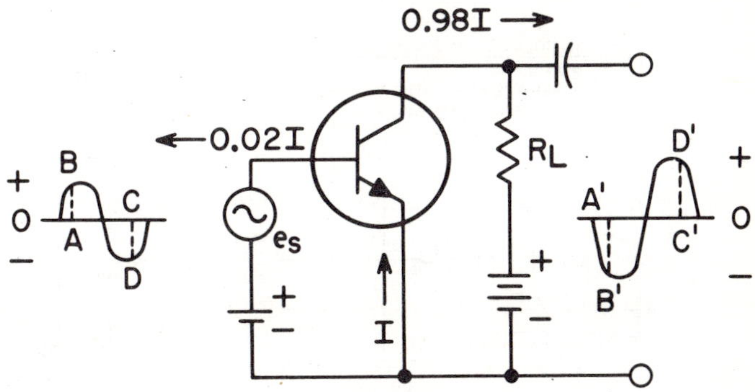

Fig. 5-36B

Figure 5-36. Basic transistor circuit configurations: A, common base; B, common emitter; C, common collector. (Courtesy of Radio Corporation of America, Harrison, N.J.)

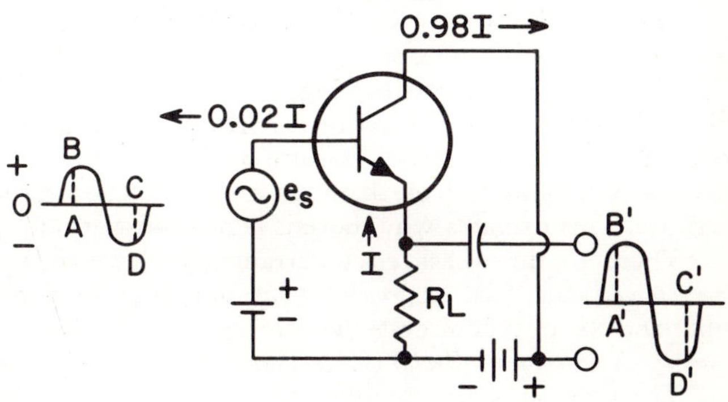

Fig. 5-36C

There is another transistor characteristic which must be kept in mind when transistors are used in instrument design. These are called breakdown voltages. When the voltage between two electrodes becomes high enough the crystal structure of the transistor changes and an increase in current occurs. When this happens localized heating may occur which will tend to destroy the junction. There are several breakdown voltages: the *first breakdown, second breakdown, punch-through,* etc. These are shown in Figure 5-35 as BV_{ceo}, BV_{cer}, BV_{ces}, and BV_{cex}; for details consult RCA Transistor Manual SC-11.

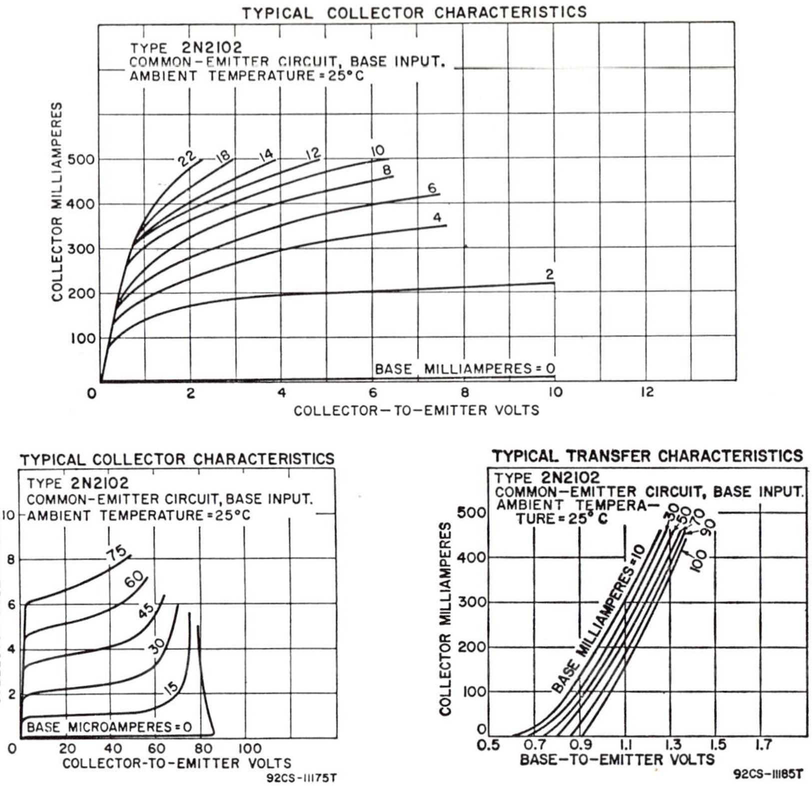

Figure 5-37. Characteristic curves for the common-emitter configuration. (Courtesy of Radio Corporation of America, Harrison, N.J.)

There is, generally speaking, no current flow in the grid-cathode circuit of a triode. There is, however, more electron flow in the transistor equivalent—emitter to collector. Vacuum tubes require a filament to heat the cathode so as to cause electron emission and electron flow. The transistor requires no filament for this purpose. This explains why there is almost no warm-up time for a transistor circuit.

There are three basic circuit arrangements which are used in transistor circuitry: the *common-base* (CB) or *grounded-base* amplifier with the base common to both the input and the output circuit; the *common-* or *grounded-emitter* (CE) amplifier, and the *common* or *grounded collector* (CC). These are shown in Figure 5-36.

The common-base amplifier has the lowest input impedance of the three types of transistor amplifiers. When $R_1 = 1\text{K}\ \Omega$ and the load resistor $R_2 = 10\text{K}\ \Omega$ and the transistor in Figure 5-36(A) is a 2N525 (bias of 5 volts and 1 ma current), then the input impedance is 35 Ω. This circuit arrangement, however, has the highest output impedance, about 1 megohm for the 2N525 and the resistor values stated above. The common-base amplifier has a low current gain and a high voltage gain, resulting in a moderate power gain.

The common-emitter circuit is shown in Figure 5-36(B). This is the arrangement most often used in transistor-amplifier circuits. The common-emitter arrangement has a moderate input impedance (using the 2N525 and the resistors mentioned above), about 1.3K Ω, and a higher but still moderate output impedance, about

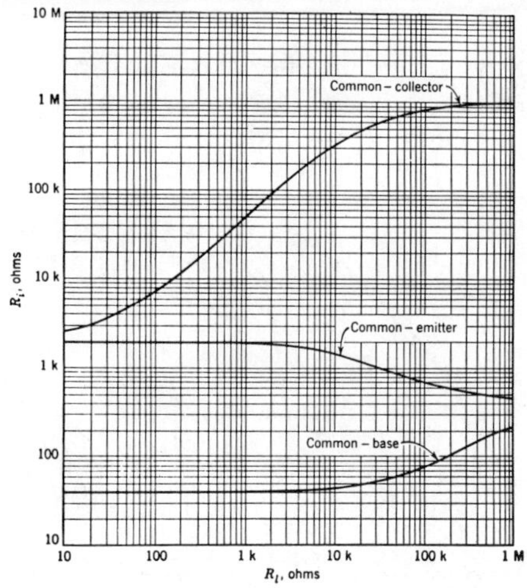

Fig. 5-38A

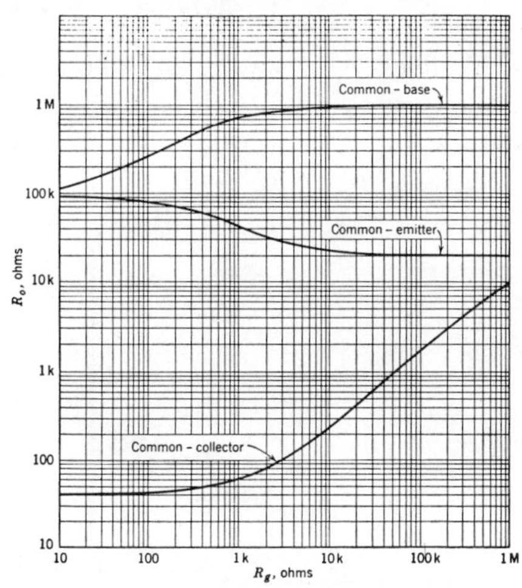

Fig. 5-38B

Figure 5-38. A, Input resistance versus load resistance. B, Output resistance versus input or generator resistance.

50K Ω. The common-emitter configuration results in a high current gain and a high voltage gain, and therefore it also has the highest power gain.

As with vacuum tubes there are both static characteristics (measured with DC signals) and dynamic characteristics (measured with AC signals) for transistors. Two dynamic characteristic curves are usually given: the first is the *collector-characteristic* and the other the *transfer-characteristic* curve. Each of these curves must be stated in terms of the transistor configuration used. Figure 5-37 shows the collector and transfer characteristics for the 2N2102 transistor. Usually only the common-emitter characteristics are published as these are the most frequently used.

The *common-collector circuit* shown in Figure 5-36(C) is also referred to as an *emitter follower.* The equivalent vacuum tube circuit is the familiar *cathode-follower* circuit as shown in Figure 5-14. The emitter follower has the highest input impedance and the lowest output impedance of the three transistor configurations (350 K Ω and 500 Ω respectively for the 2N525 circuit mentioned above). It has high current gain, but as might be expected it has a voltage gain of about unity. It has the lowest power gain of the three circuits.

Figure 5-36 illustrates the three transistor circuit arrangements using a PNP transistor. If an NPN transistor were used, the polarities of the batteries would have to be reversed. Figure 5-36(A) illustrates the current flow and voltage phase relationship between the input and the output signals in a common-base (NPN) amplifier. Notice that the output signal is not inverted but is amplified. In the common-emitter amplifier, shown in Figure 5-36(B), the output is inverted and amplified. In the common-collector or emitter-follower circuit shown in Figure 5-36(C), the output is equal to or less than the input signal. Figure 5-38(A) illustrates the variation of input resistance as a function of the load resistance in the three transistor arrangements. Figure 5-38(B) is a curve of the output resistance versus input or generator resistance for each configuration.

BIASING A TRANSISTOR CIRCUIT

Transistor amplifiers are divided into the same amplifier classes as are vacuum-tube circuits (Fig. 5-39). It is easily seen that the choice of bias voltage on the emitter junction determines the collector current which, in turn, determines the class of operation of the amplifier.

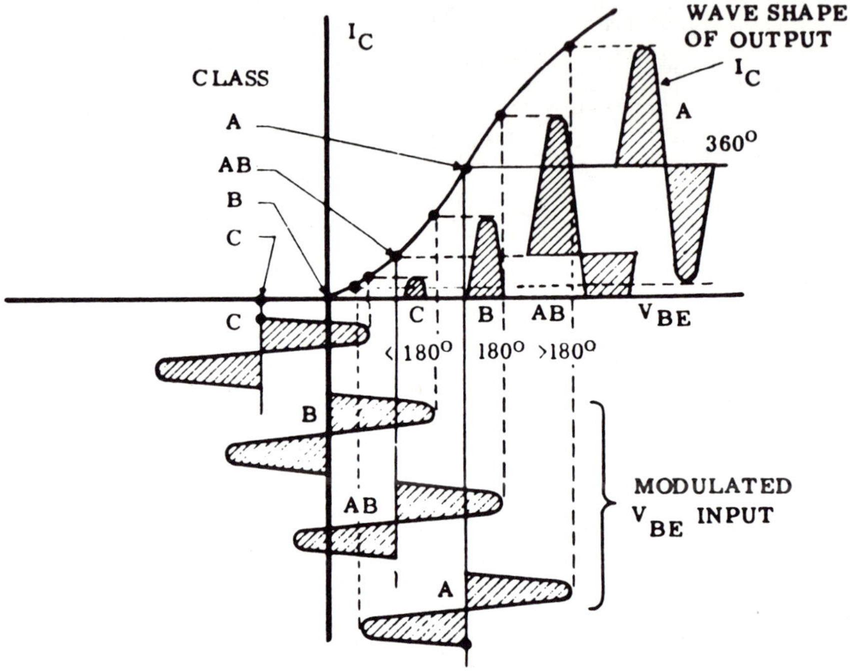

Figure 5-39. Classes of transistor amplifiers. (Courtesy of Motorola Semiconductor Products, Inc., Phoenix, Ariz.)

The bias of the base may be set, using a *fixed bias,* as shown in Figure 5-40. The resistance of the bias resistor is given by the ratio of the battery voltage to the base current. If the base current is 30 μa and a 6-volt battery is used, the resistance to provide the 30-μa current would be 200 K Ω as shown in Figure 5-40. This *fixed bias* is not desirable, however, for common-emitter circuits, because the base current is critical and parameters of transistors tend to vary.

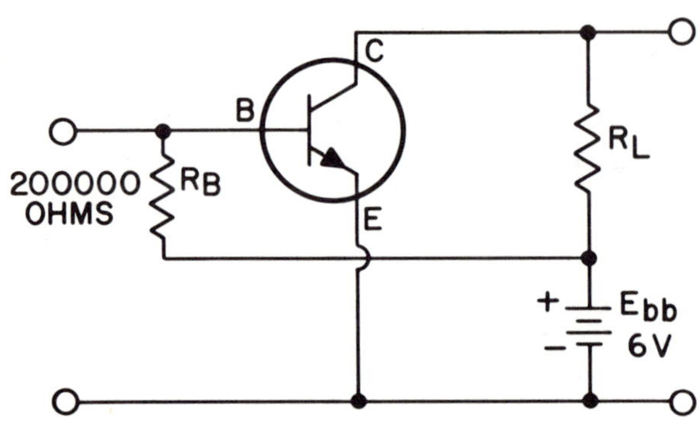

Figure 5-40. Fixed bias. (Courtesy of Radio Corporation of America, Harrison, N.J.)

A better method of bias is the *self-bias* circuit shown in Figure 5-41. Here the bias resistor R_B is given by the collector voltage divided by the base current. When the self-bias arrangement is used an increase in base current will increase the bias and maintain the operating point. An even more stable bias network is called the *voltage-divider* network. The voltage-divider network may be used with or without negative feedback. Figure 5-42(A) shows a *voltage-divider* bias circuit and Figure 5-42(B) shows essentially the same circuit but with negative feedback for increased stability. Notice the capacitor around R_3 in the emitter circuit. This capacitor tends to provide stability with a minimum of loss in gain.

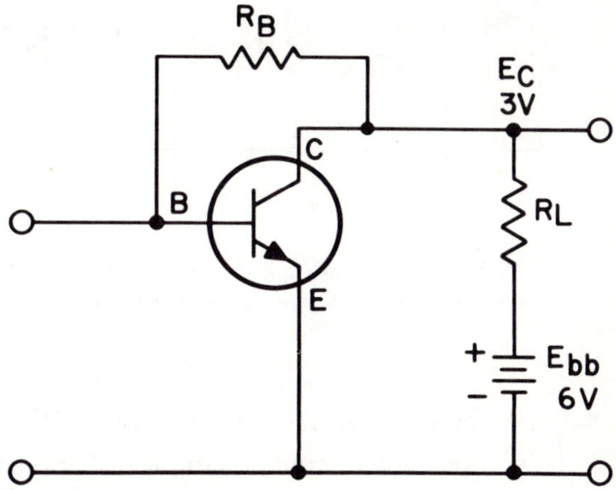

Figure 5-41. Self-bias. (Courtesy of Radio Corporation of America, Harrison, N.J.)

Later in this chapter it will be seen that special components may be used to provide transistor bias. Figure 5-43 shows two examples, namely, the thermistor and the diode. The resistance of the thermistor decreases with increased temperature. Thus, if a resistor were used to stabilize the collector current, an increase in temperature would increase the collector current and the transistor would proceed toward *thermal runaway*, but when a thermistor is used its resistance would block the increase in current due to the increase in temperature* and hence stabilize the circuit. The diode also may be used as shown in Figure 5-43(b). In this case the diode is chosen of the same crystal material as the transistor (germanium or silicon). A transistor could also be used as a diode, but this would be more expensive. Thus, changes in temperature or supply voltage affect the transistor the same as they do the diode, the bias changes accordingly to compensate for the characteristics of the transistor, and the result is increased stability. Other components, for example, the sensistor, are becoming available for this purpose.

COUPLING TRANSISTOR STAGES

Stages in a transistor amplifier are coupled, for the most part, using the same techniques as stages in a vacuum-tube amplifier. Transformer, resistance-capacitance, and direct coupling are used in transistor circuitry. Figures 5-44, 5-45, and 5-46 show examples of each form of coupling. Examination of examples of transistor amplifiers will show which forms of coupling and biasing are most commonly used and for what purpose.

*Resistance increases with temperature.

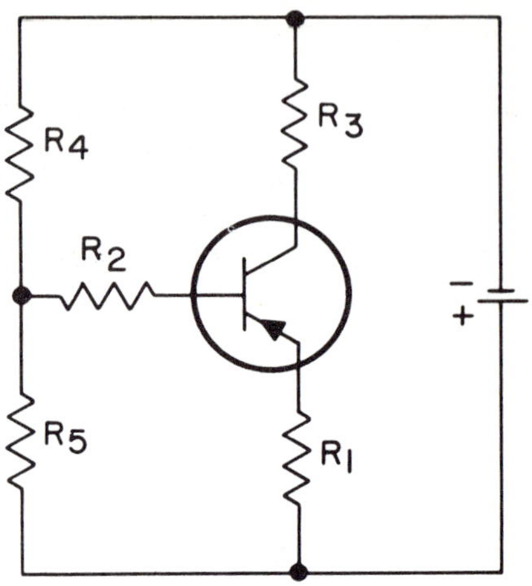

Fig. 5-42A

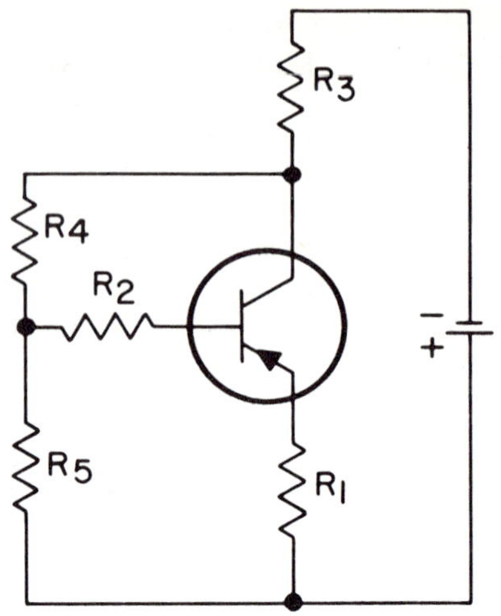

Fig. 5-42B

Figure 5-42. Bias using a voltage divider: A, without feedback; B, with feedback. (Courtesy of Radio Corporation of America, Harrison, N.J.)

EXAMPLES OF TRANSISTOR-AMPLIFIER CIRCUITS

It is important to study examples of transistor circuits which are successfully used to ascertain what characteristics the designer or purchaser of electronic instruments wishes to have incorporated into the design.

Amplifiers used in medical instruments fall into several categories. Some transistor circuits are designed to amplify small voltages such as would be the output of a transducer. These amplifiers must generate a minimum of noise and must have a large signal-to-noise ratio. It is important to note that many parts of a low-noise circuit contribute to the noise level. The resistors, for example, generate noise as a result of thermal activity. In this regard, inexpensive carbon resistors generate much more noise than the more expensive metal film resistors. Metal film resistors are also much more resistant to changes in temperature. When examining a commercial instrument, one should determine how many metal film resistors have been used. Some low-level amplifiers will contain some feedback to improve either the low-frequency or high-frequency response or both. Figure 5-47 is a circuit for a simple low-level class A amplifier. Notice the *voltage divider* bias (R_1 and R_2) and the emitter-stabilizing network (R_3 and C_1). The capacitor C_1 has a lower reactance at high frequencies and hence acts as a compensating network.

It would be advantageous to use the transistor as a push-pull amplifier. This provides increased noise elimination as well as more efficient use of available power. Transistors, unlike tubes, can be set up as shown in Figure 5-48. In this circuit very little or no DC flows through the load resistor (R_L). At the present time such *complementary* circuits (PNP and NPN transistors) are available only in low- and medium-power units. Within a year of the writing of this book NPN transistors of higher power should be available, and thus high-power complementary circuits can

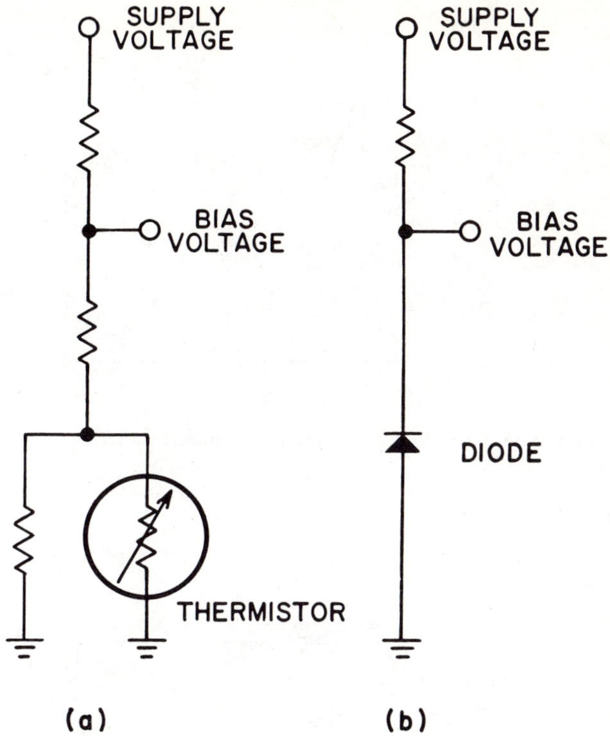

Figure 5-43. Some special components used to establish bias. (Courtesy of Radio Corporation of America, Harrison, N.J.)

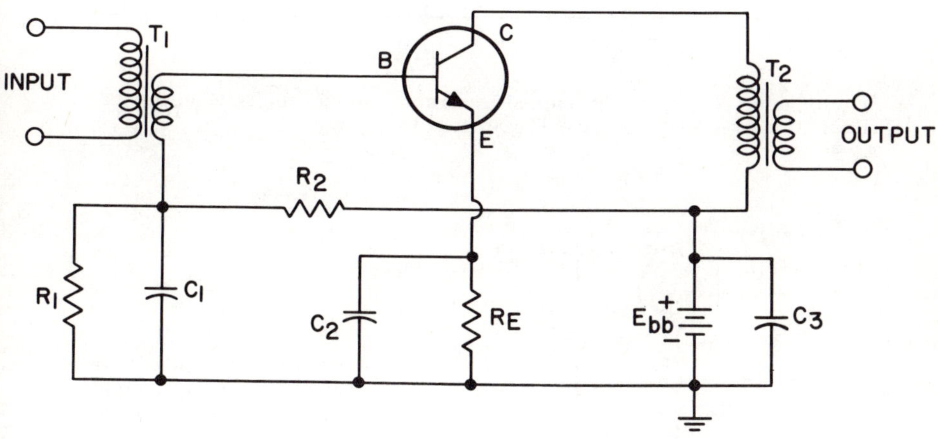

Figure 5-44. A transformer-coupled transistor stage. (Courtesy of Radio Corporation of America, Harrison, N.J.)

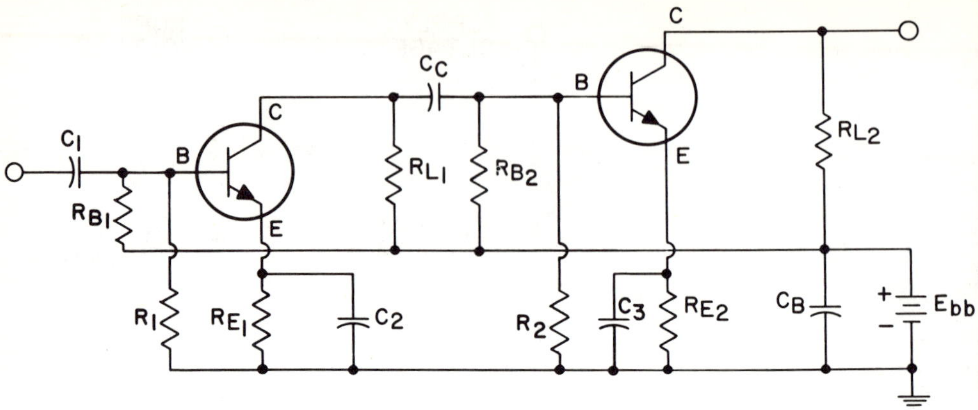

Figure 5-45. RC-coupled transistor amplifier stages. (Courtesy of Radio Corporation of America, Harrison, N.J.)

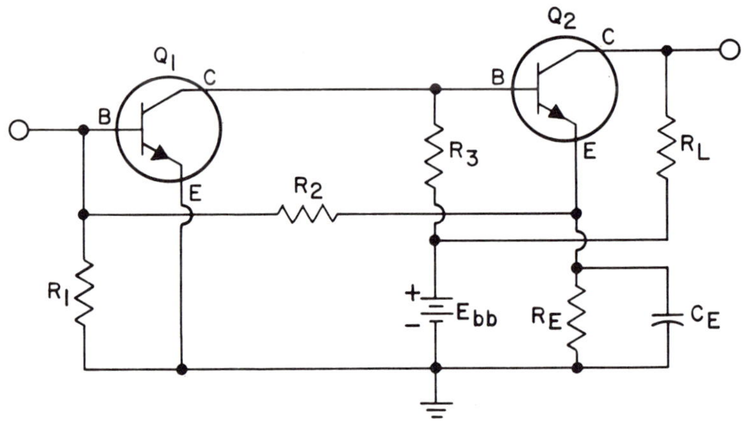

Figure 5-46. Two directly coupled transistor amplifier stages. (Courtesy of Radio Corporation of America, Harrison, N.J.)

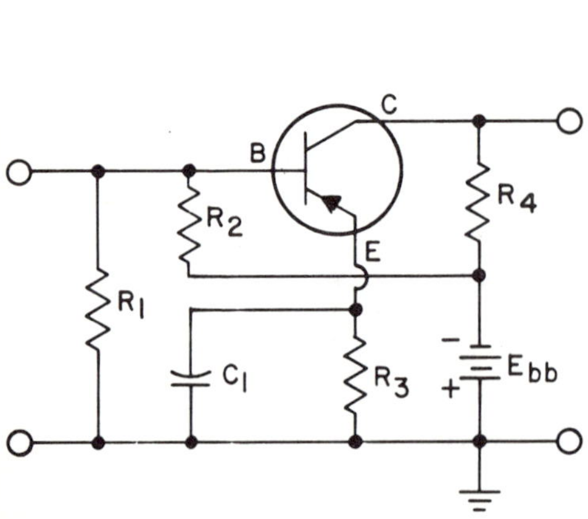

Figure 5-47. A single low-level class A amplifier. (Courtesy of Radio Corporation of America, Harrison, N.J.)

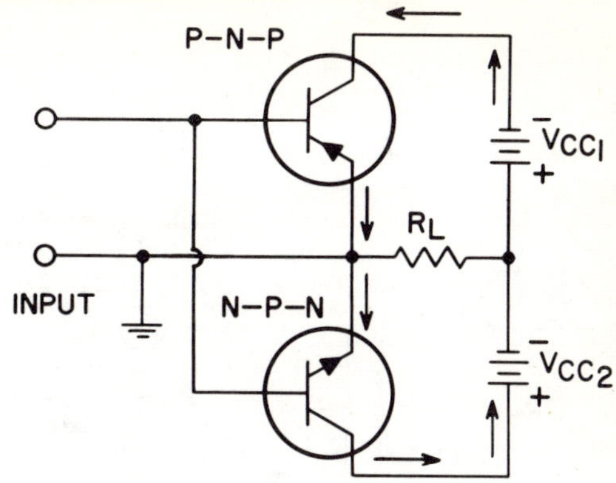

Figure 5-48. Complementary configuration—transistorized push-pull circuit. (Courtesy of Radio Corporation of America, Harrison, N.J.)

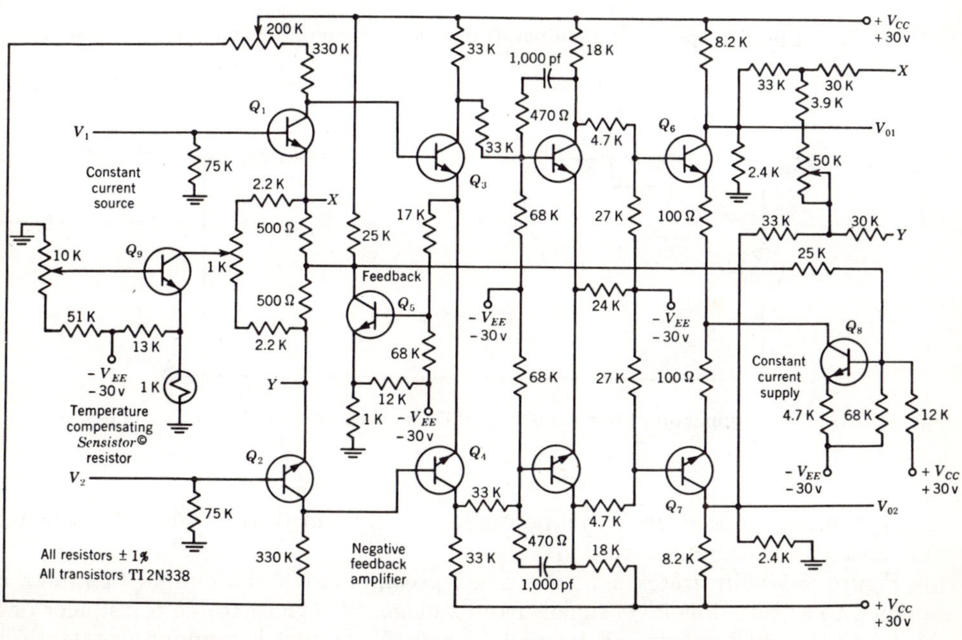

Figure 5-49. A more ideal low-noise differential amplifier. (From Transistor Circuit Design by Texas Instruments, Inc. Copyright © 1963. McGraw-Hill Book Company. Used by permission.)

be built. Transistors are now available which provide an NPN and a PNP transistor in a single combined semiconductor structure.* This, of course, provides for a more ideal push-pull circuit. Figure 5-49 illustrates a more complex low-noise differential amplifier designed by the Texas Instrument Company. This circuit incorporates series-shunt negative feedback which provides high-input impedance as well as low-output impedance. This circuit is reported to respond to a differential signal

*These semiconductor structures are called *integrated circuits.* They may consist of one or more stages and some circuit components built on a single "semiconductor chip."

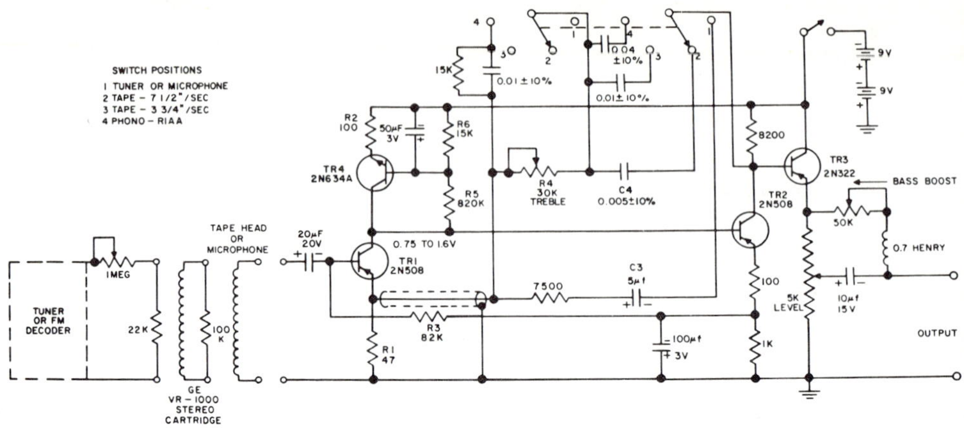

Figure 5-50. A phono-tape input amplifier. (Courtesy of General Electric Co., Syracuse, N.Y.)

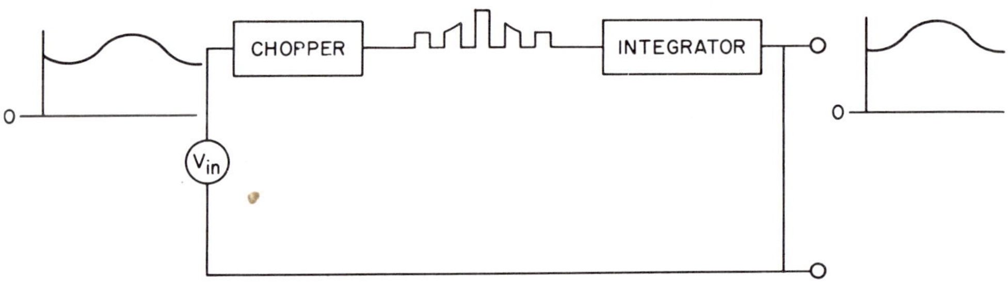

Figure 5-51. The chopper amplifier. (Courtesy of Radio Corporation of America, Harrison, N.J.)

of $25\mu V$ superimposed on a common-mode level from 0 to 5 volts. The sensitor compensates for thermal variation.

Figure 5-50 illustrates a phono-tape "preamplifier."* This input amplifier is designed to amplify low-level signals from a magnetic tape cartridge transducer or a phonograph record transducer. It employs negative feedback, temperature stabilization, and frequency compensation.

It is not uncommon to employ a circuit, as is shown in block form in Figure 5-51, to chop the incoming DC signal and treat it as an AC signal, then amplify it and integrate the amplified signal so as to present an amplified DC output signal. Since AC amplifiers are inherently more stable and have higher gain, the *chopper amplifier*† represents an advantage over the direct-coupled DC amplifier.

The chopper itself may be an electromechanical device,† such as is used in the Offner Recorder input amplifier, or it may be entirely solid state as is the Texas

*A preamplifier is an amplifier which precedes another amplifier. I do not approve of the term as it implies that the preamplifier is not an amplifier. The preamplifier should more properly be called an input amplifier.

†Choppers may now be obtained using semiconductors in module form (*e.g.,* Solid State Electronics Corporation, Sepulveda, Calif.).

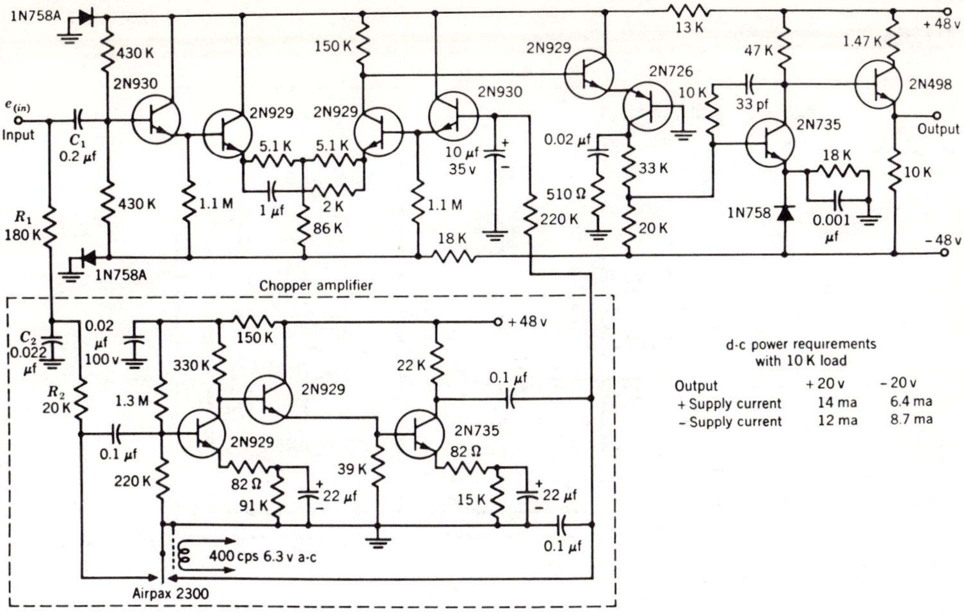

Figure 5-52. A transistorized chopper amplifier. (From Transistor Circuit Design by Texas Instruments, Inc. Copyright © 1963. McGraw-Hill Book Company. Used by permission.)

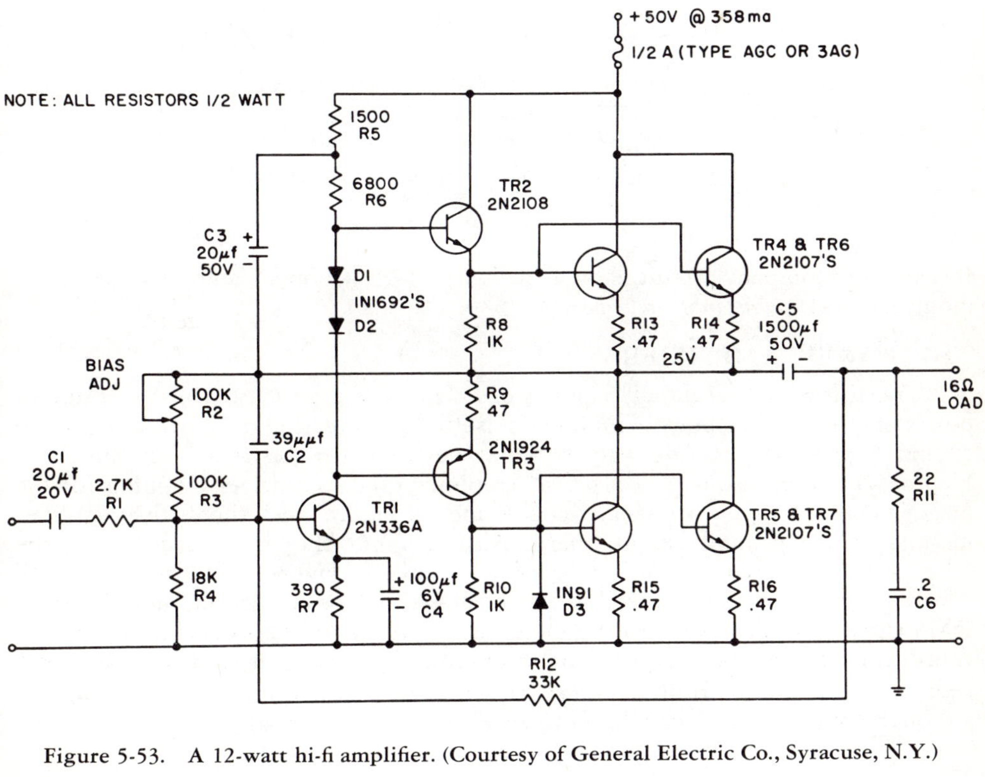

Figure 5-53. A 12-watt hi-fi amplifier. (Courtesy of General Electric Co., Syracuse, N.Y.)

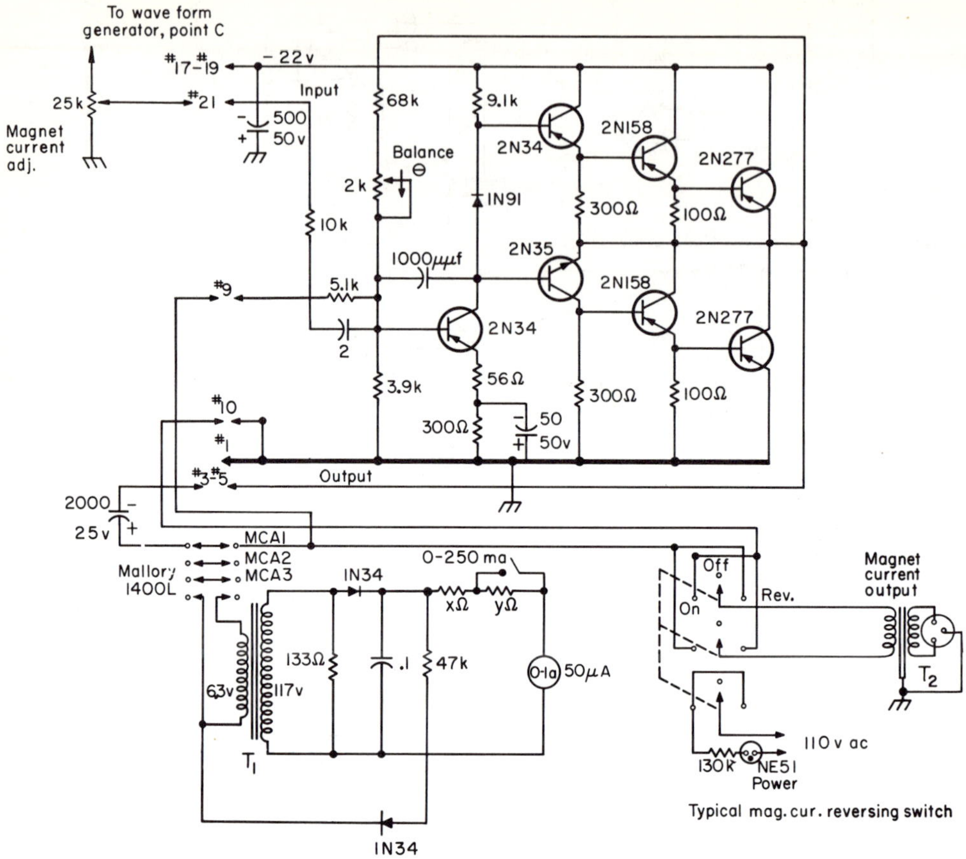

Figure 5-54. **Magnet current amplifier circuit. (Courtesy of Lawrence Radiation Laboratory, Berkeley, Calif.)**

Instrument Company circuit shown in Figure 5-52. Chopper amplifiers are commonly used as *operational amplifiers.*

THE POWER AMPLIFIER

Transistors are essentially current devices and are best used in the design of power amplifiers. The power amplifier is useful for increasing the current in a given circuit. Figure 5-53 is a 12-watt amplifier designed for use as a Hi-Fi amplifier. Figure 5-54 is an example of a power amplifier used as a driver amplifier for the magnet coils of an electromagnetic flow meter. Figure 5-55 shows this amplifier mounted on a printed circuit board. Although not shown in Figure 5-55, the schematic for the magnet current amplifier shown in Figure 5-54 also has a small separate circuit in the lower left which allows the peak current (not the usual RMS current) to be read with the meter as shown. A polarity-reversing switch and transformer to *float* the output signal (unground both sides) is shown on the lower right of Figure 5-54. Both of these circuits are useful for medical instruments although some variation may be necessary for special application.

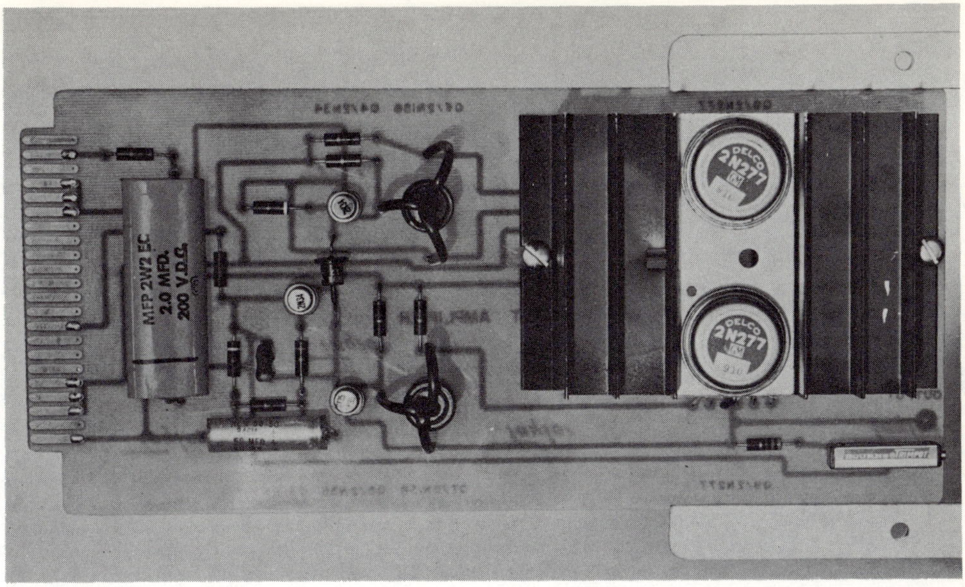

Figure 5-55. Magnet current amplifier on a printed circuit board. (Courtesy of Lawrence Radiation Laboratory, Berkeley, Calif.)

REFERENCES:

Basic Theory and Application of Electron Tubes (TM 11-662 to 16-1-255, Department of the Army, 1952).

Basic Theory and Application of Transistors (TM 11-690, Department of the Army, 1959).

GE Transistor Manual, Ed. 6 (General Electric Company, 1962).

Hall Effect Manual, Ed. 2 (Helipot Division of Beckman Instruments, Inc., Fullerton, Calif., 1964).

Handbook of Operational Amplifiers (Burr-Brown Research Corporation, Tucson, Ariz.).

Motorola Power Transistor Handbook (Motorola Semiconductor Products Division, 1960).

Pulse and Digital Circuits, Jacob Millman and H. Taub (McGraw-Hill Book Co., New York, 1956).

Radio Amateur's Handbook, revised annually (American Radio Relay League, Newington, Conn.).

RCA Receiving Tube Manual RC-22 (Radio Corporation of America, 1963).

RCA Transistor Manual SC-11 (Radio Corporation of America, 1964).

Transistor Circuit Design, Texas Instruments Inc. (McGraw-Hill Book Co., New York, 1963).

6

The Oscillator

the sinusoidal oscillator

non-sinusoidal oscillators

references

There are many ways in which a voltage or current that varies sinusoidally with time can be used in all types of electronic instrumentation and, because of usefulness, a large group of circuits has been classified under the heading of *oscillator circuits.* This chapter is devoted to discussing the principles of operation of the oscillator circuit; a variety of examples are given so that the student will understand the circuitry found in medical instruments. The class of circuits called *oscillator circuits* also includes non-sinusoidal generators such as the square wave and sawtooth generator as well as special circuits.

Oscillators are described not only by the wave shape of the signal but also by the frequency of oscillation. The names given to various frequencies of sinusoidal oscillation are listed in Table 6-1.

TABLE 6-1

Subaudio	less than 20 cps
Audio	20-20,000 cps
Ultrasonic	20,000 cps-1 MC
Radio Frequency (Broadcast)	150 KC-1.5 MC
RF (high frequency)	1.5 MC-40 MC
VHF (very high frequency)	40 MC-100 MC
UHF (ultra high frequency)	over 100 MC

Perhaps the most important application of the oscillator circuit is in providing the carrier for the transmission of radio and television signals. The oscillator is useful in a variety of medical electronic instruments, such as the stimulator, and in providing a carrier signal for use with bridge detectors, such as the strain gauge.

In addition to these specific applications, oscillator circuits of one sort or another are used in oscilloscope sweep circuits and in diverse calibrating and time-indicating devices.

When an oscillating signal is required, it should be a pure oscillation (one frequency). It should not have multiples of that frequency (harmonics) in its output. The frequency of oscillation chosen should be maintained within definite limits. It should not wander or drift. The voltage or current output must also be maintained at a constant amplitude over its useful range. In some instances it may be necessary to provide a *current* of constant amplitude; in other applications it is necessary to provide an oscillation *voltage* of constant amplitude. The oscillator should be stable over a wide range of temperatures and under any special conditions that might be anticipated such as acceleration, impact, and operation inside an animal or in a bag of salt water.

THE SINUSOIDAL OSCILLATOR

There are many ways in which a sinusoidal oscillator can be designed using both vacuum tube and transistor circuitry. Mechanical devices also exist which may act as oscillator elements, some of which have characteristics superior to electronic circuitry, although, in general, the frequency of these units is limited. Examples of such oscillating components are tuning forks and electromechanical devices similar to a doorbell.

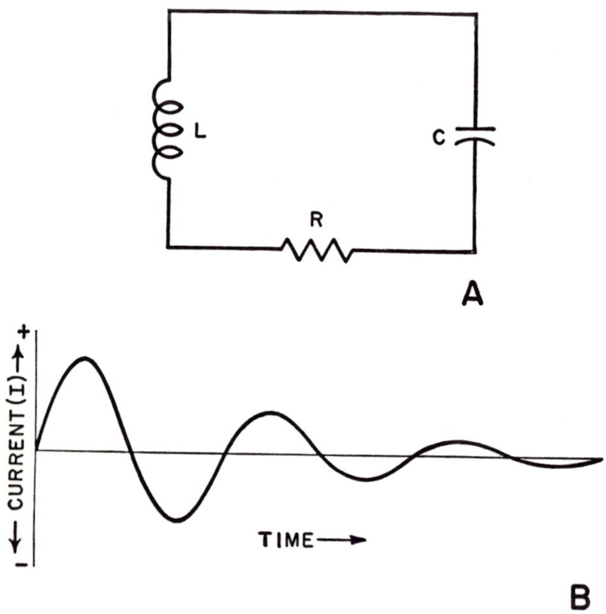

Figure 6-1. Damped oscillatory current in an RLC circuit. (Reproduced by permission of the Department of the Army.)

If the capacitor in the RLC network shown in Figure 6-1(A) were suddenly energized, the current would flow in the network as indicated in Figure 6-1(B). The student should at this time review the discussion of the RLC network in Chapter 2 and re-examine Figures 2-11 and 2-12. Notice that damped oscillations may be generated in an RLC network. Were the resistance of the network shown in Figure 6-1(A) to be eliminated as shown in Figure 6-2, oscillation of the current to

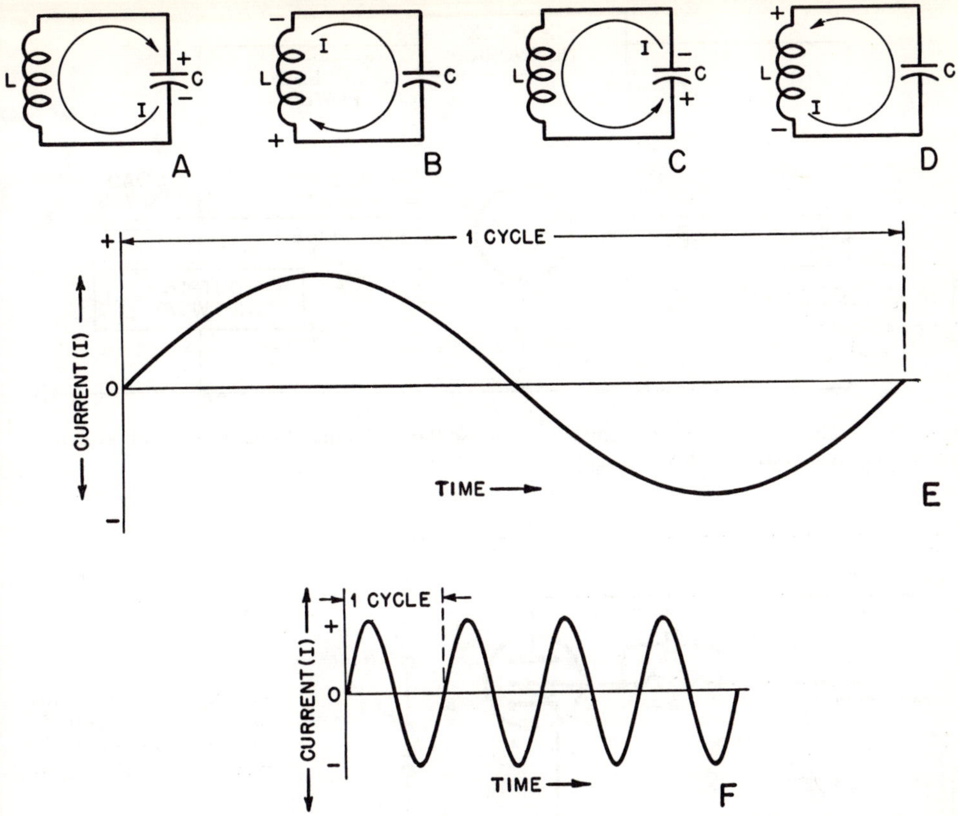

Figure 6-2. Oscillatory current in an LC circuit. (Reproduced by permission of the Department of the Army.)

and fro in the network would be produced by suddenly charging the capacitor and would continue as graphed in Figure 6-2(F). This is, of course, an impossible physical situation because all components must have some resistance associated with them.

How then can we produce a continuing oscillation in a "real" circuit? The answer is that the signal shown in Figure 6-1(B) must be applied as signal to the input of an amplifier, and a portion of the amplified output must be then used as positive feedback power to prevent degeneration of the signal (Fig. 6-3). That portion of the output power which is not used as positive feedback may be used to drive a load and is thus a useful oscillator signal.

Figure 6-4 is a schematic for the *tuned-grid* oscillator. When the switch is closed, electrons travel from the cathode to the plate. This current through the tube causes a small amount of grid current to flow. The grid current charges capacitor C_1, and the LC network begins to oscillate. Oscillation of the LC network (L_1 and C_1), called a *tank circuit,* causes variation of the signal to the grid, which assists in generating the sinusoidal oscillation. The time constant of the coupling RC network (C_g and R_g) is chosen to keep the grid at the proper bias so as to allow the tube to be in Class A operation and prevent distortion. The flow of current through the

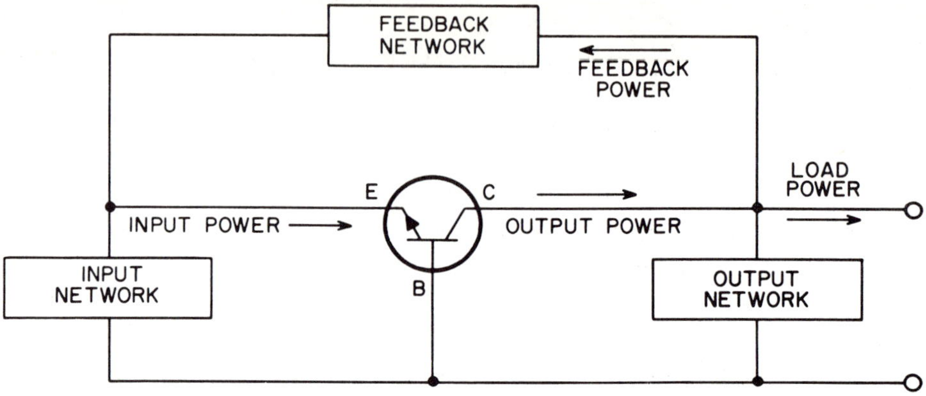

Figure 6-3. The use of positive feedback in an oscillator circuit. (Courtesy of Radio Corporation of America, Harrison, N.J.)

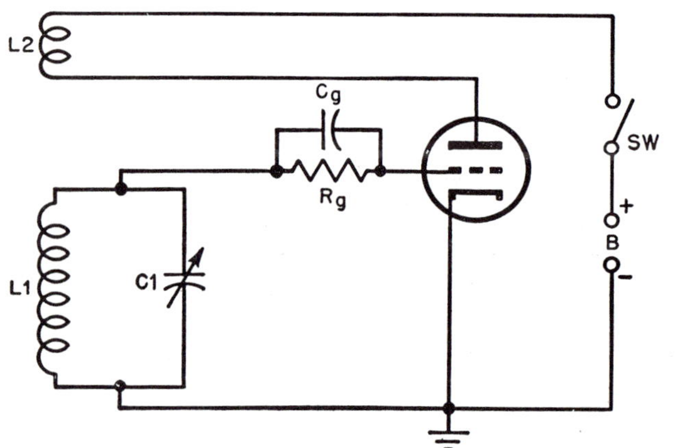

Figure 6-4. Tuned-grid oscillator. (Reproduced by permission of the Department of the Army.)

inductor L_2 in the plate circuit, when the inductor is wound correctly on the same core as L_1, causes regenerative feedback which maintains amplitude of oscillation. It is important to point out that if this oscillator is required to supply much current, the tube will be loaded down. This is the equivalent of adding resistive components to the LC network and thus increasing the damping. To avoid this a buffer stage, which is an amplifier with a low gain, is used to isolate the oscillator from the remaining circuitry in the instrument.

The transistor equivalent of the tuned-grid oscillator is the *tuned-base oscillator* and is shown in Figure 6-5. The amplifier portion of the oscillator in Figure 6-5 is shown within the dotted lines, and resistors R_B, R_F, and R_C supply the proper DC bias to forward bias the emitter-base junction and apply reverse bias to the collector-base junction and thus maintain the transistor in Class A operation. The feedback is developed across windings 1 and 2. The tank circuit consists of the inductance as associated with windings 3 and 4 of transformer T_1 and capacitor C_1. Capacitor C_c prevents the DC bias from affecting the tank circuit. The frequency at which the tank circuit oscillates can be adjusted within limits. Notice that the resistor

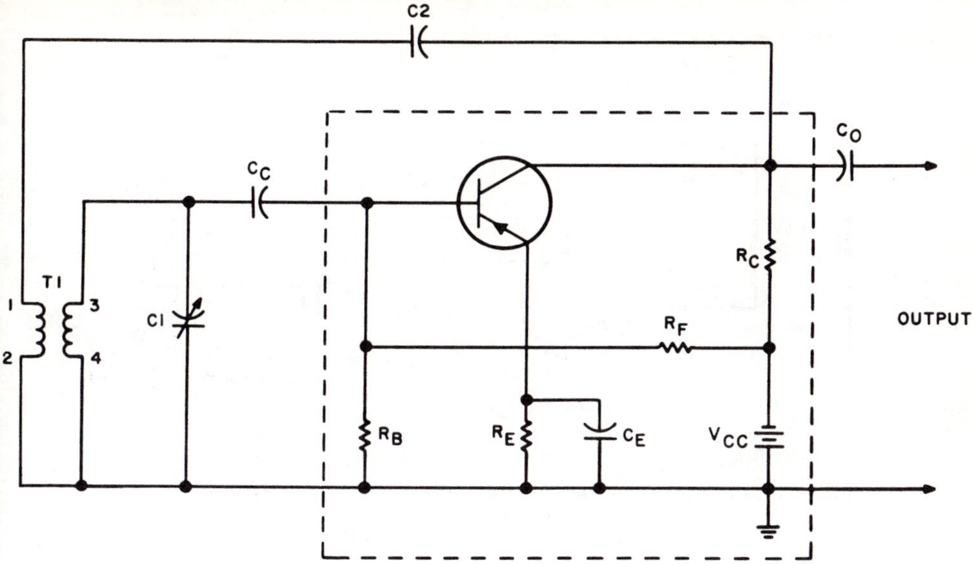

Figure 6-5. Tuned-base oscillator. (Reproduced by permission of the Department of the Army.)

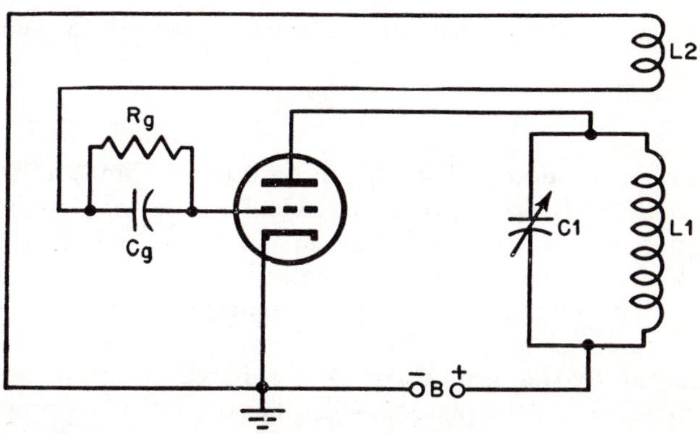

Figure 6-6. Tuned-plate oscillator. (Reproduced by permission of the Department of the Army.)

in the emitter leg of the circuit prevents an excess of DC current from destroying the transistor, while the capacitor C_E allows the AC signal to bypass R_E and prevent degeneration of the signal at high frequencies.

It is also possible to place the LC tank circuit in the plate circuit of the vacuum-tube oscillator or equivalent in the collector circuit of the transistor oscillator. Then the positive feedback would have to be in the grid circuit or in the emitter circuit. The vacuum tube and its transistor equivalent for this type of oscillator are shown in Figures 6-5 and 6-6.

In the oscillator circuits described above the tank circuit is located in one section of the oscillator, either the grid circuit or the plate circuit, and the positive feedback is generated in the other section. Although not shown, it is possible to *tune* both the plate and the grid circuits. In this circuit, called the *tuned-grid-tuned-plate* oscillator, there are two separate tank circuits. In another variety of

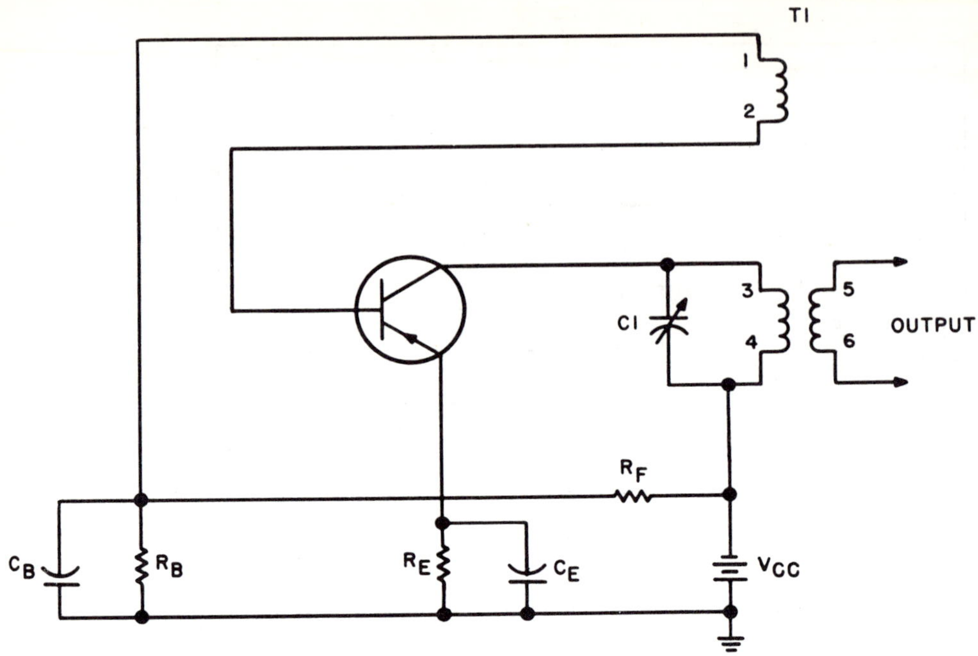

Figure 6-7. Tuned-collector oscillator. (Reproduced by permission of the Department of the Army.)

oscillator, one tank circuit can be used in common for both the plate and the grid circuit. This arrangement is called a *split-tank* oscillator, and there are two basic types, the *Hartley* oscillator, in which the inductance is split, and the *Colpitts,* in which the capacitance of the tank circuit is split.

The Hartley and Colpitts oscillators are shown in Figures 6-8 and 6-9. In the Hartley oscillator one end of the variable capacitance C_1 of the tank circuit is connected to the grid through a suitable coupling network and the other end is connected to the plate. Notice that the DC plate power supply is in series with the plate circuit. This is a *series-fed* Hartley oscillator. The power supply could be in parallel with the plate circuit. If this were so, it would be a *shunt-fed* Hartley oscillator. The shunt-fed oscillator reduces the effect of power supply variation (ripple in a DC power supply; see Chapter 7). Notice that in the Hartley oscillator the inductance is center tapped, and the center goes to the cathode. In this case, an inductive voltage divider has been used. The transistor version of the Hartley oscillator is shown in Figure 6-9. The Hartley oscillator can also be modified to be operated push-pull for increased power output.

Figure 6-8(B) is the circuit for the Colpitts oscillator. In this circuit the cathode is supplied by a capacitive voltage divider, but the tank circuit is still common between the plate and the grid circuits. The Colpitts circuit in Figure 6-8(B) shows the *shunt-fed* plate voltage while Figure 6-8(A) illustrates the *series-fed* arrangement.

The transistorized Colpitts oscillator is shown in Figure 6-10. In this circuit, as in Figure 6-9, R_B and R_F set the base bias, R_C is the collector load resistor, and the positive feedback is applied to the emitter through resistor R_F. The input signal

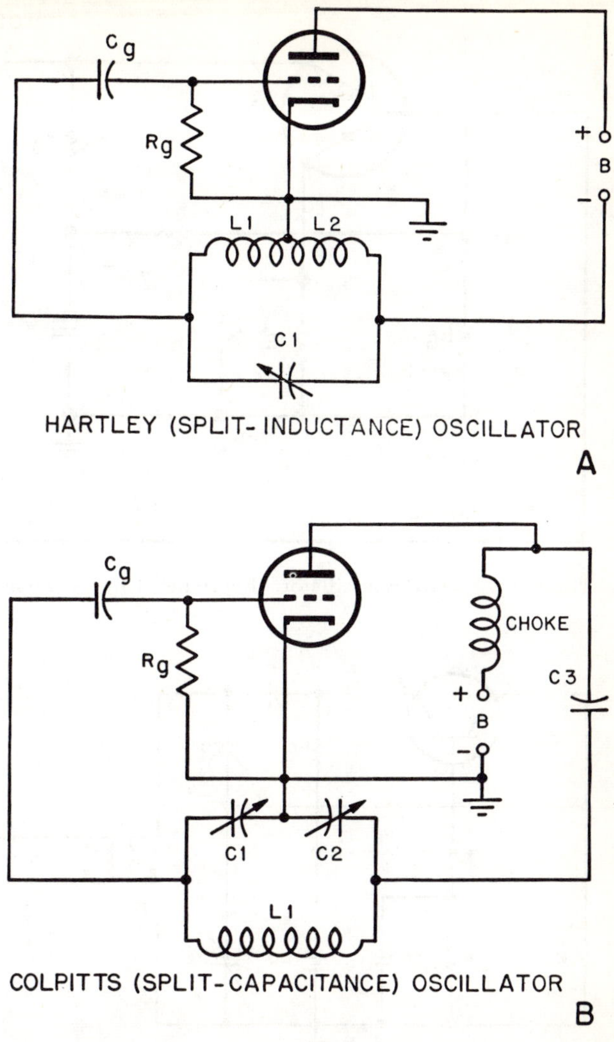

Figure 6-8. Split-tank oscillators. (Reproduced by permission of the Department of the Army.)

is developed across the resistance in the emitter leg R_F, and capacitor C_E prevents degeneration of the signal. Notice that the Colpitts oscillator is a *common-base* configuration and the feedback signal developed between C_1 and C_2 is in phase with the collector. A modification of the Colpitts circuit is called the *Clapp* oscillator. The oscillator circuit is shown in Figure 6-11. The Clapp oscillator is operated in the *common-emitter* configuration, and the feedback signal must be shifted 180° in phase in order to provide positive feedback. This is done with a capacitor in series with the inductance in the tank circuit. In this arrangement, when the series capacitor C is large in comparison with tank capacitors C_1 and C_2, then the operating frequency is determined by the inductance of the transformer windings 1 and 2 of T_1 and is thus controlled with the series capacitor C.

Thus far, all the oscillators which have been described have employed an LC network to determine the frequency of oscillation. The LC tank circuit is frequently

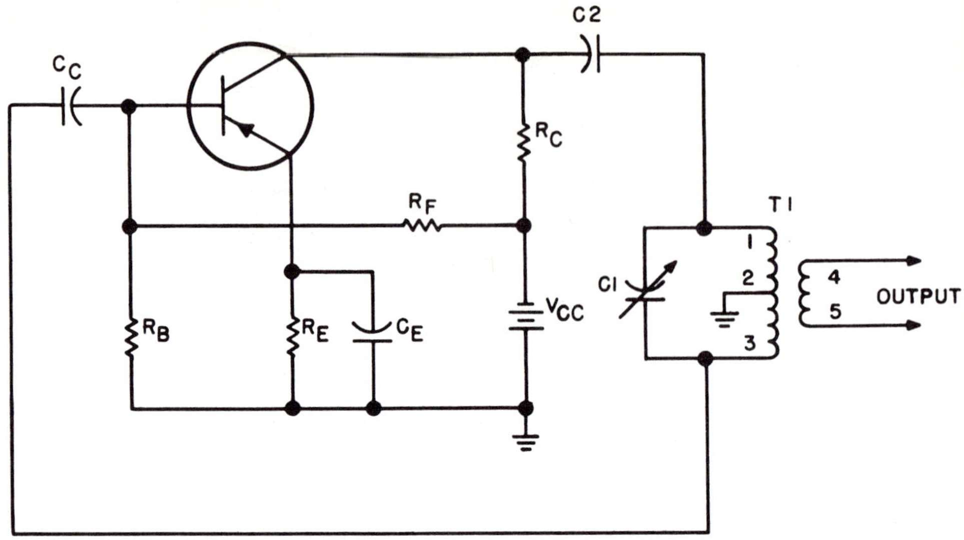

Figure 6-9. Hartley oscillator. (Reproduced by permission of the Department of the Army.)

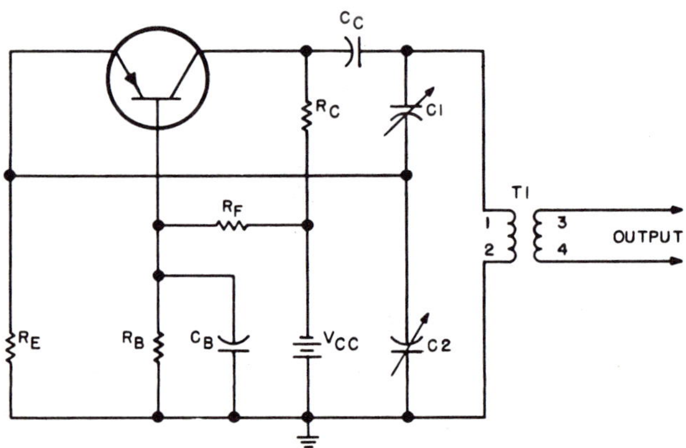

Figure 6-10. Colpitts oscillator. (Reproduced by permission of the Department of the Army.)

used in communication electronics as it is possible to generate higher frequency oscillations. It is, however, more common to use an oscillating RC network rather than the LC network in medical instruments. The RC network can be made smaller since it does not need large, heavy, and expensive inductors, and is less expensive. Figure 6-12 is an example of an *RC phase-shift oscillator.* Again in this circuit, R_B, R_F, and R_C determine the bias and should be chosen so that the transistor is in Class A operation. Resistor R_E limits the amount of current flowing through the emitter circuit, and capacitor C_E allows the alternating signal to bypass R_E, thus preventing high-frequency degeneration.

In the common-emitter arrangement the signal at the base is 180° out of phase with the signal at the collector. The RC network usually consists of three or more RC pairs or sections. When three sections are used, the time constant of each RC pair shifts the signal 60° for a total of the required 180°. In Figure 6-12 these RC

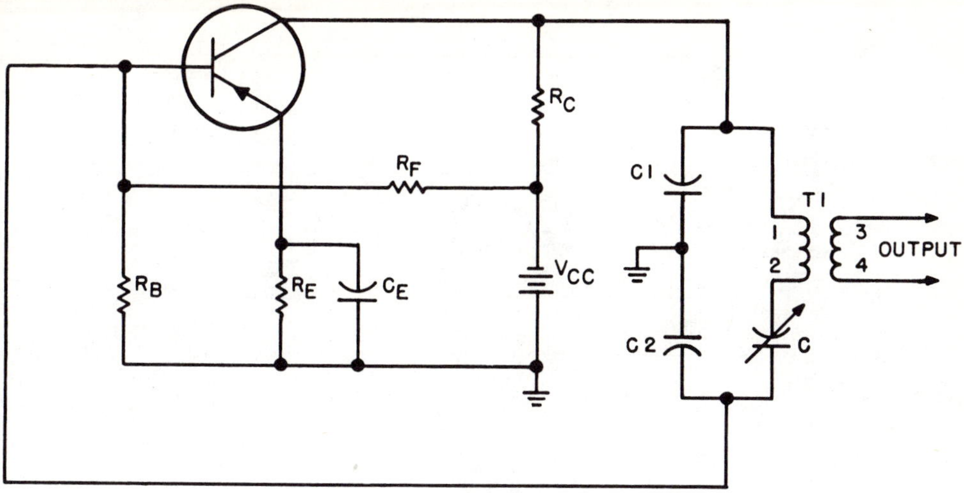

Figure 6-11. Clapp oscillator. (Reproduced by permission of the Department of the Army.)

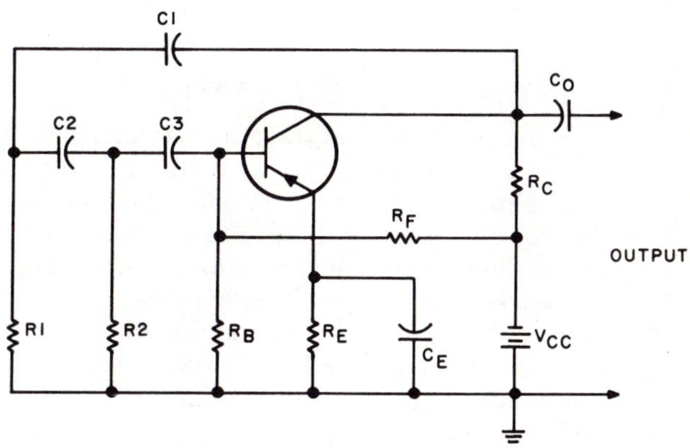

Figure 6-12. The RC phase shift oscillator. (Reproduced by permission of the Department of the Army.)

pairs are R_1C_1, R_2C_2, and R_3C_3. Since a given time constant for an RC network is valid for only one frequency, at other frequencies the phase shift does not total 180° and distortion occurs. The RC phase-shift oscillator requires a transistor or tube with a much higher gain than does an oscillator employing an LC tank circuit. This is so because there is considerable loss of signal across the RC pair compared to losses in LC network. The amount of loss, however, decreases as the number of RC sections is increased.

As previously mentioned, many other devices may be used to establish the oscillating frequency. Figure 6-13 is an example of a commercially available plug-in module. This particular circuit operates from 10 to 75 kilocycles and its output is 8 volts peak-to-peak. It requires a positive and a negative 12-volt power supply. The crystal acts as a coupling impedance between the output of the second stage and the input of the first stage. Since a full 360° phase shift occurs through two cascaded common-emitter transistor amplifiers, the circuit will oscillate at a frequency which experiences neither phase shift nor appreciable attenuation through the crystal.

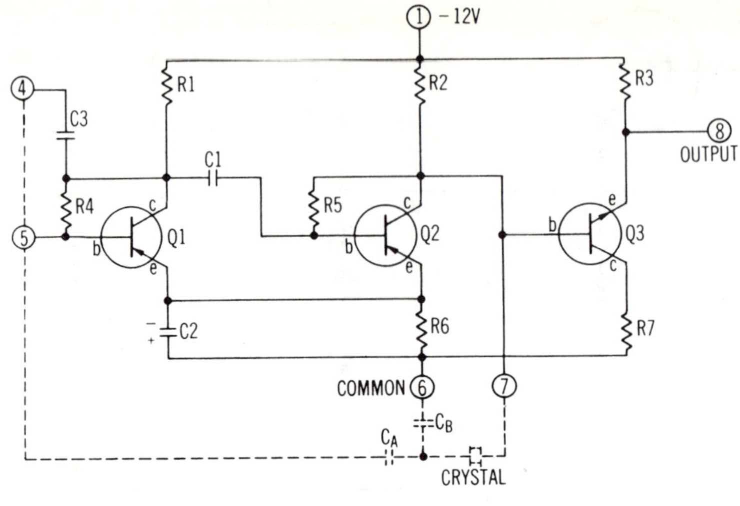

Figure 6-13. A crystal oscillator, module T-107. (Courtesy of Engineered Electronics Co., Santa Ana, Calif.)

NON-SINUSOIDAL OSCILLATORS

Both vacuum tube and transistor circuits have been designed to serve a number of specialized applications other than amplifiers, *per se,* or sinusoidal oscillators. Many of these circuits are quite important in the design of medical instruments. Included for discussion in this group are the multivibrator, the one-shot, the Schmitt trigger, the flip-flop, and the squaring circuit. All of these circuits are basically alike. The circuit has two *states.* The first state occurs when one of the components (transistor or tube) is turned on and the other is off. The other state occurs when the second component (transistor or tube) is turned off while the first is on. This amounts to a two-stage amplifier with positive feedback, the feedback always being supplied by the other stage.

The circuit described above may be called a *multivibrator.* If the multivibrator oscillates regularly between the two states, the multivibrator is called an *astable multivibrator.* This circuit may be used as a square-wave generator. Figure 6-14 is a commercially available multivibrator module (E.E. Co.* Z-90036). The bias applied to the grids of both stages will vary according to the fixed power supply bias (B−) and the dynamic state of each of the triode stages. When the grid is at zero bias the other stage V_2 is driven to cut off. When V_2 is cut off, capacitor C_1 will discharge through resistor R_4. This causes the voltage on the grid of V_2 to approach a range where V_2 will conduct. When V_2 does conduct, it drives V_1 to cut off. The transfer of conduction from V_1 to V_2 is extremely fast, depending on the circuit components. The circuit described above is a free-running multivibrator which continuously generates a square wave when power is applied. The frequency of the square wave depends on the values of C_1 and C_2 and the bias applied to R_3 and R_4.

Figure 6-15 is a transistorized multivibrator; it consists of two RC coupled common-emitter stages. The output of the first stage, Q_1, is coupled to the base of

*Engineered Electronics Co., Santa Ana, California.

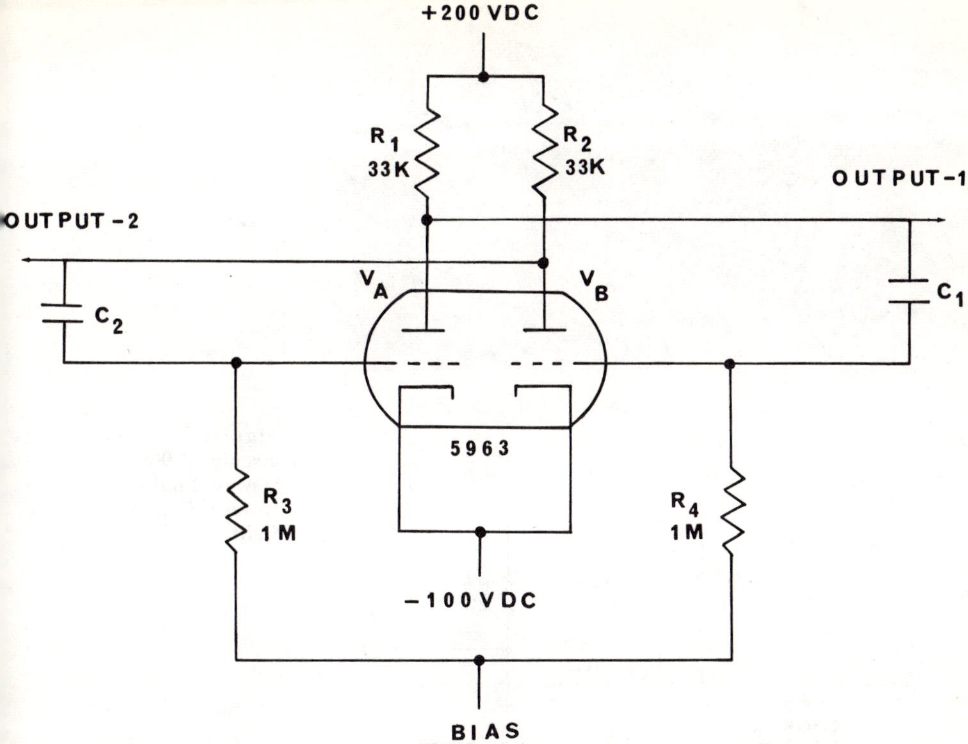

Figure 6-14. A vacuum-tube multivibrator, module Z-90036. (Courtesy of Engineered Electronics Co., Santa Ana, Calif.)

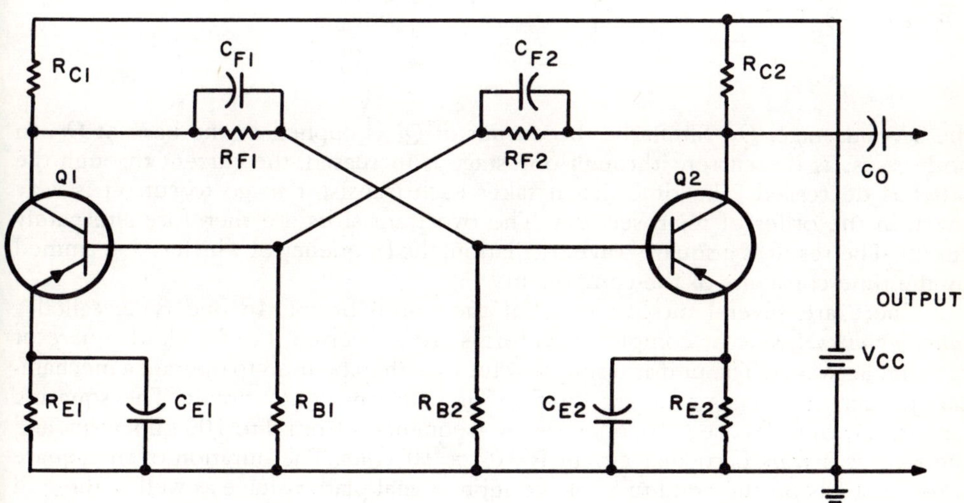

Figure 6-15. RC-coupled common-emitter multivibrator. (Reproduced by permission of the Department of the Army.)

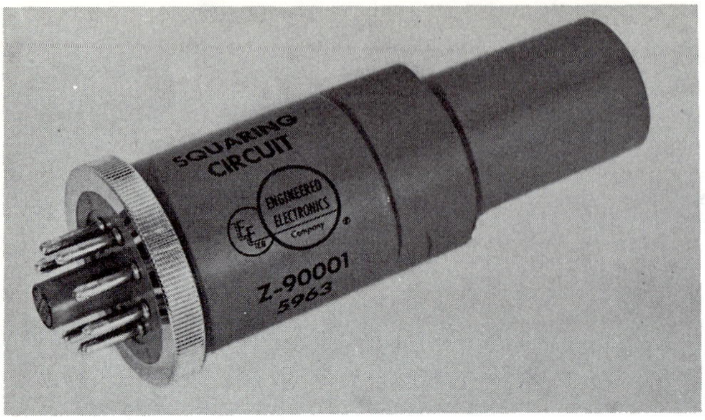

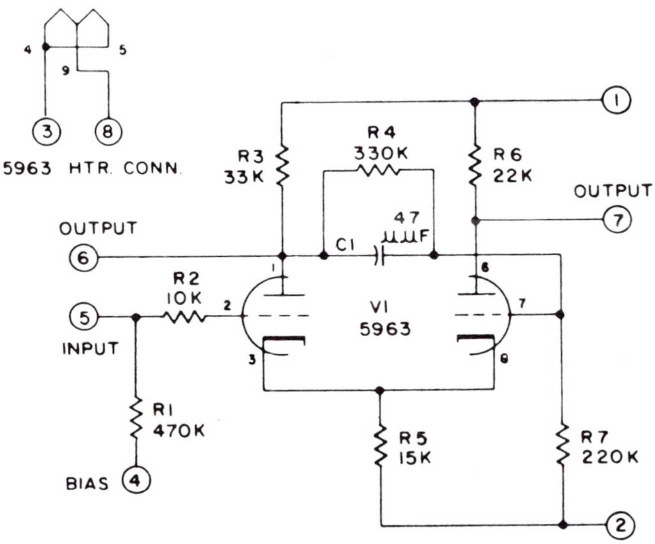

Figure 6-16. Squaring circuit Z-90001. (Courtesy of Engineered Electronics Co., Santa Ana, Calif.)

the second stage, Q_2. Similarly, the output of Q_2 is applied to the base of Q_1. In both cases, as the current through one stage is increased, the current through the other is decreased. The time that it takes each transistor to go to cut off is very short, in the order of microseconds. The two transistors are therefore alternately cut off. The result is a square-wave oscillation, the frequency of which is determined by the time constants of the components.

There are several modifications of the multivibrator. In one type, called a *squaring circuit,* sine or complex waveforms are converted into standard square or rectangular waves. The uniform square wave may then be used to operate a mechanical counter, to trigger a binary flip-flop, or to open a gate circuit. The squaring circuit, shown in Figure 6-16, operates on frequencies from 0 to 100 kilocycles, and the output of this particular circuit is 100 ± 10 volts. The duration of the square wave depends on the amplitude of the input signal plate voltage as well as the grid bias. Figure 6-17 is an example of a transistorized squaring amplifier.

It is often necessary to use the output of a transducer to trigger some action or activate some circuit or system. This may be done with still another modification

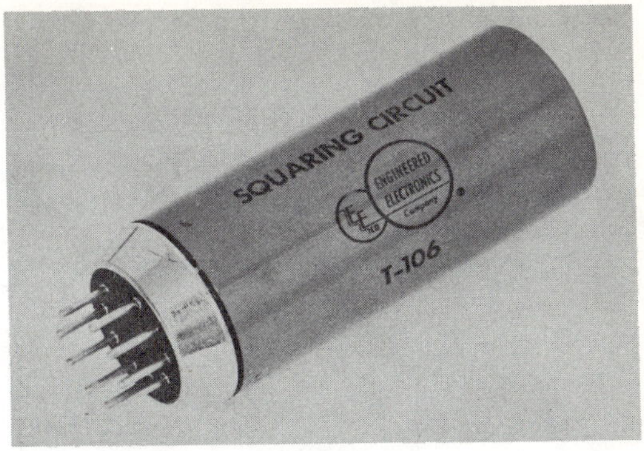

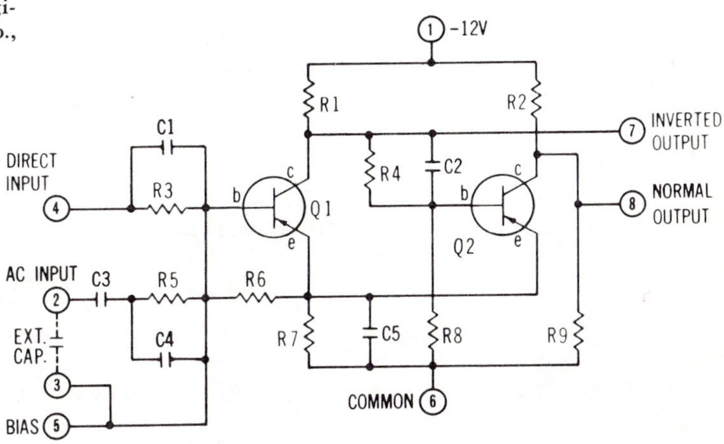

Figure 6-17. Transistorized squaring circuit T-106. (Courtesy of Engineered Electronics Co., Santa Ana, Calif.)

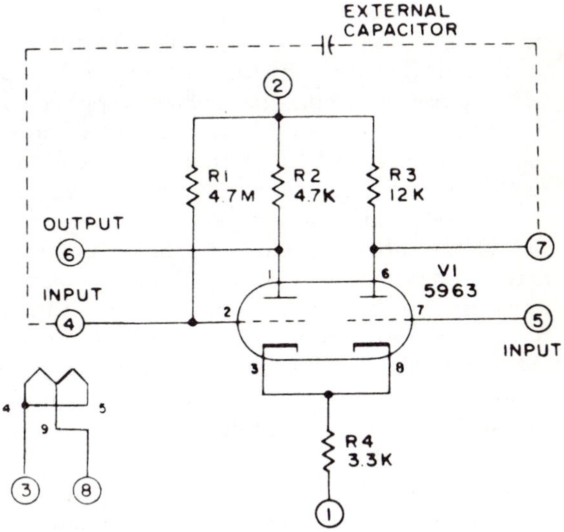

Figure 6-18. Vacuum tube one-shot, Z-8889. (Courtesy of Engineered Electronics Co., Santa Ana, Calif.)

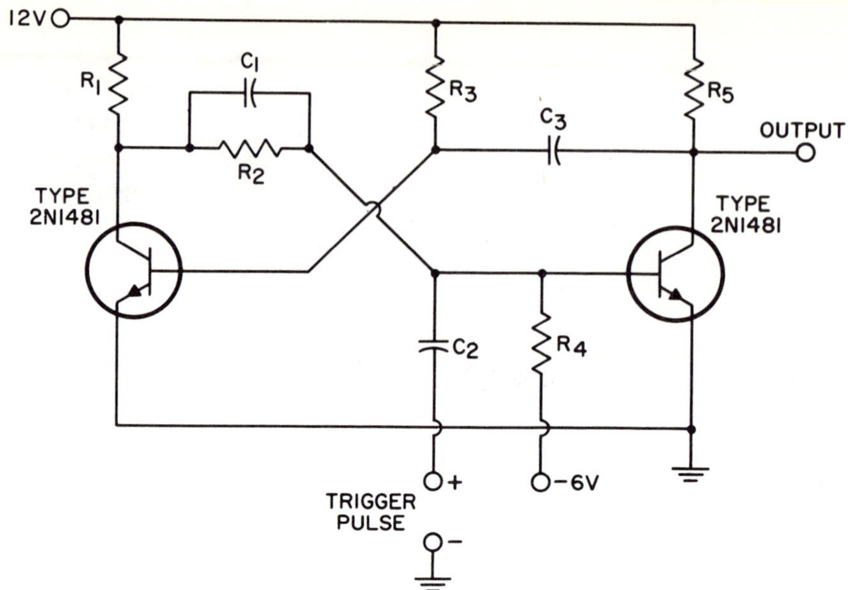

Figure 6-19. Transistorized monostable multivibrator or one-shot. (Courtesy of Radio Corporation of America, Harrison, N.J.)

of the multivibrator, called a *one-shot*, shown in Figure 6-18. A minimum signal, in this case +27 volts, applied to pin 7 will cause the tube to *bootstrap** to cut off, which in this case is +60 volts, and stay there for a period depending on the time constant of the feedback capacitor between the plate of the second stage and grid of the first. Figure 6-19 is the circuit for a basic transistorized multivibrator or one shot. This circuit, like its vacuum-tube equivalent, operates for a set period when triggered with a pulse of sufficient amplitude.

The multivibrator shown in Figure 6-20 is called a *bistable* or *flip-flop*. The output, which is the collector of Q_2, will remain at a steady voltage until input A is pulsed. The output will then change to another voltage (or state) and remain there until input *A* is pulsed with a voltage of the opposite polarity or input *B* is pulsed with a pulse of the original polarity (applied to input *A*). The multivibrator circuits described in this section have many uses appropriate to medicine and biology. The most important use for these circuits is for digital computer modules. It may be seen now that this circuit is the basis for the digital computer which operates using the binary number system. One state represents the "1" and the other represents the "0" as shown in another flip-flop circuit in Figure 6-21. These flip-flop units may be combined in a variety of ways to be used as binary adding and subtracting units (reverse polarity), as well as multiplication and division. The flip-flop may also be used in a variety of circuits representing logic equations, such as OR, NOR and AND/OR circuits. Figure 6-22 is a transistorized flip-flop microcircuit made with welded wiring. In a few years even this unit may appear large as the electronic industry develops smaller and smaller microcircuits built on a single "semiconductor chip."

*Pull itself up by its bootstraps—positive feedback forces the output to the maximum.

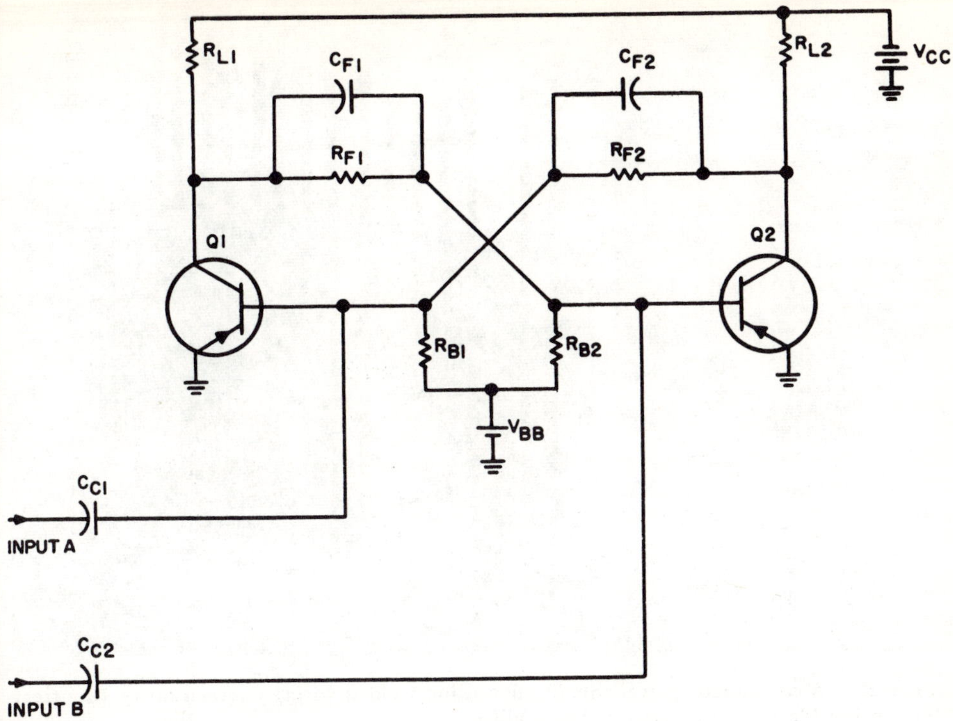

Figure 6-20. The bistable multivibrator. (Reproduced by permission of the Department of the Army.)

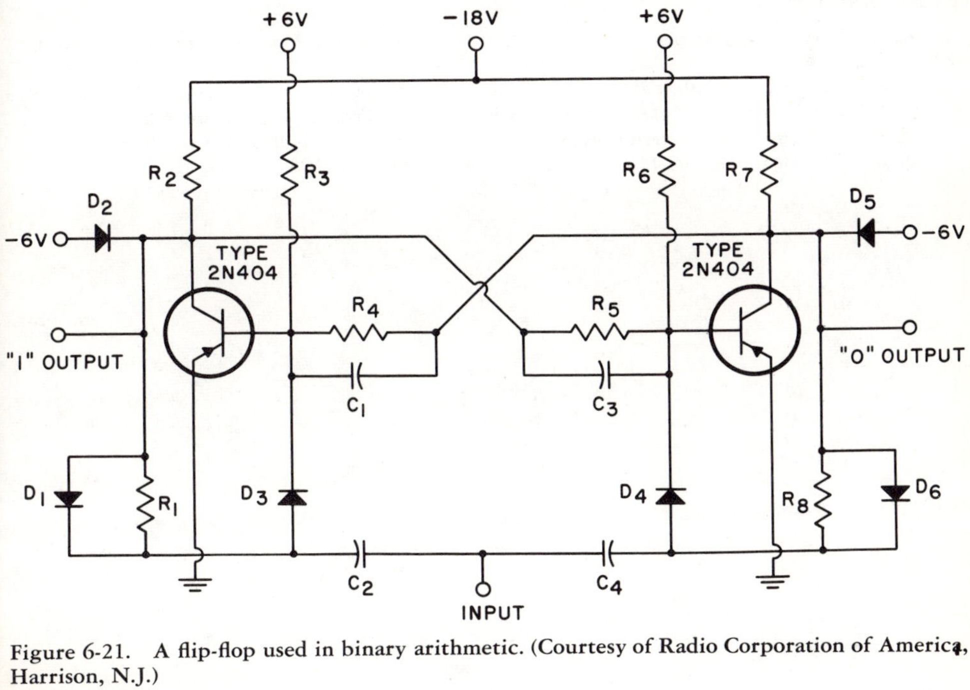

Figure 6-21. A flip-flop used in binary arithmetic. (Courtesy of Radio Corporation of America, Harrison, N.J.)

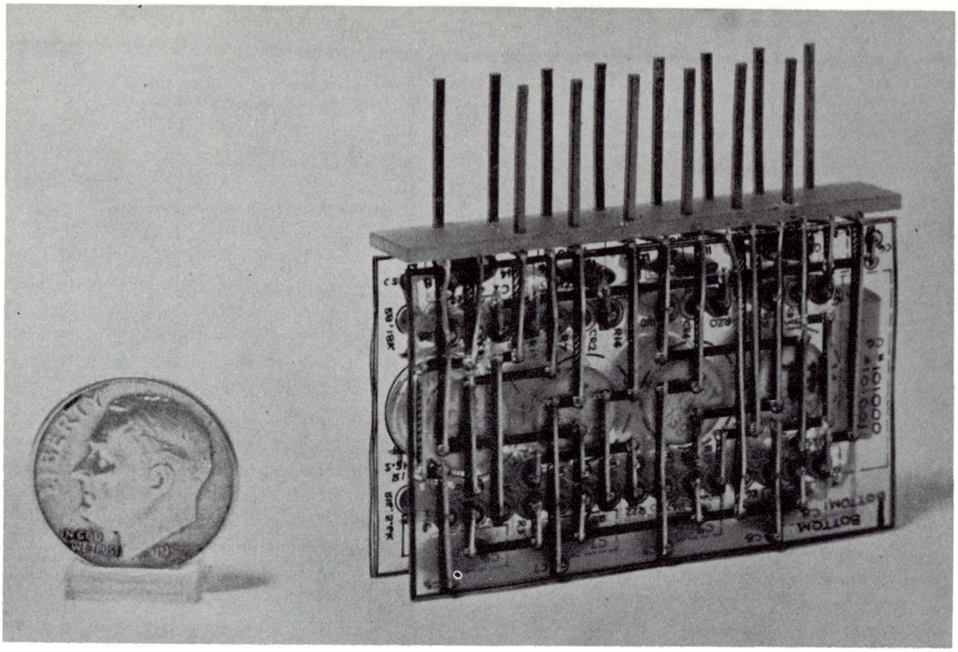

Figure 6-22. A commercially available flip-flop using welded wiring microcircuitry. (Courtesy of Engineered Electronics Co., Santa Ana, Calif.)

REFERENCES:

Basic Theory and Application of Electron Tubes TM 11-662 to 16-1-255 (Department of the Army, 1952).

Basic Theory and Application of Transistors TM 11-690 (Department of the Army, 1959).

Handbook of Operational Amplifiers (Burr-Brown Research Corporation, Tucson, Ariz.).

Pulse and Digital Circuits, Jacob Millman and H. Taub (McGraw-Hill Book Co., 1956).

RCA Transistor manual SC-11 (Radio Corporation of America, 1964).

7

The Power Supply

ripple

regulation

references

All electronic instruments must have some source of power in the form of one or more DC voltages. Some AC voltage is also required. In this chapter various methods for obtaining power will be described, and the criteria will be established for judging a power supply. For most biomedical applications it is less expensive, and a better power supply is obtained, when the instrument is purchased rather than made in the laboratory.

The power supply is, indeed, one of the most important parts of an instrument. The reader should understand that when the plate voltage of a vacuum-tube amplifier changes, the gain of the amplifier also changes. The result is instability, nonlinearity, and distortion of the signal. To avoid changes in power supply voltage, the power supply must be regulated so that a constant DC voltage is supplied—in this case, to the plate of a vacuum tube. Most power supplies draw power from the *house current,* which in the United States is usually about 110 volts AC with a frequency of 60 cps. The house voltage and frequency vary considerably in different countries. The power supply of an instrument will first rectify the 60-cps voltage, usually by means of full-wave rectification (see Chapter 4), then filter the signal to remove the 60-cps variation left after rectification. This remaining 60-cps variation is called *ripple.* A good power supply will regulate its DC output so that it remains constant *within* 10 per cent when the load it is supplying varies within limits. A good supply also remains constant when the line voltage varies, *e.g.,* from 100 to 125 volts AC. An extremely good power supply may regulate to within 0.01 per cent or at least 0.1 per cent with less than 10 mv RMS (root-mean-square) of ripple in several hundred volts of DC output.

In order to achieve good regulation and low ripple, the power supply must be constructed with a number of *regulator stages,* large capacitors, and inductors (chokes). All of these components are physically heavy, and therefore a good power supply must be heavy. In general, one can assume that the heavier a power supply, the better it is. It is not fair, however, to compare a vacuum-tube power supply and a transistorized supply.

In general, a good power supply should also provide a fuse or other protection against extreme load changes and internal failure. It should provide a DC output which is isolated from ground so that it may be used *floating,* that is, as either a positive or negative voltage with respect to ground. The power supply should also provide some unregulated power for tube filaments, *i.e.,* 6.3 volts and/or 12 volts AC. The power supply will, by its nature, liberate heat. *It is important in any instrument to be certain that this heat is removed from the vicinity of the components of the instrument. This is important because electronic components are sensitive to heat, and their useful life is considerably shortened by continuous exposure to it.*

RIPPLE

Rectification of the house current produces an output which may be called a pulsating DC signal, that is, it has a certain mean voltage (DC component) with alternating pulses, which are referred to as *ripple,* superimposed on it. The number of pulses per cycle of house current or, said in a different way, the frequency of the ripple, depends on whether the house current is supplied as single phase or three phase and whether half-wave or full-wave rectification is employed (see Chapter 4).

The removal of the pulsations is accomplished, for the most part, by means of capacitors and inductors, although sometimes special components such as a transistor might be used for a special effect. The filter used becomes a part of the load supplied by the power supply. The *load,* in ohms, is determined by dividing the output DC voltage by the output DC current. The amount of ripple is the ratio of the RMS value of the ripple in the output to the DC component expressed as a percentage. The amount of reduction (expressed as percentage) in the ripple which results from the addition of a filter is an estimate of the efficiency of the filter.

Figure 7-1 shows examples of the basic types of filter networks used in power supplies. In Figure 7-1(A) the network is a *capacitor-input* filter. Notice that the first component is the capacitor, which is in parallel with the load (R_L). When the pulsating wave goes to a maximum it charges the capacitor, then, while the ripple pulse is decreased, the capacitor injects some of the energy it has stored back into the circuit and tends to smooth out the ripple. Capacitors, especially the large electrolytic ones used in power supply design, have a definite upper limit with

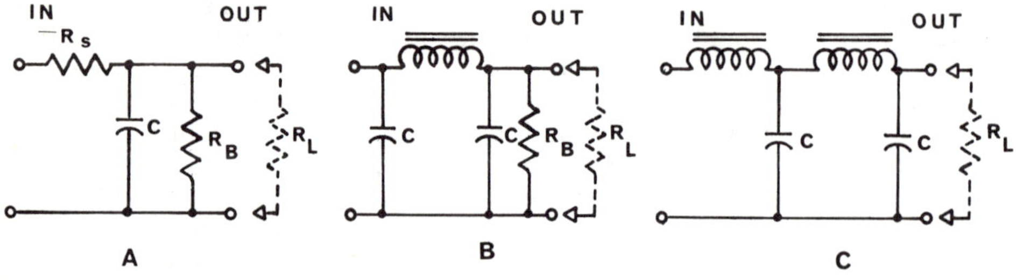

Figure 7-1. Filter networks: A, capacitor-input; B, capacitor-input type with an additional LC section; C, two-section choke input filter.

regard to the peak input voltage. Therefore the capacitor must withstand at least 1.4 times the RMS voltage as measured with a voltmeter. Usually a safety factor of at least two is used, and sometimes a surge resistor must be used. The resistor (R_B) in Figure 7-1 is a *bleeder* resistor which discharges the filter capacitor when the power supply is not in use. Otherwise, if the power supply is turned on exactly at peak voltage and there is already a charge on the capacitor, the capacitor may be destroyed. Capacitor-input filters allow higher DC output voltages but provide poor regulation. The reduction in ripple is only efficient with this type of smoothing filter when the load resistance (R_L) is high. Ripple reduction is considerably improved when additional sections of series inductance and parallel capacitance sections are added, as shown in Figure 7-1(B). When only one such section is added the filter is called a π-filter.

The *choke-input* filter provides good ripple reduction when the load resistance is small. The choke-input filter offers a much higher reactance to the ripple than it does to the DC voltage; this lowers the available DC voltage but effectively reduces the ripple. The filter, shown in Figure 7-1(C), combines the effective features of the capacitive-input filter with those of the choke-input filter. The input choke must have a minimum amount of inductance or the filter will act as a capacitor-input filter. This minimum is determined by the load resistance. (The load resistance is equal to the *critical inductance* in henrys when the output voltage is expressed in volts and the current is expressed in milliamperes.) The maximum inductance may be determined from the maximum current that the rectifier diodes can withstand. The iron-core choke is available with an inductance which varies with current. Such an inductance is called a *swinging choke* and is commonly used in power supply circuits. Thus, when swinging chokes and multiple LC sections are used, ripple can be eliminated almost completely with good regulation even in the face of a variable load.

As stated earlier, a transistor can be used as a rectifier diode in all of the rectifier circuits described in Chapter 4, including the half-wave, the full-wave, and the full-wave bridge circuits as well as the doubler and tripler circuits. When a transistor is used as a diode the emitter is not employed, and forward conduction occurs when the collector is the anode and the base is the cathode of the diode. These transistor-diodes can handle much larger currents than the usual diode.

One use of the transistor-diode is the *capacitance multiplier* shown in Figure 7-2. This circuit increases the capacitance of the base capacitor (500 μf) by a factor equal to the gain of the transistor. In Figure 7-2, the effective capacitance in parallel with the load resistor (R_L) is about 25,000 μf. Figure 7-3 is an example of an

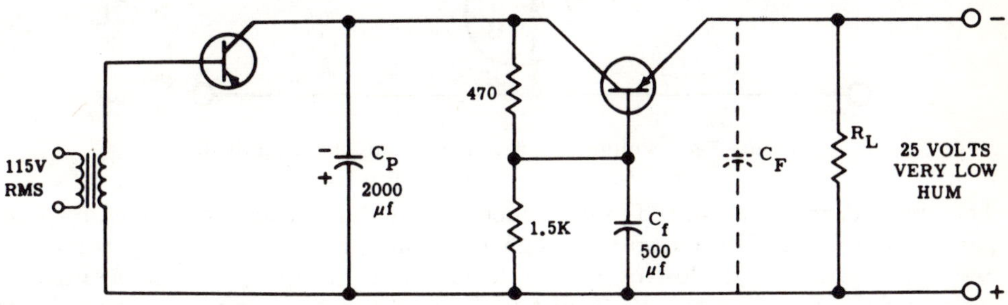

Figure 7-2. A capacitance multiplier circuit (Courtesy of Motorola Semiconductor Products, Inc., Phoenix, Ariz.)

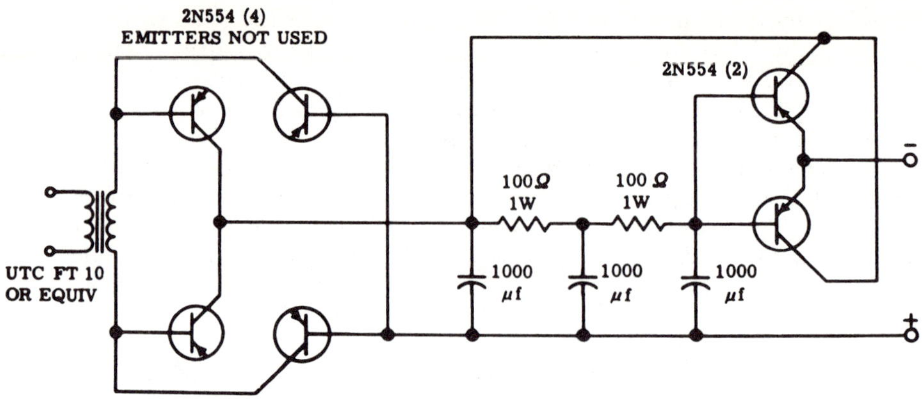

Figure 7-3. An electronic filter. (Courtesy of Motorola Semiconductor Products, Inc., Phoenix, Ariz.)

electronic filter. This filter considerably decreases the ripple voltage in spite of high current loads. It would take large, heavy, and expensive inductor and capacitor components to provide the same ripple attenuation. Such an electronic filter can be applied to any DC power supply if the collector-emitter voltage is kept low.

REGULATION

It is important to regulate the output of a power supply with respect to two variables—changes in the input voltage, usually the house line voltage, and changes in the output load resistance, which cause more or less current to be drawn from the power supply. The manufacturer will state in the specifications of a particular power supply that the regulation of the DC voltage is better than 1 per cent, 0.1 per cent or 0.01 per cent for loads which draw some specific amount of current, and better than 1 per cent, 0.1 per cent, or 0.01 per cent with line voltage variation, for example, from 100 to 125 volts.

The simplest form of voltage regulator is a zener diode used in parallel with the load as shown in Figure 7-4.

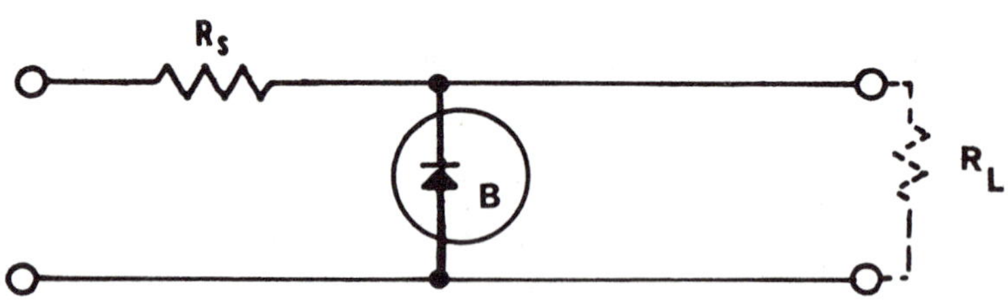

Figure 7-4. Voltage regulation by means of a zener diode.

The zener diode is "cultured" in production to exhibit avalanche breakdown at a particular voltage. The current through the junction is controlled by the circuit resistance so as not to destroy the diode junction. In Figure 7-4, this controlling resistor is the series resistor R_s. Zener diodes are available at a great many voltages, from about 1 volt to over 100 volts (at the time this book was being written). Ordinary zener diodes are subject to thermal variation, but a special zener diode,

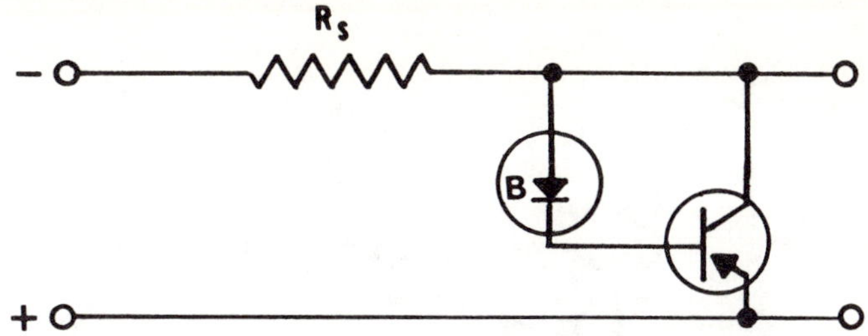

Fig. 7-5A

Figure 7-5. A, Shunt regulator; B, cascade shunt regulator. (Courtesy of Radio Corporation of America, Harrison, N.J.)

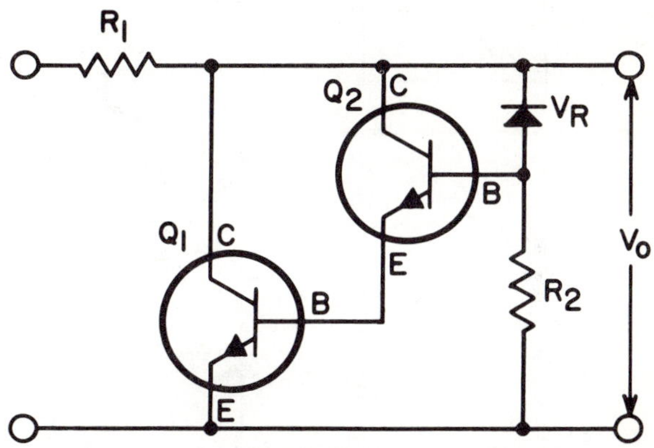

Fig. 7-5B

called a *reference zener*, is available which is made of more than one zener diode in series, chosen so that some of the individual elements have a positive temperature coefficient (resistance increases with temperature) and at least one element has a negative coefficient. The individual zener elements are packaged together to form the *reference zener.*

The maximum current that a zener diode can handle is limited by the temperature at which its junction breaks down. The zener diode can be used to provide more current if it is used in combination with a transistor as shown in the simple *shunt regulator* circuit shown in Figure 7-5(A). Figure 7-5(B) is a *cascade shunt regulator.* The addition of more transistors in parallel with Q_1, as in Figure 7-5(B), will increase the current capabilities of the shunt regulator.

Figure 7-6(A) is an example of a series transistor regulator. Here all of the current must flow through transistors Q_2 and Q_3, which are in parallel. The series transistors absorb some of the voltage fluctuations that may occur, and their inherent dynamic impedance also decreases the ripple as it did when the transistor was used as a capacitance multiplier. Transistor Q_1 compares the reference voltage developed across the diode V_R (a zener diode or a reference diode) with some portion of the output voltage. The difference or *error* between the reference diode and the output is amplified and supplies the input to the regulating transistors Q_2 and Q_3, which are working to keep the output *voltage* constant.

There are some circuits which require a *constant current* power supply rather than a constant voltage supply as all the circuits described thus far have been. A simple constant current regulator can be designed which uses a series transistor as a resistance which varies with changes in the load resistance in such a way as to

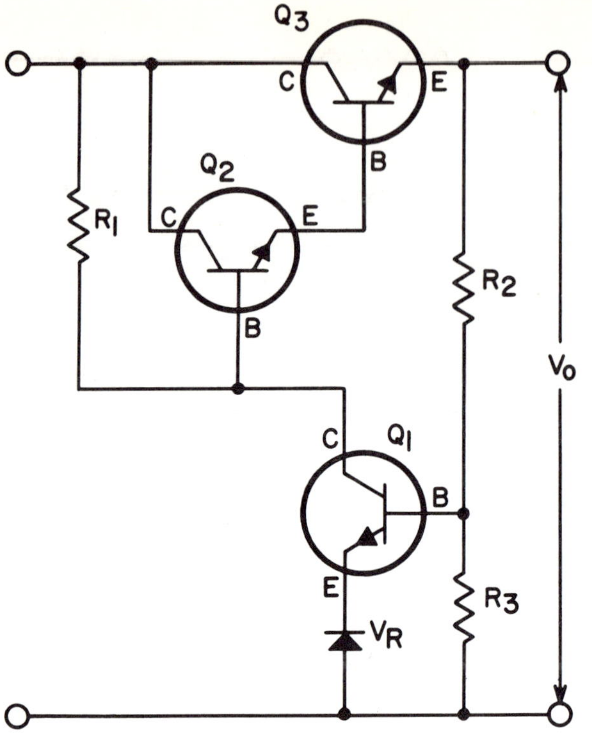

Fig. 7-6A

Figure 7-6. A, Series regulator for constant voltage regulation; B, series regulator for constant current regulation. (Courtesy of Radio Corporation of America, Harrison, N.J.)

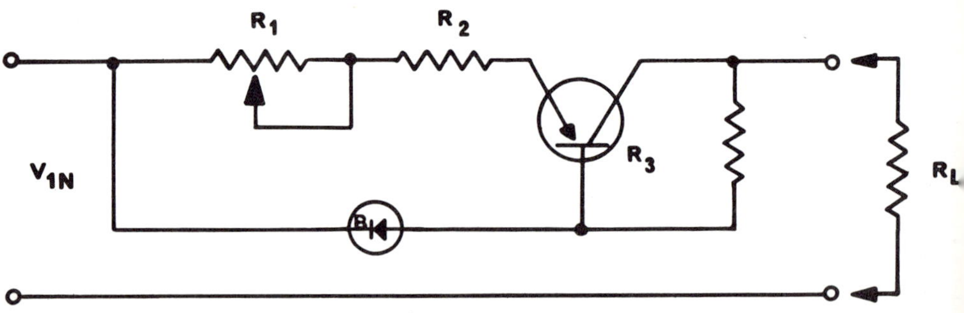

Fig. 7-6B

keep the current constant. Figure 7-6(B) is a constant current supply which allows two paths for the current: the first, through the zener diode and resistor R_3, and the second, in parallel with the first, through R_1, R_2, and the series transistor. Any change in current through R_3 causes a change in the base bias of the transistor which causes a compensating change in the junction resistance. The over-all result is that the current is kept constant. It should be clear from Ohm's law, however, that both the current and the voltage cannot be held constant at the same time.

Figure 7-7 is the schematic of a transistorized power supply which can be used to supply $\pm$ 12 volts, $\pm$ 22 volts, or as shown, one of each. Each section of this

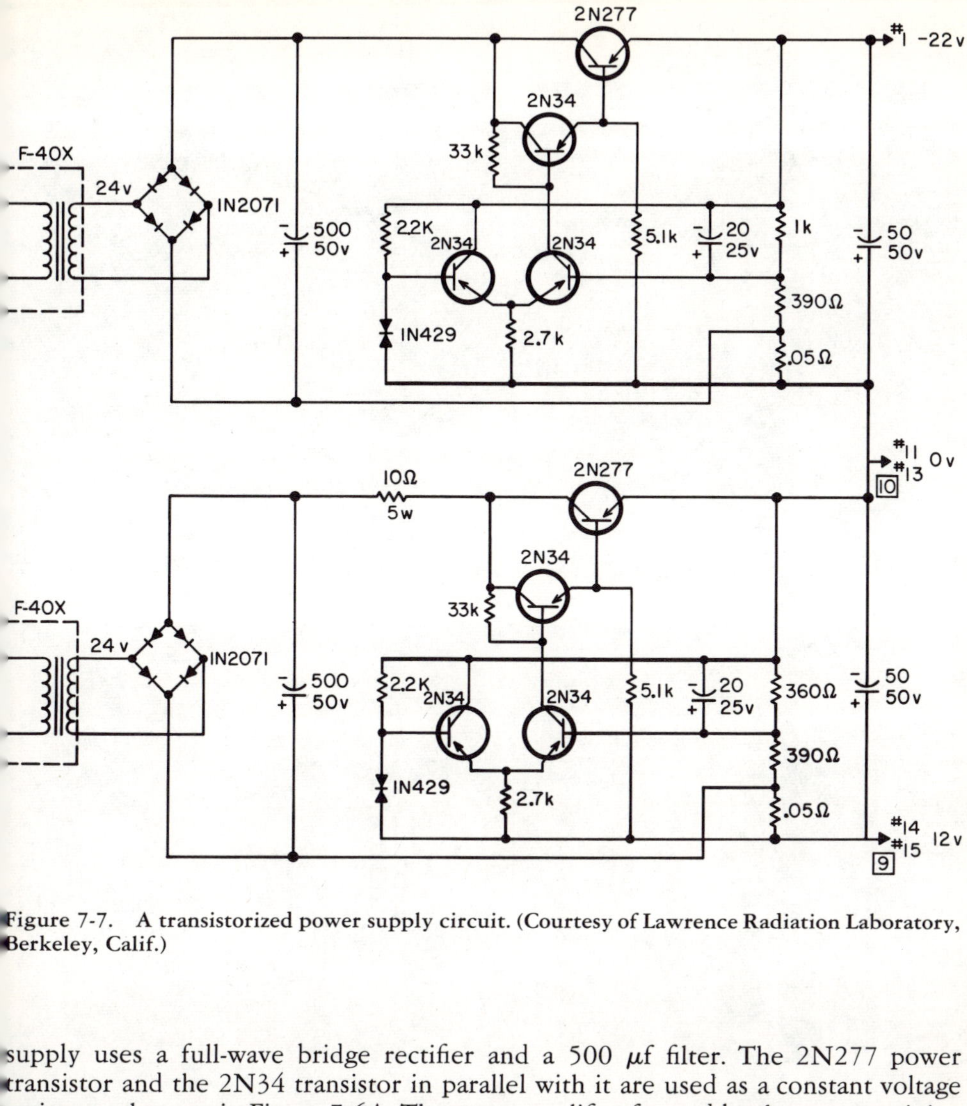

Figure 7-7. A transistorized power supply circuit. (Courtesy of Lawrence Radiation Laboratory, Berkeley, Calif.)

supply uses a full-wave bridge rectifier and a 500 μf filter. The 2N277 power transistor and the 2N34 transistor in parallel with it are used as a constant voltage series regulator as in Figure 7-6A. The error amplifier formed by the two remaining 2N34 transistors is a *differential amplifier.* Notice that the zener diode 1N429 maintains the base of one 2N34 at the reference potential and a portion of the output is supplied to the other base, and that the difference between the top and the bottom circuit is the voltage divider supplying the differential amplifier. This error signal controls the series regulator stages. Figure 7-8 shows the circuit in Figure 7-7 mounted on a printed circuit board.

Another simple form of regulating element is the *gaseous regulator tube*, which is essentially a gas-filled rectifier. When the voltage across a gaseous regulator tube, sometimes called a *glow tube*, reaches a threshold or starting voltage, the gas ionizes and conducts, and current flows through the tube. After starting, the tube will maintain a constant but somewhat lower voltage drop across itself. This is called its working voltage. These tubes do not require filaments, and, hence, the nomenclature

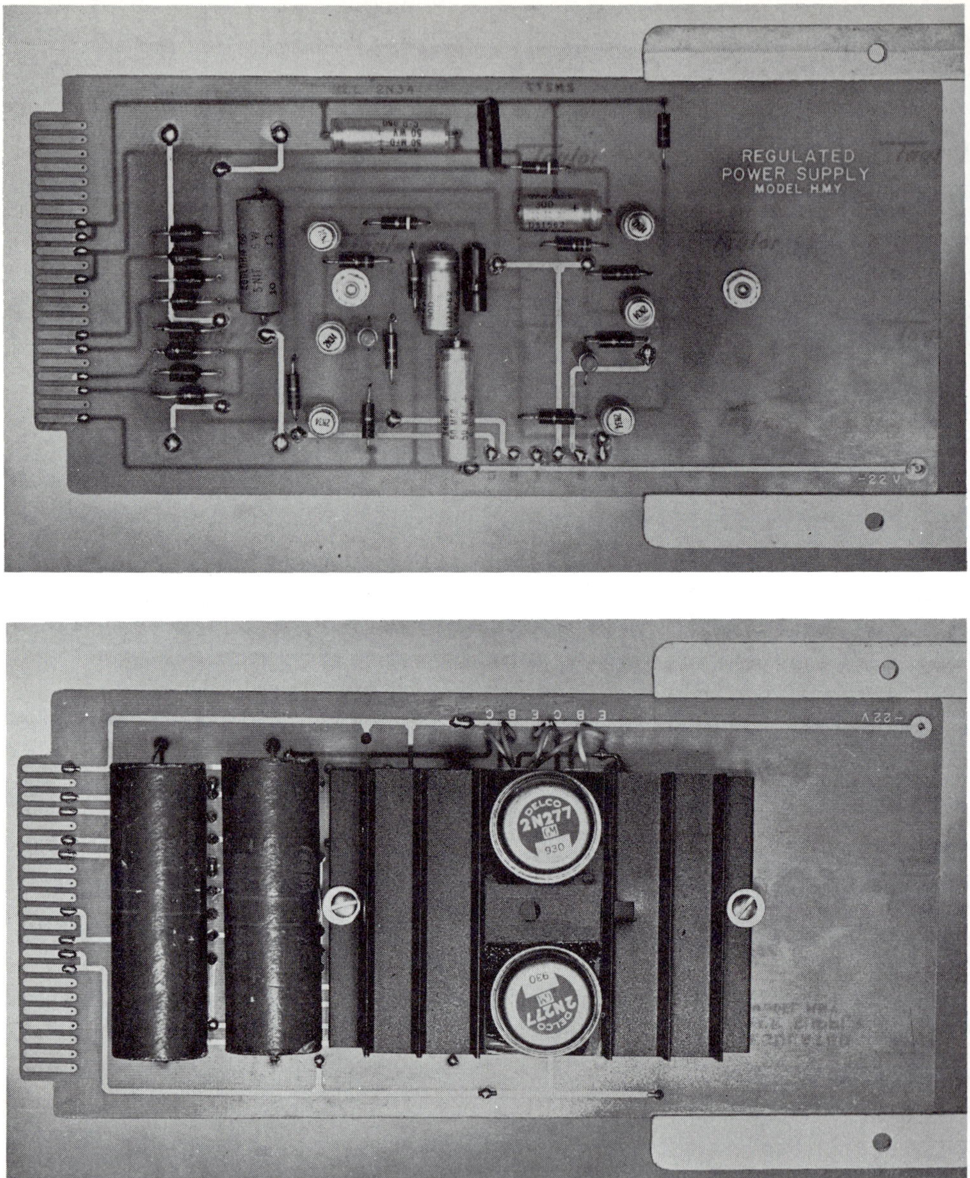

Figure 7-8. The power supply shown in Figure 7-7. (Courtesy of Lawrence Radiation Laboratory, Berkeley, Calif.)

describing them begins with 0, *e.g.*, 0A2, 0B2, 0G3. One exception to this nomenclature is the 5651 used in Figure 7-12. Each tube is designed to have a particular starting and working voltage, and each tube has a definite current capacity. Several gaseous regulator tubes are shown in Figure 7-12. In recent years, the glow tube has almost been replaced with its semiconductor equivalents.

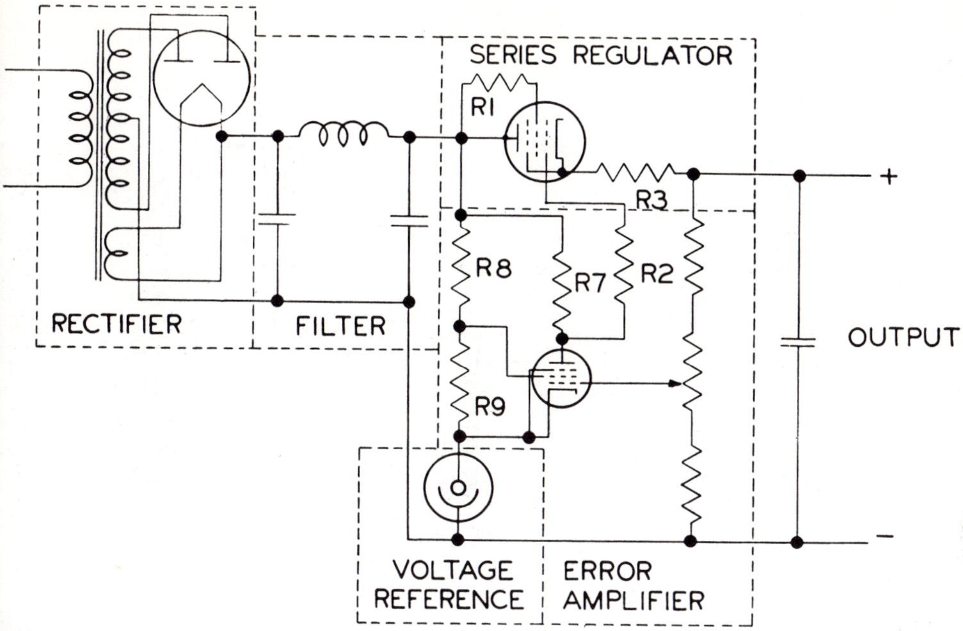

Figure 7-9. **The elements of a vacuum-tube power supply. (Courtesy of Associated Specialties Co., Orefield, Pa.)**

Generally speaking, power supplies that provide the best regulation and are best suited to supply circuits requiring 100 to 300 volts employ a negative feedback system (Fig. 7-9). A full-wave rectifier using center-tapped transformer (see Fig. 4-7) provides *pulsed DC* to the filter section. The peak voltage supplied by the rectifier must be high enough to allow for losses in the *filter* and in the regulating circuit and still have the required output voltage. The smoothing filter is the capacitor-input type (Fig. 7-1[B]). The *reference voltage* is supplied by the voltage drop across the gaseous regulator tube which is chosen to provide a minimum of current change over the useful range of the power supply.

An amplifier which senses changes in the output of the circuit and provides a signal to correct the output voltage is called the *error amplifier.* The grid to cathode voltage on this tube is the difference between the reference voltage (voltage across the gaseous regulator tube) and a fixed portion of the power supply output voltage. This difference or error voltage then determines the voltage drop across R_7 which controls the grid of the *series-regulator* stage. The screen grid to cathode voltage on the regulator will increase if the output voltage from the filter network increases. This occurs when the line voltage increases. The increase in screen grid voltage will, in turn, increase the current through R_7 which lowers the grid to cathode voltage on the series regulator. In this way, the screen grid voltage makes a correction for line voltage changes. The series regulator consists of a heavy duty pentode stage, and often two such pentodes in parallel are used to handle the power. The voltage drop across the series regulator tube is then adjusted by means of the amplified error signal applied to the grid of the series regulator. The output voltage of the power supply is held constant by varying the plate impedance of the series

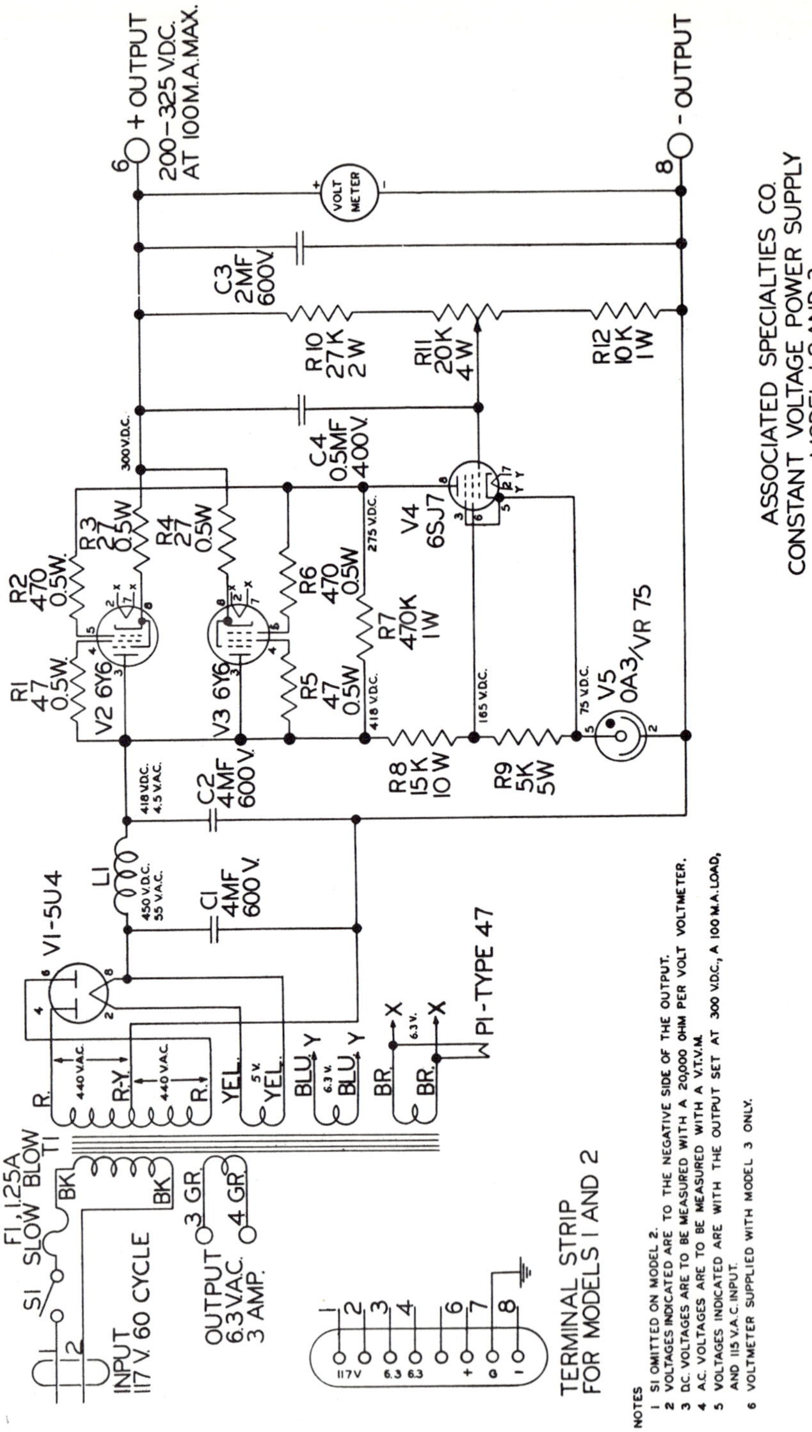

Figure 7-10. A complete constant-voltage power-supply circuit. (Courtesy of Associated Specialties Co., Orefield, Pa.)

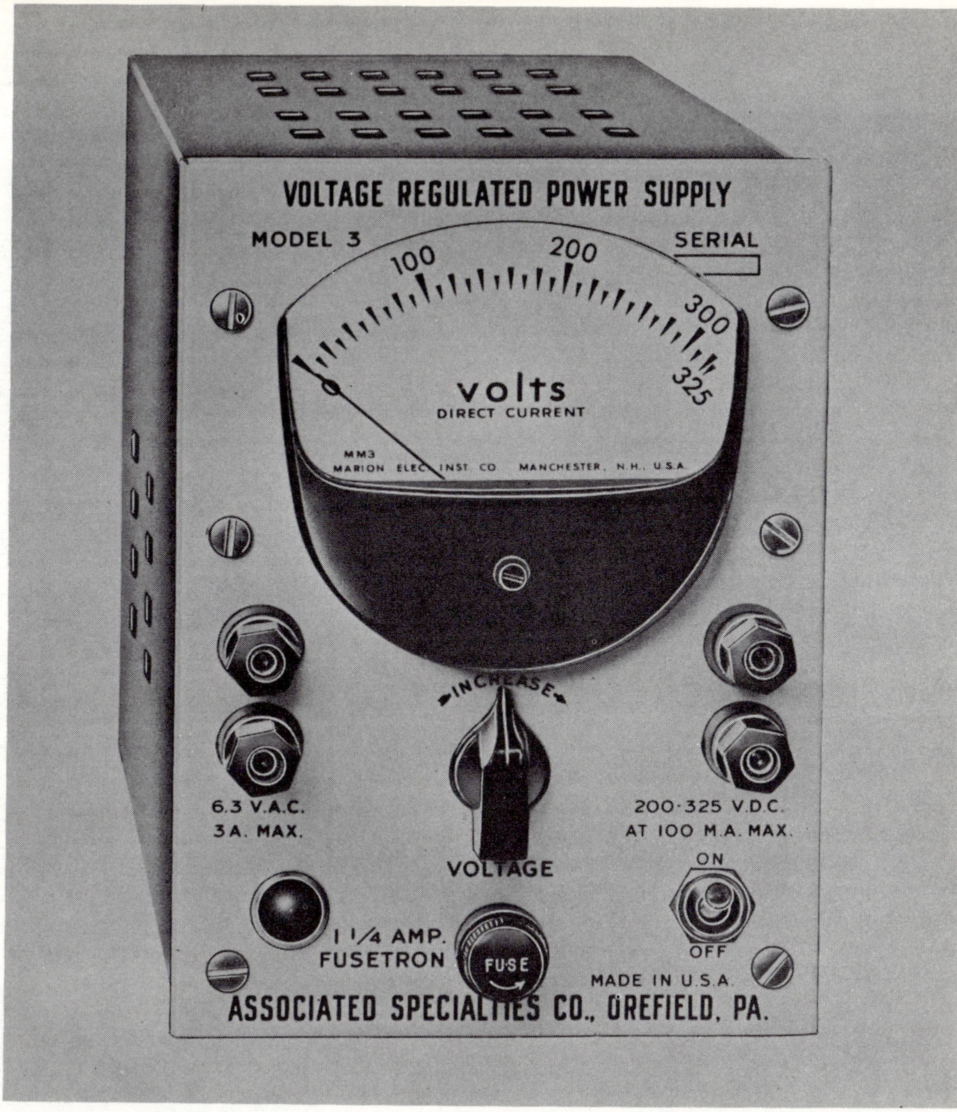

Figure 7-11. A photograph of the unit shown in Figure 7-10. (Courtesy of Associated Specialties Co., Orefield, Pa.)

regulator tube. The resistor R_1 is used to prevent oscillation, called *parasitic oscillation,* in the pentode which would prevent it from operating as a regulator. Figure 7-10 shows the complete circuit for a commercially available power supply which supplies 100 ma of current with a constant voltage which is adjustable from 200 to 325 volts. It regulates better than 1 per cent for loads which draw from 0 to 100 ma current, and it maintains the desired voltage within 1 per cent with line variations from 105 to 125 volts AC. Figure 7-11 is a photograph of the instrument as supplied by the manufacturer.

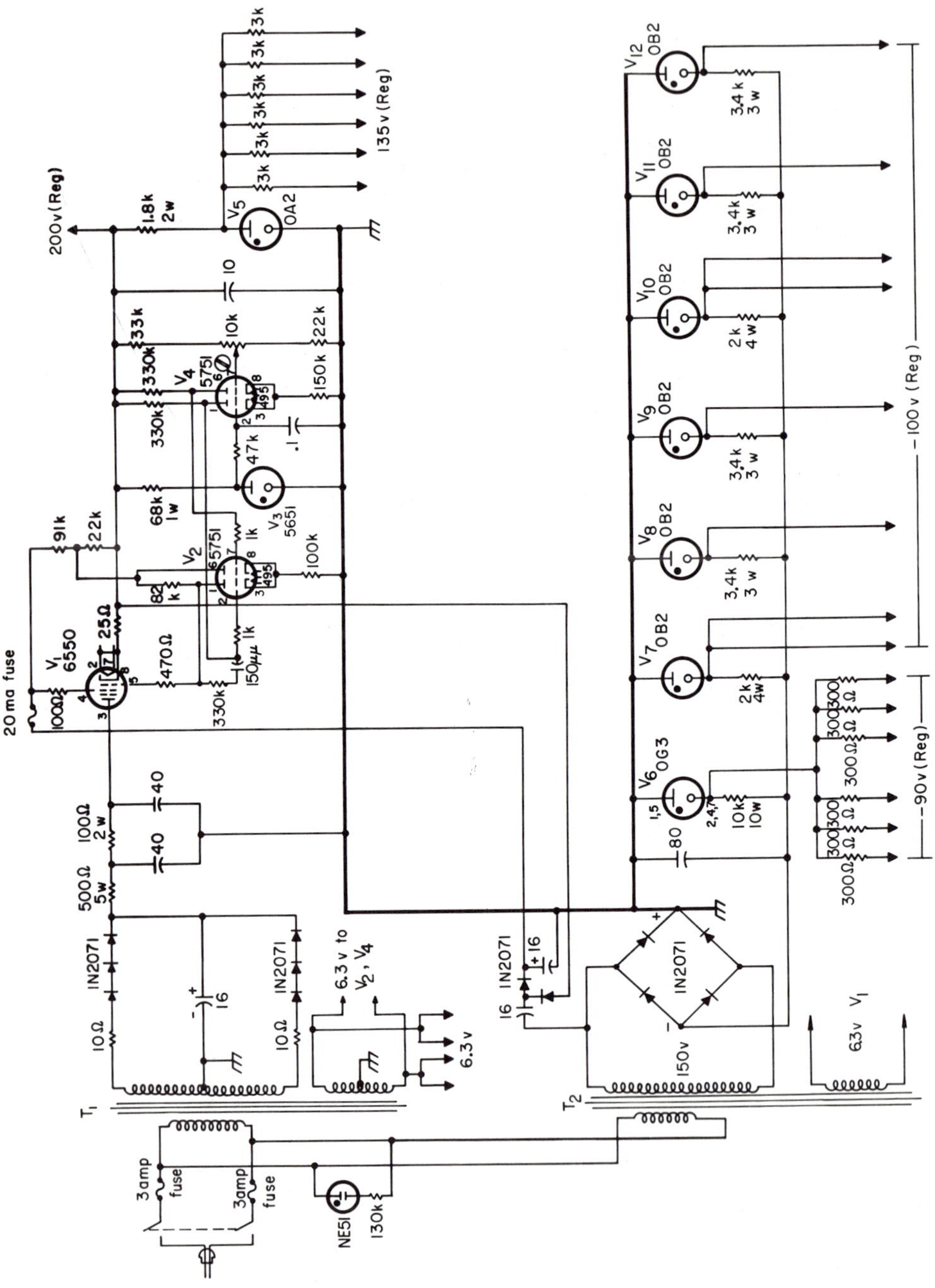

Figure 7-12. A power supply designed for a specific biomedical instrument. (Courtesy of the Lawrence Radiation Laboratory, Berkeley, Calif.)

Figure 7-12 is a circuit for a power supply used by the author to provide +200 and +135 volts DC regulated to ±0.1 per cent and −90 and −100 volts regulated to ±1 per cent. In this power supply the line voltage is rectified with a center-tapped transformer and a full-wire rectifier. Notice that the primary of the transformer is fused, and a 110-volt neon indicator lamp is provided to show when the circuit is turned on. The inexpensive 1N2071 diodes are in series to share the voltage drop. This eliminates the need for a heavy duty more expensive diode. The filter is of the capacitor-input type. The voltage reference is provided by the 5651 gaseous regulator tube. The reference voltage is supplied to one grid of the differential amplifier (V_5) and is compared to a fixed portion of the DC voltage output. The difference or error is then amplified, using V_2, and subsequently fed to the grid of the series regulator V_1 to maintain the output at 200 volts DC. If the load resistance should suddenly decrease and draw too much current from the power supply the 20-ma fuse would blow, and conduction through the series regulator would stop, protecting the power supply. The OA2 in combination with the 3-K Ω resistor provides a regulated +135 volts DC.

Transformer T_2 in the lower circuit in Figure 7-12 increases the line voltage to 150 volts AC which is then rectified by means of a full-wave bridge rectifier and filtered with an 80-μf capacitor. The choice of the gaseous regulator tube and its series resistance is determined from the required regulated voltage and the amount of current to be drawn from the supply. In this example, each arrow represents a separate place in the over-all circuit (not shown) that required a source of power. The 6.3 volts AC secondary coil on transformer T_2 heats the filament of the main series regulator tube, while the additional winding on transformer T_1 supplies 6.3 volts AC to heat the two 5751 error-amplifier tubes. This power supply was designed to provide power for a rather large instrument that required much better regulation for the +200-volt and the +135-volt supplies than it did for the −90-volt and −100-volt supplies.

REFERENCES:

Radio Amateur's Handbook, revised annually (American Radio Relay League, Newington, Conn.).
Radiotron Designers Handbook (Radio Corporation of America, 1964).
RCA Transistor Manual SC-11 (Radio Corporation of America, 1962).
Zener Diode Handbook, No. 2 (International Rectifier Corporation, 1960).

SECTION 2

Introduction to Biomedical Instruments

8

General Remarks

general remarks

calibration

references

what is a biomedical instrument?

the parts of a biomedical instrument

characteristics of some biomedical parameters

This textbook has become necessary because of the relatively recent marriage of the oldest science, medicine, with perhaps one of the most recent sciences, electronic instrumentation. As early as 1791, the Italian scientist, Luigi Galvani, discovered that a frog muscle would twitch when touched by a metal instrument (a knife) which happened to be near an electric generator. The early nineteenth century produced many scientists with interests in physics, and about 1850, Helmholtz, a German physician, was the first to study the transmission time of the "electrical signal" from the nerve to a muscle in the human. Also in that century Johannes Müller reported that an electric current is generated by a frog's heart during contraction. Augustus D. Waller published a paper in 1887 which was the first to describe the electromotive changes which accompanied each heart beat. Electrical activity present on the surface of the brain had already been demonstrated by Richard Caton in 1875; thus, the basis for an interest in physical physiological measurements was launched.

In the early twentieth century great steps were taken in laying the groundwork for what today is known as biomedical instrumentation with the invention of the string galvanometer by Johannes S. C. Schweigger. Willem Einthoven, in 1903, introduced one of the most important advances, the perfection and physiologic application of the string galvanometer and the subsequent study of the electrocardiogram with it. The string galvanometer was perhaps the first instrument to have clinical significance, and it is fitting that Einthoven won the Nobel Prize for his contribution. The biomedical world had been bitten by a bug which was to almost completely change the practice of medicine.

By the second World War a variety of investigators had begun using electronic equipment to study the genesis of the bioelectric potentials, i.e., the action potential, the electroencephalogram, and the electromyogram, as well as to study blood flow, blood pressure, and many other phenomena. In 1937 Erlanger and Gasser introduced the oscilloscope as a biomedical device for the investigation of electrophysiological events. The war, however, had a profound effect on the advancement of electronic instrumentation. Large amounts of money were spent on the development of instruments for a variety of wartime needs. After the war, medical and physiological laboratories began to accumulate and use this equipment. Since research funds were meager, the low-cost *war surplus* provided many of the components, amplifiers, recorders, and instruments.

Even in the late 1940's and early 1950's it was unusual to find a physician or even a biologist or physiologist who could design more than an elementary electronic instrument. In the past 20 years, however, the use of electronics in medicine and biology has become so important that no physician, surgeon, or research worker can afford not to at least understand the electronic instruments which are in evidence in every hospital and laboratory. It is only now, in this decade, that the marriage of electronics and the biological sciences has been consummated. The space age is again significantly changing the character of the electronic instruments found in hospitals and laboratories. These instruments may be expected to become smaller and more reliable than those in use when this book was written.

WHAT IS A BIOMEDICAL INSTRUMENT?

There are many types of instruments which could be classified as biomedical instruments. These instruments have many uses, but some of them are biomedical instruments only insofar as they are used in a biological or medical laboratory. An example of an instrument used in many fields is the strain gauge for measuring blood pressure. It is also commonly used to measure strain in such things as concrete bridges and railroad rails, and to measure gas and air pressure. Many other examples could be given of instruments which are utilized in almost every scientific field.

Some instruments are used only for measurement of a parameter, such as the electrocardiograph or a pH meter, while others are used only for administration of therapy, such as a diathermy machine which is used to administer heat to deep tissues. The "stimulator" used to electrically stimulate a nerve or muscle is a nonclinical example of an instrument of this general type. Still other instruments generate voltages or currents of various amplitudes and waveforms which are used specifically for calibration of other instruments.

Another class of instruments may be used to control an entire system. In this case, a group of instruments in which one instrument is in control of another is called a servo-system. A very important class of instruments, known as a *read-out* device, accepts voltage or current output signals from other instruments and registers or records the signal by writing directly on paper, by controlling a light beam which registers on photosensitive paper, by recording information on magnetic tape, or by displaying the data on the face of a cathode-ray oscilloscope.

There are many other classifications of instruments. Every author and, in fact, almost every instructor of instrumentation has his own classification. The reader is encouraged to examine the books listed in the reference section at the end of this chapter.

The instruments with which this book is most concerned are those which enable the investigator to extend his capacity for making measurements. The instrument enables one to quantitatively determine and record minute and gross environmental changes. The instrument must almost always convert some physical or, in a few cases, chemical parameter into an electronic signal. These electronic or electromechanical devices are an intimate part of any experimental design. The device may monitor a parameter such as blood flow or it may, as in the case of the stimulator, alter a parameter such as the heart rate. In any case, the experimenter must know not only what effect the experiment has on the instrument, which is the measured entity, *but he must also know what effect the instrument will have on the experiment.* Ideally, the instrument should have no effect on the system being measured. It often does, however, and the experimenter is responsible for evaluating whether this effect is significant. This is most important in physiological experiments involving animal measurements.

Every instrument has limitations. Sometimes the limitation is the range, sometimes it is the sensitivity, and at other times it could be the frequency response. The reader has already been exposed to a discussion of the theory of measurement in Chapter 3, and this material should be reviewed at this time. It seems appropriate to list here the qualities which the ideal instrument should possess:

1. sufficient but not excessive range
2. sufficient but not excessive sensitivity
3. linearity
4. a high signal-to-noise ratio
5. complete isolation from the system being measured and from the environment
6. stability
7. sufficient output which is easily read (see discussion of scale error and parallax in Chapter 3)
8. adequate but not excessive frequency response
9. the simplest design that is necessary to accomplish the purpose of the instrument.

THE PARTS OF A BIOMEDICAL INSTRUMENT

Almost all biomedical measuring instruments have the following parts: (1) a sensing element or transducer to detect the *primary signal;* (2) a signal modifier or processer, which is an amplifier or combination of amplifiers; and (3) a device which allows the user to read and/or record the amount of output signal. The transducer element converts the magnitude or change in magnitude of the parameter to be measured into a usable electronic signal. The parameter to be measured could be almost any physical entity and many chemical entities, and there are, therefore, an almost infinite number of electromechanical devices which may be called transducers. A more complete discussion of typical transducers will follow in the next chapter. The design of the signal processor or modifier portion of the instrument will depend on what is to be done with the data. In some cases, it is a straightforward amplifier, while in other cases, it is extremely complicated and requires sophisticated combinations of specialized circuitry.

The output of the instrument may be indicated by means of a meter, or it may be made available at the terminals of an output connector. There are many types of devices which may be used to present or record the output signal, and, in general, they are referred to as *read-out devices.* The most common read-out device is the cathode-ray oscilloscope, which may be obtained in almost any size and shape and with a variety of sensitivities and sweep times. Another common form of read-out device is the direct-writing recorder. There are also a great many varieties of recorders, some with a slow response time and others as fast as 200 cps. Other recorders use a light beam instead of a pen or stylus, and these are at least an order of magnitude faster than the direct-writing instrument. The output of the instrument may also be recorded on magnetic tape or on punched paper tape or cards for eventual computer use. Modern computers also operate *on-line*, that is, they accept voltage signals directly for storage and calculation. Read-out devices are more thoroughly discussed in Chapter 10.

CALIBRATION

Prior to using any instrument, one should know quantitatively whether the instrument will do what is demanded of it. When the physician wants to know a patient's oral temperature, he pops an inexpensive thermometer into his mouth for about 3 minutes, removes it, and reads it. Usually he believes the markings on the glass capillary. In doing so, the physician is assuming that the factory calibration was sufficiently accurate for his purpose. It may or may not be correct, but the physician cannot know this unless he calibrates the instrument. Many instruments, especially electronic instruments, must be calibrated by the user. Sometimes this is not difficult, as in the case of the thermometer mentioned above. At least it is not difficult to attain any reasonable degree of accuracy.

Quantitative calibration of some instruments, however, will not be easy, and it will often strain the ingenuity of the user. When an instrument is being calibrated, the following questions must be asked:

1. Does the instrument measure *only* the desired parameter?

2. Does the instrument affect the parameter being measured? In the case of an animal measurement does the instrument change the physiology of the animal?

3. Does it have enough *range* to include all possible quantities of parameter, large and small, that are of interest to the user?

4. Does the instrument have enough *sensitivity* throughout the desired range?

5. Does the output of the instrument increase *linearly* with equal increments of the desired parameter? In some cases, it is desirable to have the output increase as the log of the input, but this should be necessary only when the range is very large or for special applications.

6. Does the instrument provide a sufficient signal to distinguish it from its inherent noise? In other words, does the instrument have a high enough *signal-to-noise* ratio?

7. Can the instrument respond fast enough to faithfully reproduce changes in the parameter being measured? Another way of asking this question is to ask whether the instrument has an adequate *frequency response.*

The above list of questions may be separated into two classes: those which require specific electronic techniques and those which require an intimate knowledge of the parameter to be measured, including the relation of this parameter to the system in which it exists.

Let us take, for example, an instrument which measures blood pressure. Without bothering with the details at this time, let us assume that the instrument complex consists of a transducer, an amplifier, and a recorder such as is shown in Figure 8-1.

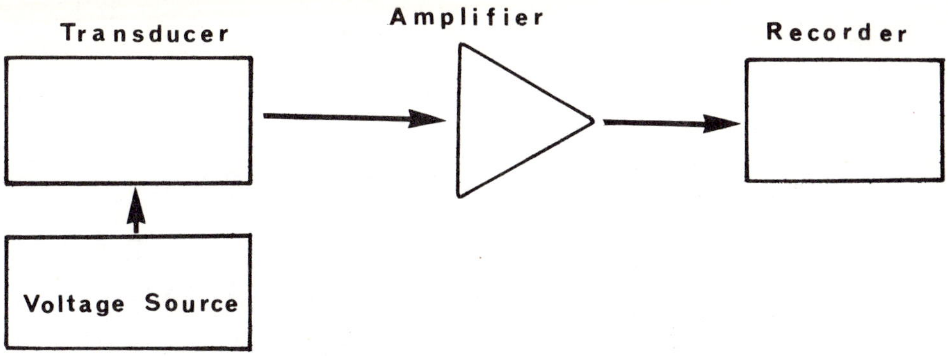

Fig. 8-1

The transducer, in this case, is a Wheatstone bridge such as the one shown in Figure 3-11, except that the resistance of one arm of the bridge varies as a function of the pressure applied to the stiff membrane to which it is attached. The voltage source is necessary to supply a voltage across the legs of the bridge (A and D); this is called the *excitation voltage* of the bridge. In a commercial pressure strain gauge from my laboratory, for example, the voltage across the opposite legs of the bridge (B and C) will change 400 microvolts per centimeter of mercury pressure for each volt of bridge excitation voltage. The amplifier boosts the microvolt *primary signal* from the transducer to a signal with sufficient voltage and current to drive the galvanometer of the recorder.

The first step in the calibration of the instrument described above is to determine the electronic characteristics of the amplifier and the recorder. Each must be checked to determine whether the output of the instrument is linear and has the proper sensitivity and range for the primary signal, which in this case is the output of the pressure manometer. This is done by connecting a battery across a voltage divider which will provide the proper microvolt or millivolt signals. The calibrating voltage signal is then applied to the input terminals of the amplifier, and the output is registered on the recorder. Enough values of calibrating voltages must be used to determine the linearity and sensitivity of the electronic system. The investigator should also determine the output signal when the input terminals are short-circuited. This output signal, when divided by the gain of the amplifier or amplifiers, is referred to as the *shorted input noise.* The lowest permissible value will depend on the magnitude of the signal. This is why the *signal-to-noise* ratio is more often quoted and in the final analysis is the more important of the two parameters.

The second step is to determine the *frequency response* of the electronic system. This is done by applying a sinusoidal voltage (using a voltage divider to obtain a microvolt signal) to the input of the instrument. As the frequency is increased the amplitude of the output should not decrease. The investigator should carry out this analysis to a high enough frequency to insure that the highest frequency component in the data to be measured is not attenuated more than the usual allowance of 5 per cent. *The frequency response of the instrument is one of the most important parameters in the calibration of an instrument whenever that instrument is to be used to measure quantities which vary in time.* Every laboratory should own and use a sinusoidal generator of high quality for this purpose. Should the fre-

TABLE 8-1
Characteristics of Some Biomedical Parameters
CARDIOVASCULAR SYSTEM

Parameter		*Parameter Range*	*Required Frequency Response (cps)*	*Transducer or Type of Transducer Usually Employed*
Blood pressure (Direct cannulation)	arterial venous	30-300 mm Hg 0-15 mm Hg	DC to 50 cps DC to 20 cps	Strain gauge, differential transformer, capacitance manometer, Bourdon tube, Pitot tube
Blood pressure (Indirect)	{continuous systolic}	50-300 mm Hg $P = 0\text{-}6 \frac{\text{mm Hg}}{\text{sec}}$	DC to 60 cps	Pulse pickup element, piezoelectric crystal, or photocell transducer
Blood pressure	diastolic and systolic		.1 to 50 cps .1 to 50 cps	Pressure gauge, mechanical displacement, or pressure (palpatory signal)
Heart sounds			30 to 150 cps	Microphone
Pulse sounds		Variable	.1 to 50 cps	Photoelectric, piezoelectric mercury resistance gauge, strain gauge, etc.
Phonocardiogram			5 to 2000 cps	Crystal, condenser, or magnetic microphone
Plethysmogram (volume measurements)			DC to 30 cps	Plethysmograph (with strain gauge or other pressure-sensing device)
Ballistocardiogram			DC to 40 cps	Special platform with accelerometer
Heart rate (Tachometer)	human animal	.75 to 3 beats/sec .75 to 9 beats/sec	.75 to 3 cps .75 to 9 cps	Signal electronically derived from ECG, pulse wave, etc.
Blood flow		.05 to 1650 ml/sec	DC to 50 cps	Electromagnetic, sonic, bristle, turbine-type, thermal dilution
Oximetry			0 to 5 cps	Photoelectric ear or finger pickup
Electrocardiogram		$10\mu V$ to 5 mV	0 to 5 cps	Direct recording from electrodes

RESPIRATORY SYSTEM

Parameter	*Parameter Range*	*Required Frequency Response (cps)*	*Transducer or Type of Transducer Usually Employed*
Flow rate	250-500 ml/sec	DC to 40 cps	Pneumograph with strain gauge Pneumograph with differential transformer Impedance pneumograph, mercury-filled rubber tube, turbine type, thermistor
Breathing rate (human) (animal)	0.2-0.4 breaths/sec .1-1	0.2 to .4 cps 0.1 to 1 cps	Strain gauge belt, mercury-filled rubber tube, thermocouple, thermistor Impedance pneumograph
CO_2 O_2	0-10% 0-20%	0.5 to 2 cps 0.5 to 2 cps	Mass spectograph, Severinghaus electrode, light absorption Mass spectograph, Clark electrode, light absorption
Dissolved Gases, Ions, and pH pH (in vitro)	Signal range 0 to ± 700 mV	0.5 to 2 cps	Glass electrode
Na^+, Cl^-	0 to 600 mV	0.5 to 2 cps	Membrane electrode, light absorption

BIOELECTRIC POTENTIALS

Parameter	*Parameter Range*	*Required Frequency Response (cps)*	*Transducer or Type of Transducer Usually Employed*
Electroencephalogram	15 to 100 μV	DC to 100 cps	Surface electrodes
Cerebral potentials (intracranially recorded)	10 to 100 μV	Pulse duration (0.6 msec to .1 sec)	Needle or microelectrodes
Electromyogram (primary signal)	50 to 200 μV	10 to 2500 cps Pulse duration (0.6 msec to .1 sec)	Surface or needle electrodes
Smooth muscle potential (*e.g.*, electrogastrogram)	.5 to 80 mV	DC to 0.6 cps	Surface or needle electrodes
Electroretinogram Electrocardiogram	.1 mV .1 mV	DC to 20 cps (see Cardiovascular System)	Electrodes

TABLE 8-1 (continued)

PHYSICAL QUANTITIES

Parameter	*Required Frequency Response (cps)*	*Transducer or Type of Transducer Usually Employed*
Temperature	Variable	Thermistor, thermocouple, resistance thermometer
Voice	20 to 20,000 cps	Microphone with output proportional to: amplitude, log amplitude, average energy
Skin resistance (GSR)	Resistance: 1 KΩ to 500 KΩ	Direct recording from electrode
Displacement	Variable	Differential transformer, mutual inductance, strain gauge, photoelectric transducer, etc., variable capacitance, variable inductance, variable resistance, photoelectric, potentiometric: usually large displacement on weak spring
Acceleration	Variable	Displacement (small displacement—stiff spring)
Velocity	Variable	Voltage induced in a conductor moving in a magnetic field, magnet moving past coil, etc.
Drop counter	Variable	Strain gauge, photoelectric transducer, conductivity

quency response not be sufficiently high for the application, the investigator should analyze the amplifier and the recorder separately to determine which is the culprit. Adequate frequency response must be designed into an amplifier or a recorder, but there are sometimes adjustments that the user can make for highest frequency response as per the manufacturer's directions. In general, the higher the frequency response of an instrument, the higher the cost of the instrument.

The third step is to consider the transducer element in combination with the rest of the system, assuming that the range, sensitivity, linearity, signal-to-noise ratio, and frequency response of the amplifier and recorder are adequate. A series of mean pressures should be applied to the transducer to determine the sensitivity and linearity of the entire system. The user should be careful not to exceed the manufacturer's specifications regarding the usable range of the manometer. The investigator or user should now apply sinusoidal pressure to the transducer element in a closed fluid-filled system and study the frequency response of the entire system, including any catheter or cannula that is to used with the manometer. Although the application of sinusoidal pressure is, in my opinion, the most satisfactory procedure for determining the frequency response, there are other procedures that may be used. These are discussed in Chapter 11.

The last step is to determine all of the factors which may have some effect on the system. In the case of the pressure measurement, air bubbles in the transducer or cannula considerably decrease the response of the manometer with respect to frequency response. The length and diameter of cannula are also important. Actually, there are many different factors that have an effect on the measurement of blood pressure, and these must be determined and individually evaluated. Every biomedical instrument, particularly the transducer element, must be studied from all sides before it can be used. The manufacturer's calibration should be discounted, and a calibration study of the type described above should be done by the investigator himself. Whenever possible this should be done in his own laboratory. In the chapters to follow, some calibration techniques will be described in more detail.

CHARACTERISTICS OF SOME BIOMEDICAL PARAMETERS

There are, of course, many biomedical parameters to measure. Table 8-1 lists some of them along with the range of the parameter that may be expected, the frequency response required to accurately evaluate the measurement, and some of the transducers or types of transducers that might be employed for this purpose.

Almost all biomedical parameters are difficult to *isolate* from their environment. Placing an electrode on the inside of a nerve fiber may easily alter the potential difference between the inside and the outside. Inserting a needle into the anterior chamber of the eye alters the intraocular pressure it is to measure. Almost any measurement made in the vicinity of the heart will pick up unwanted electrocardiographic signal. These few examples are the kind of problems that make biomedical instrumentation interesting and challenging.

An animal is essentially *a bag of salt water* as far as biomedical instrumentation is concerned. Many of the instruments now made are designed to fit into the animal, and some to broadcast the signal to an external receiver. In other cases the transducer is implanted in the animal and wired to a connector which is sewed into the skin of the animal. The transducer is then connected to the remainder of the instrument with a suitable cable. In both of these cases, the parts of the instrument which are left inside the animal must be encapsulated in a salt-water proof material. The material chosen must be as non-toxic as possible, such as Teflon or a variety of silastic rubbers or some of the epoxy resins. All such materials eventually deteriorate when left inside an animal. Sometimes they cause the animal to deteriorate as well. The instrumentalist should be prepared to spend a considerable amount of time, barring some unforeseen development, choosing and testing the proper "potting" or encapsulation material. The leakage of salt water and tissue fluids into transducer elements or cables causes many problems, such as increasing the capacitive coupling to ground and losing signal or even a direct short or low-resistance short between two parts or wires which should have an infinite resistance between them.

With modern materials and the "state of the art" in electronics, the biomedical investigator should be able to measure almost any parameter. The instrumentalist should be prepared to take advantage of each development in electronic engineering. The instruments and techniques developed for use in space vehicles should be examined closely for their application to common biomedical instrumentation problems.

REFERENCES:

An Introduction to Electronics for Physiological Workers, Ed. 2. I. C. Whitfield (Macmillan, New York, 1960).

Biological and Medical Electronics, Ralph W. Stacy (McGraw-Hill Book Co., New York, 1960).

Biophysical Research Methods, Fred M. Uber (Interscience Publishers, New York, 1950).

Electronic Apparatus for Biological Research, P. E. K. Donaldson (Butterworth, London, 1958).

Instrumentation Methods for Physiological Studies, vol. 1, Norman N. Goldstein, Jr. (University of California Berkeley Extension, Berkeley, Calif., 1964).

Medical Physics, vol. 1, 2, 3, Otto Glasser, ed. (Yearbook Publishers, Chicago, 1944, 1950, 1960).

Technology of Instrumentation, Eric B. Pearson (D. Van Nostrand Co., Princeton, N.J., 1957).

The Response of Physical Systems, J. D. Trimmer (John Wiley and Sons, New York, 1950).

9

The Transducer

the measurement of displacement

light sensors

physiological gas analysis

references

The transducer may be defined as a device which converts one type of energy into another. It may be considered an extension of the sensory facility of man. It allows measurement of phenomena which are not electrical in nature and therefore cannot be amplified or modified by means of electronic circuitry and ultimately recorded. There are many forms and kinds of transducers. Almost every relationship involving one physical entity in terms of another may be used as the basis for a transducer. For example, the resistance of a wire varies with its length, so energy used in stretching the wire, as when measuring displacement or pressure, can be converted into resistance changes, which when in the proper circuit can be converted to a usable voltage signal, which is, in turn, proportional to the original displacement or pressure. A magnet moving past a coil induces a current in the coil proportional to the velocity of the magnet.

Transducers are commonly used to convert light intensity, color, concentrations in liquids and gas, liquid or gas flow, sound intensity or frequency, velocity and acceleration, as well as pressure and force. As new materials and new electronic components are invented new transducers will also be invented. Many new transducers have been described with the appearance of the semiconductor such as a new species of semiconductor pressure-transducer, new radiation-sensitive transducers, and light-sensitive elements. An entire measurement procedure may be considerably enhanced merely by improving or simplifying the transducer. The more sophisticated measurements almost always utilize simple but unique transducers.

Often a transducer is easier to design than it is to fabricate. The investigator may decide to measure some biomedical parameter, and he may decide on a fairly simple plan. In most cases, he will find that the actual building of the device requires the talents of a jeweler, a tool and die maker, a watch maker, and an expert in plastics and potting materials as well as an electronic enthusiast. It is not extremely difficult to build a device such as those used in an elementary physiology laboratory, some of which will be described in this chapter. The difficulty lies in refining the transducer for use in a research situation, for which the instrument must be accurately calibrated. It must also be reliable and in all probability must be able to withstand animal implantation.

Simply stated, calibration of the transducer involves controlled variation of the amplitude of the parameter being measured to determine the amplitude of the primary signal. The parameter being measured must also be varied with time to a frequency at least as high as that expected to be found in an adequate Fourier analysis of the data itself. It is important to calibrate the transducer in the exact situation in which it is to be used. If it is to be used inside an animal it must be calibrated inside the animal. The transducer must not overload the system or parameter being measured and cause an alteration in the data itself. Otherwise the investigator may be fooling himself.

Calibration of a transducer is no different in principle from calibration of any instrument, as discussed in the last chapter. The basic principles of measurement discussed in Chapter 3 also apply to the transducer. The biomedical instrumentalist should be aware that it is the transducer that is most often the weakest part of the instrument design.

A transducer that has mechanical parts must be calibrated by moving the mechanical parts. A pressure transducer is an example of this. Most amplifiers designed for use with strain-gauge manometers have one or more resistors which can be put in parallel with the active strain element. These resistors will mimic a change in pressure. Use of the shunting resistor does not constitute a calibration, *per se.* The best way to calibrate a pressure transducer statically and dynamically is with a pressure source.

The remainder of this chapter will be devoted to presentation of a variety of transducer elements. There are many others which could have been included, and still others which will be invented while this book is in press and thereafter. The references at the end of this chapter list some monographs devoted solely to transducer elements. The beginning instrumentalist should also examine The Review of Scientific Instruments, the special communication section at the back of the Journal of Applied Physiology, Medical Electronic News, and other publications in which instruments are described. The beginner will soon have a personal catalog of transducers available to do a specific task.

THE MEASUREMENT OF DISPLACEMENT

One of the easiest physical parameters to measure is displacement. A search through any elementary or intermediate level physics text will reveal that the term "displacement" (distance or length) appears in many physical relationships. Because displacement is easy to measure there are many transducer designs for this purpose.

There is no way to dissociate the measurement of displacement from that of pressure or force. It is possible, however, to minimize the force necessary to move the actuating part and hence to more nearly measure pure displacement. One of the easiest ways to do this is to vary the resistance in a straight wire as shown in Figure 9-1(A). Here a wiper (C) is arranged to short out a portion of the resistance as it moves along the wire.

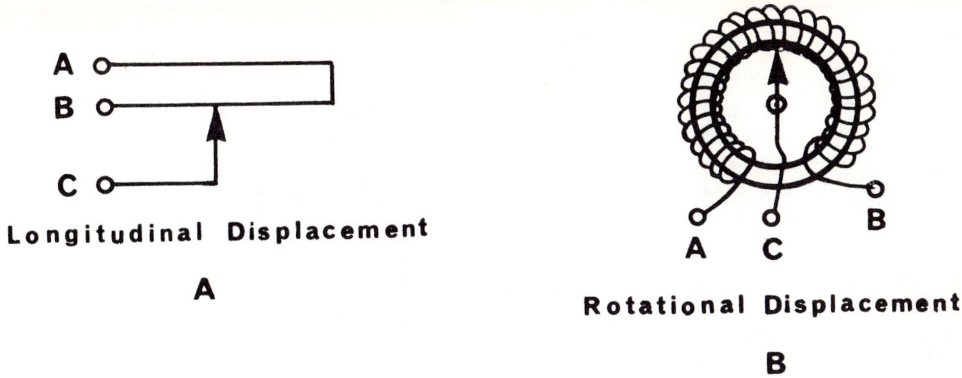

Figure 9-1. Displacement transducers.

Approximately the same procedure can be used to measure rotational displacements (Fig. 9-1 [B]). Applications for these devices are rather limited, however, because of the friction encountered by the wiper arm. The slide-wire device is adequate for measuring displacements when good resolution is not needed and the velocity is not high.

Another device, described by Goldstein, is a *rheostatic displacement transducer.* In Figure 9-2, the transducer is being used to study the contraction of muscle. A silver wire is attached by means of a cotton applicator to a muscle lever. The wires in the beaker are silver, and the connecting wire on the applicator stick is 40-gauge insulated magnet wire. If there is too much voltage across the input terminals of the oscilloscope, or whatever display or recording device is employed, a resistor can be used in series with the battery.

Perhaps the most common transducer for the measurement of displacement is the *strain gauge.* A strain gauge is based on the principle that the resistance (R) of a wire varies directly with the length (l) and inversely with the cross-sectional area (a).

$$R = \frac{\rho l}{a} \tag{9.1}$$

where ρ is the resistivity, which is defined as the resistance per unit length and unit diameter. When a wire is stretched, it not only gets longer but it also decreases in cross-sectional area; there may also be changes in the crystalline structure of the wire which may affect the resistivity. If a stress is applied to a length of wire, the strain may be measured in terms of the change in resistance of the wire. Equation 9.2 represents the sum of all factors on the change in resistance in a wire strain gauge.

$$\Delta L = \frac{FL}{R} \Delta R \tag{9.2}$$

where ΔL and ΔR are small changes in L and R respectively. In equation 9.2, F is called the *gauge factor* and is defined as the unit change in resistance per unit change of length. The gauge factor may be expressed as follows:

$$F = \frac{\Delta R}{R} \frac{\Delta L}{L} \tag{9.3}$$

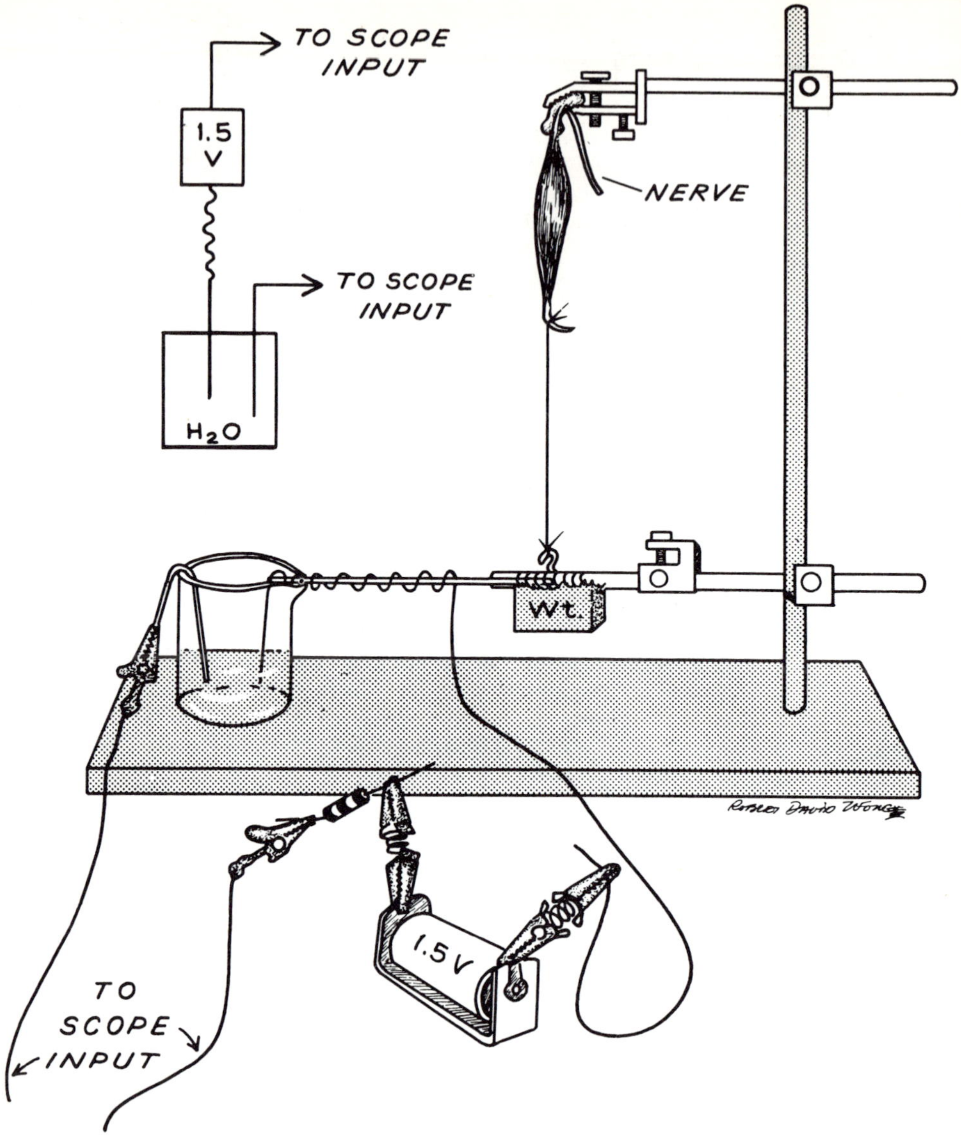

Figure 9-2. Rheostatic displacement transducer. (Courtesy of Norman N. Goldstein, Director Project IMPS, University of California Extension, Berkeley, Calif.)

Gauge factors are quite variable: nickel has a gauge factor of 12.1, platinum is 6.1, constantin is about 2, and manganin is about 0.4.

The strain-gauge element may be bonded directly to the material on which the stress is applied. Such a gauge is called a *bonded strain gauge.* The strain wire is usually arranged as shown in Figure 9-3 and cemented in place. It may be cemented onto a steel part to measure bending on a large I beam in a building or onto a rail to measure vibration. It may also be bonded onto a supporting structure in a concrete foundation to measure stress developed in the wall.

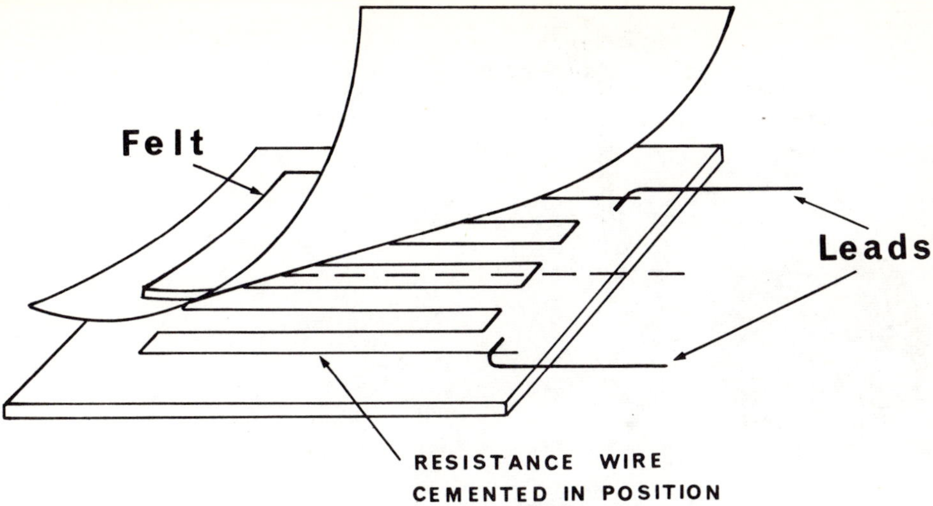

Figure 9-3. The bonded strain gauge.

Any strain-gauge transducer may be designed to measure displacement, force, pressure, or acceleration, depending on the nature of the device to which the gauge is bonded. The difference between a transducer designed to measure displacement and one designed to measure force is merely the amount of bend in the supporting structure. Figure 9-4(A) is a force transducer designed by the Minneapolis Honeywell Company to measure the mechanical activity of the heart. The basic unit, shown in the insert, may be sewn directly into the myocardium. It is $35 \times 9 \times 2.5$ mm, weighs 2 grams, and is completely watertight (potted) by using epoxy resin. When excited, using 10 volts, the gauge has an output of 300 mv for its maximum displacement of 0.04 inches, which represents a force of ± 200 grams. It can resolve a signal representing less than 100 mg. When attached to a ring stand the transducer is a *force transducer* with exactly the same characteristics as previously described. If, however, a 3-inch bronze extension is added (Fig. 9-4[B]), a displacement of 0.5 inches may be measured. The above device has a resonant frequency of 450 cps, which represents, of course, a much higher frequency response and much higher resolution than the transducer in Figure 9-2.

The strain gauge described above is the bonded strain gauge. The *unbonded strain gauge* is an electromechanical device which transduces minute movements into proportional changes in resistance. An armature is arranged so that it can move longitudinally. The armature is mounted in a hole in the frame and secured by a crossbar. The unit is shown in Figure 9-5 (A) and (B). Four resistance wires are used, each of equal length. When the armature is moved longitudinally by an external force, two of the wires stretch and the other two relax. The wires are arranged as in a Wheatstone bridge. The wires that are stretched are opposite the wires that relax. By use of an arrangement of this sort, temperature effects on the strain wires are minimized. Mechanical stops are inserted to avoid damage to the strain wires.

The unbonded gauge described above may also be used as a displacement transducer, a pressure transducer, or an accelerometer. When a mechanical arm is attached to the armature (Fig. 9-6) the device becomes a force transducer. Figure

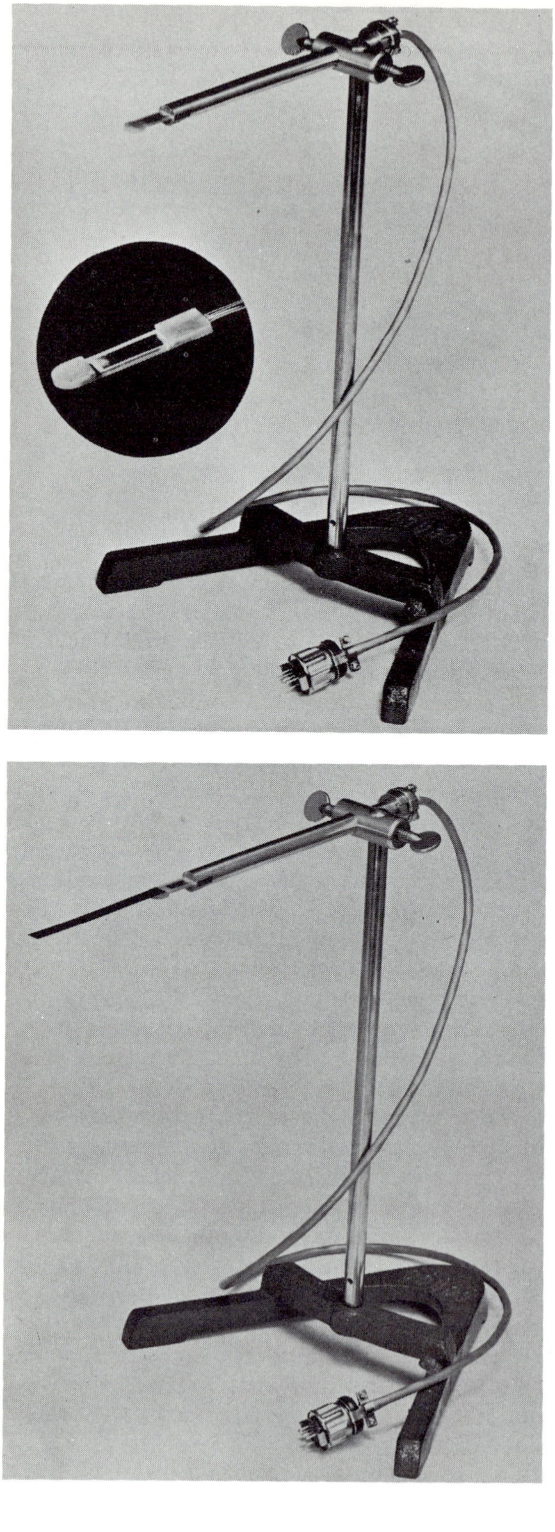

Figure 9-4. Force transducers made with a bonded strain gauge. (Courtesy of the Minneapolis-Honeywell Regulator Co.)

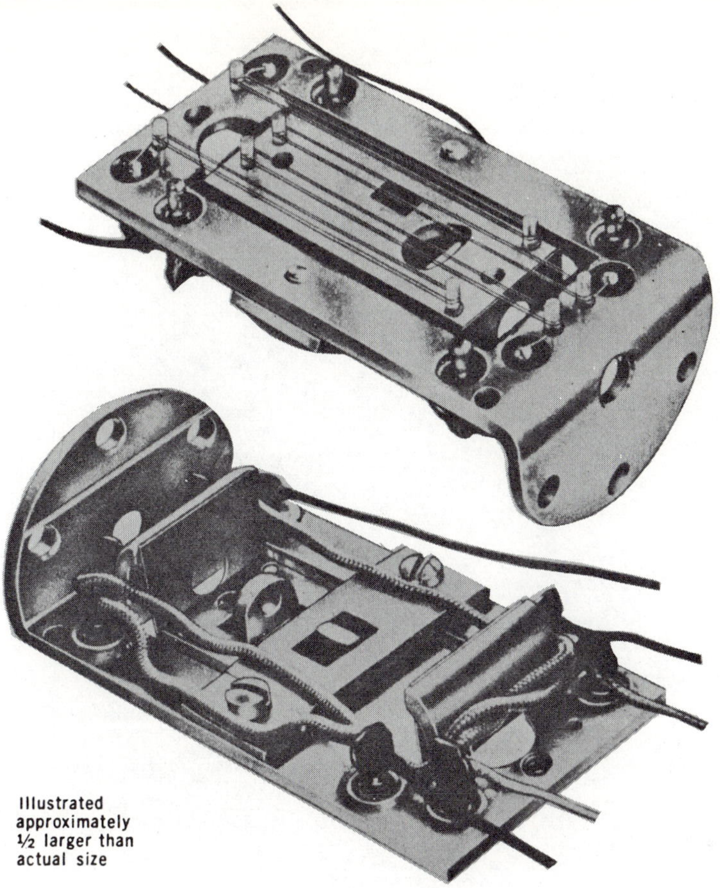

Figure 9-5. The unbonded strain gauge: top view; bottom view. (Courtesy of Statham Physiological Transducers, Inc., Puerto Rico.)

9-7 is the unit described above arranged as an accelerometer. Notice that a weight has been added. The weight is attached so that its mass lies directly on the sensitive wires. Since the mass is constant, the acceleration is proportional to the force (Newton's second law), and the force transducer becomes an accelerometer.

When the armature is rigidly connected to the center of a metallic bellows or diaphragm, the unit becomes a pressure transducer. The stiffer the diaphragm the higher the frequency response. Figure 9-8(A) is another arrangement for the unbonded strain gauge designed by Statham Physiological Transducers, Inc., for their miniature pressure transducer. In this arrangement, travel of the armature is less than 0.0003 inches. Figure 9-8(B) is the completed pressure manometer. Measurement of blood pressure is considered in more detail in a later chapter.

Another form of displacement transducer employing the *photoelectric principle* is shown in Figure 9-9(top). This transducer, manufactured by the E & M Instrument Company, uses an electric lamp to activate a photocell with a shutter assembly between them as shown in block diagram form in Figure 9-9. Notice that the

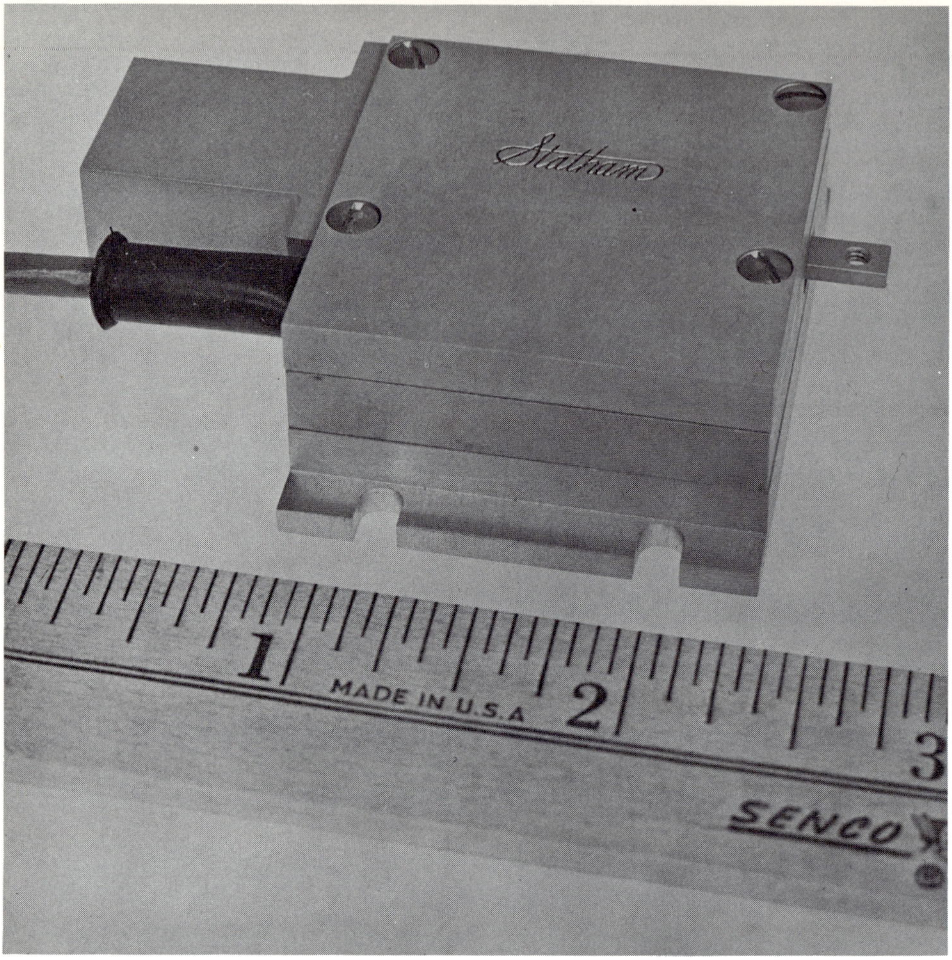

Figure 9-6. The unbonded strain gauge used as force transducer. (Courtesy of Statham Physiological Transducers, Inc., Puerto Rico.)

hook, which is restrained by a spring, is attached to the shutter. Movement of the hook, within limits, causes the shutter to move, allowing more of the surface of the photocell to be exposed to the light. An electrical signal may then be derived from the photocell which is proportional to linear displacement. The unit is designed to allow adjustment of the spring tension. Figure 9-9(bottom) is the completed transducer. This instrument may be used as shown in Figure 9-10 to study the properties of skeletal muscle or as in Figure 9-11 to study the properties of cardiac muscle.

The photoelectric principle may also be applied to the pressure transducer. Figure 9-12(top) is a cut-away diagram of a pressure transducer also manufactured by the E & M Instrument Company. A photograph of the completed transducer is shown in Figure 9-12(bottom). This transducer consists of a Bourdon tube, which is a hollow tube with an elliptical cross-section, bent as shown in the illustra-

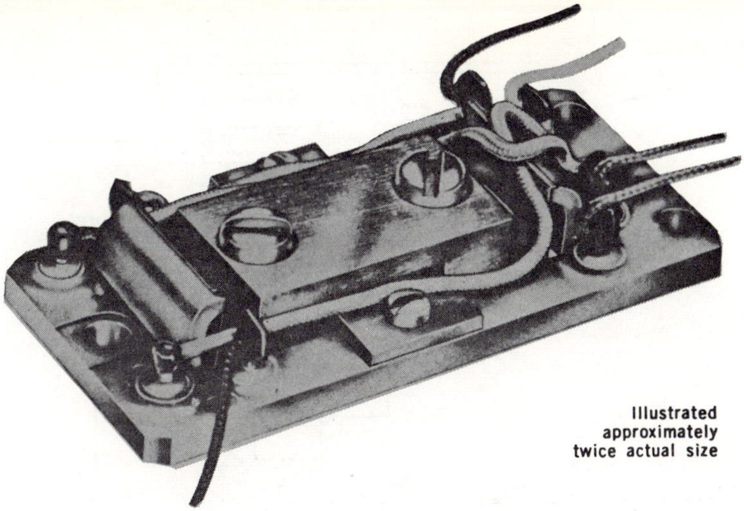

Figure 9-7. The unbonded strain gauge used as an accelerometer. (Courtesy of Statham Physiological Transducers, Inc., Puerto Rico.)

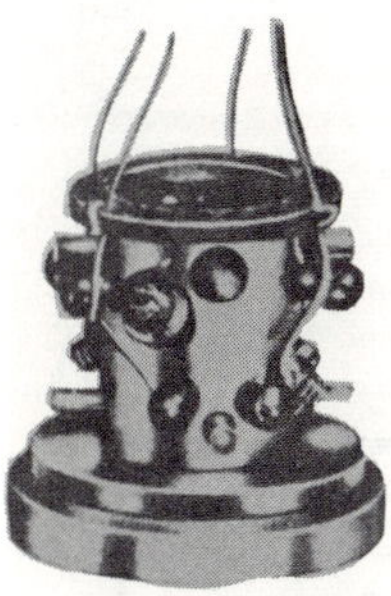

Illustrated approximately twice actual size

Fig. 9-8A

Figure 9-8. The miniature Statham pressure transducer. (Courtesy of Statham Physiological Transducers, Inc., Puerto Rico.)

Fig. 9-8B

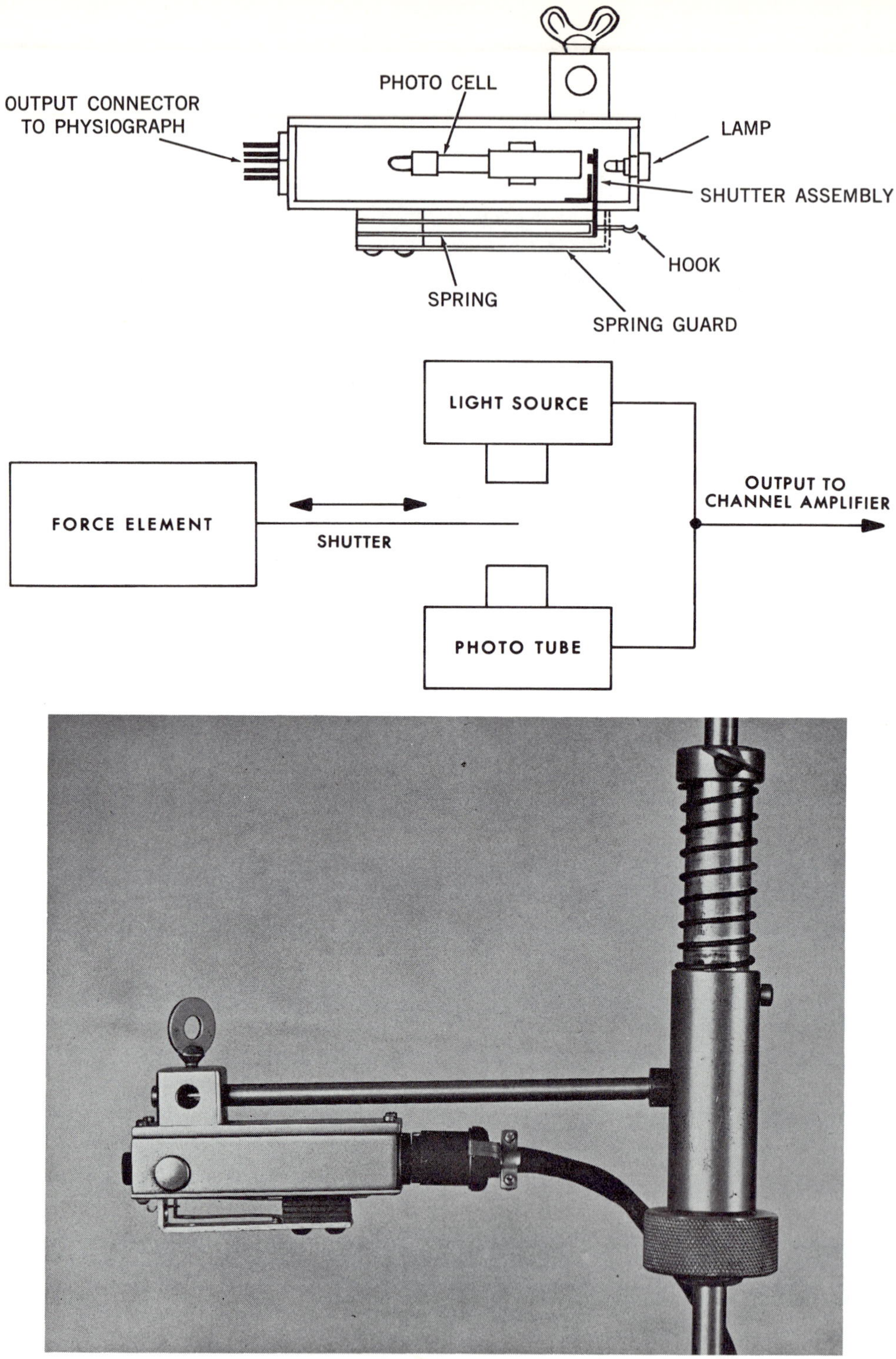

Figure 9-9. The photoelectric displacement transducer: *top,* cut away view; *center,* block diagram; *bottom,* completed transducer (Courtesy of E & M Instrument Co., Inc., Houston, Tex.)

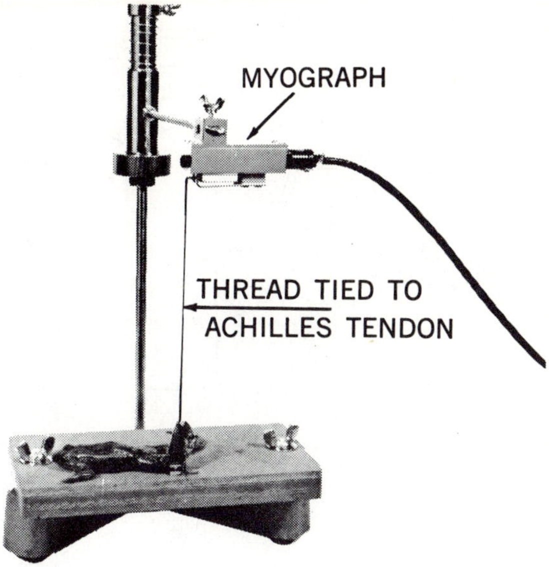

Figure 9-10. Using the photoelectric displacement transducer to study the properties of skeletal muscle. (Courtesy of E & M Instrument Co., Inc., Houston, Tex.)

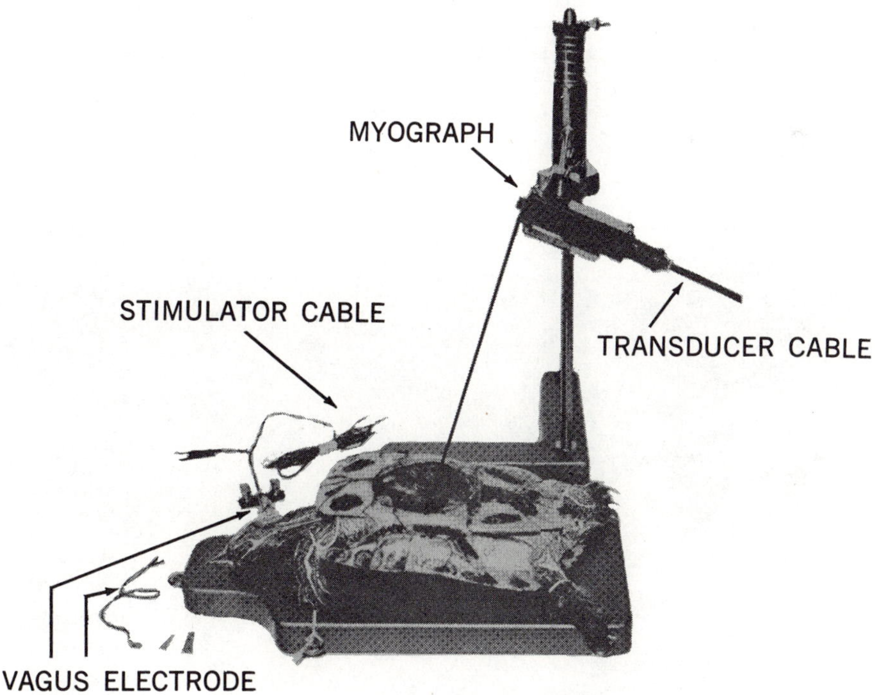

Figure 9-11. Using the photoelectric displacement transducer to study the properties of cardiac muscle. (Courtesy of E & M Instrument Co., Inc., Houston, Tex.)

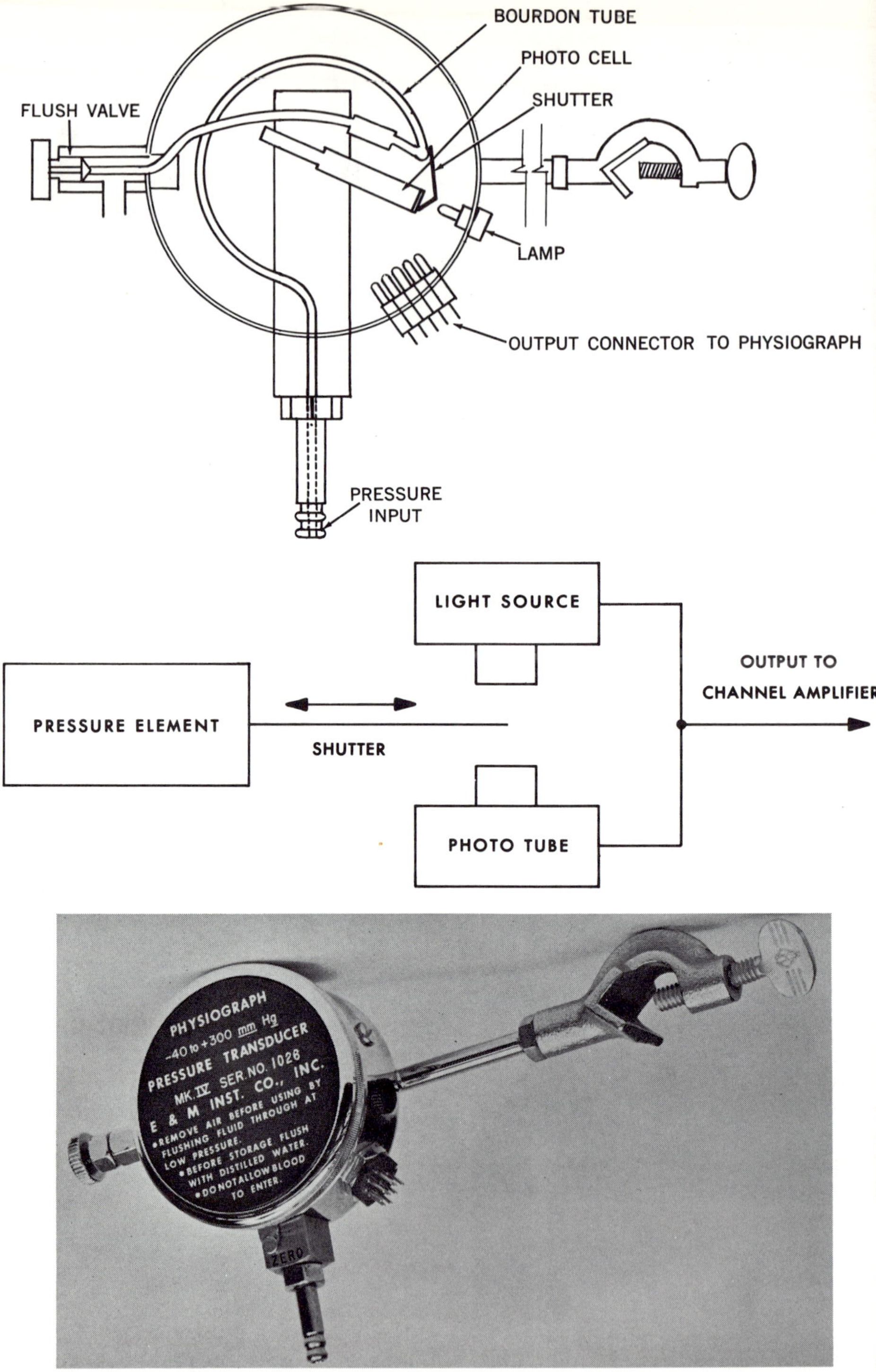

Figure 9-12. The photoelectric Bourdon pressure transducer: *top,* cutaway diagram; *center,* block diagram; *bottom,* completed instrument.

tion. The tube uncurls when the pressure inside is increased, and the end of the Bourdon tube, which is attached to a shutter assembly located between the lamp and the photocell, alters the amount of light permitted to strike the photocell (Fig. 9-12[center]). This particular unit has a rather large volume displacement, as 15 mm^3 of fluid must move into the Bourdon tube to cause an increase of 100 mm mercury pressure. The strain gauge in Figure 9-8(B) can have a volume displacement as small as 0.1 mm^3 for every 100 mm mercury pressure. This means that the Bourdon tube is considerably less sensitive than the strain-gauge manometer. It also has a lower frequency response. It is, however, adequate for measurement of blood pressure in large laboratory animals, when a high-frequency response is not necessary. The photoelectric pressure transducer costs about one-fourth as much as the strain-gauge manometer mentioned above.

Another form of the photoelectric transducer, also manufactured by the E & M Instrument Company, is the *bellows pneumograph* shown in Figure 9-13. This transducer is very similar to the photoelectric displacement transducer described above. The valve on the left is opened, and the unit is strapped around the chest. The bellows are then compressed slightly and the valve closed. As the subject breathes, the shutter assembly moves back and forth, exposing more or less light to the photocell. This transducer cannot be calibrated to measure the actual volume of air going into and out of the lungs, but it does indicate the rate, various time sequences, and patterns of respiration. The completed transducer is illustrated in Figure 9-13(bottom).

A device which transduces air flow rate to a differential pressure which can be measured, using two pressure transducers or a differential pressure transducer, is shown in Figure 9-14. The transducer is called a *double cone pneumotach.* As the air passes through the double cone arrangement, in which there are three No. 200 mesh air flow screens, the pressure on one side is somewhat greater than that on the other side. This type of transducer can be calibrated from 0 to 200 liters per minute. An air flow of 100 liters per minute generates a pressure differential of 1.5 cm of water in the transducer shown above with three screens.

There are several varieties of transducers that measure displacement which are based on the use of electromagnetic coupling as in a transformer. Such a transducer is referred to as a *differential transformer.* In principle, the transducer consists of three parts: a primary coil, a secondary coil, and a movable magnetic core. The core is attached to a stiff connecting rod which is used to transmit the motion being studied to the transducer.

The circuit diagram (Fig. 9-15[center]) shows a primary coil which is energized with a suitable alternating current. As a rule of thumb, the frequency of this alternating current should be at least four times the highest frequency component expected in the data. The secondary coil is center tapped and arranged, as shown in Figure 9-15(top), on either side of the primary. The core is made of a material with high magnetic permeability such as iron or a special alloy. The secondary coil is arranged differentially, that is, the voltage induced in one half of the secondary is 180° out of phase with the voltage induced across the other half of the secondary coil, and, hence, at one pre-set position of the core with respect to the primary and secondary coils, there is no voltage induced across the secondary.

As the high-permeability core moves back and forth from its null position, making no mechanical contact with the coils, it changes the amount of magnetic flux energizing the secondary coil; hence, a voltage is developed across the sec-

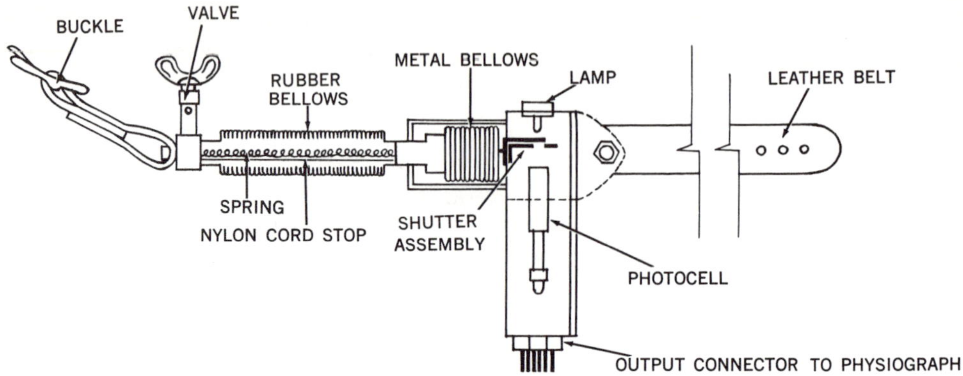

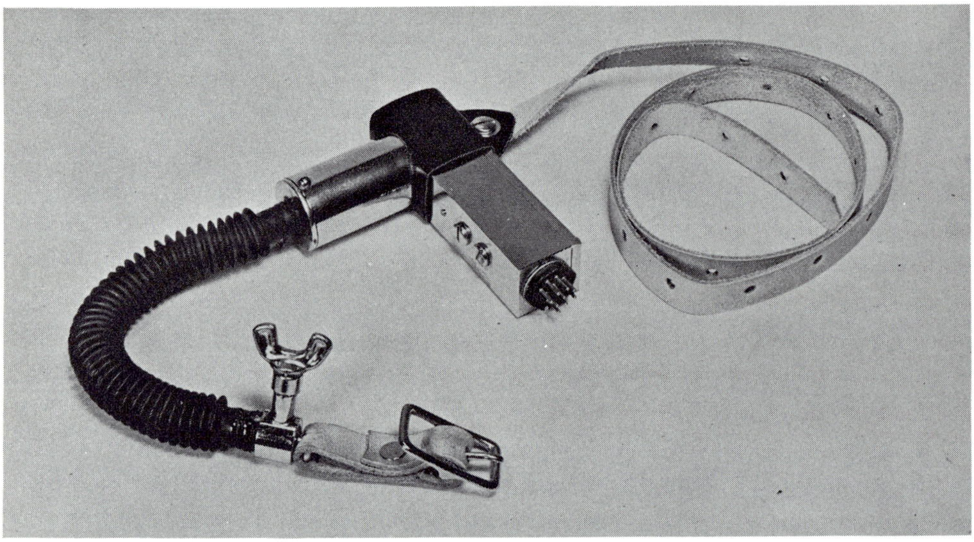

Figure 9-13. The photoelectric bellows pneumograph: *top,* cutaway view; *bottom,* complete transducer. (Courtesy of E & M Instrument Co., Inc., Houston, Tex.)

ondary which is proportional to displacement of the core. As the core gets close to the edges of the coils, however, the magnetic flux density is not uniform as the output signal across the secondary becomes non-linear. A transducer of this design can be used only within a specific range of displacement to insure linearity. It does, however, have a high resolution, being capable of detecting a displacement of a millionth of an inch. The calibration curve of the Sanborn transducer, illustrating the usable range, is shown in Figure 9-15(bottom). A transducer which operates as described above, manufactured in many sizes by the Sanborn Company, is shown in Figure 9-16.

Another variety of the *differential transformer* displacement transducer is shown schematically in Figure 9-17(top). Notice that this transducer is essentially the same as the one described in Figure 9-15 except that the alternating current is generated within the transducer, thus conserving space. The unit also demodulates the alter-

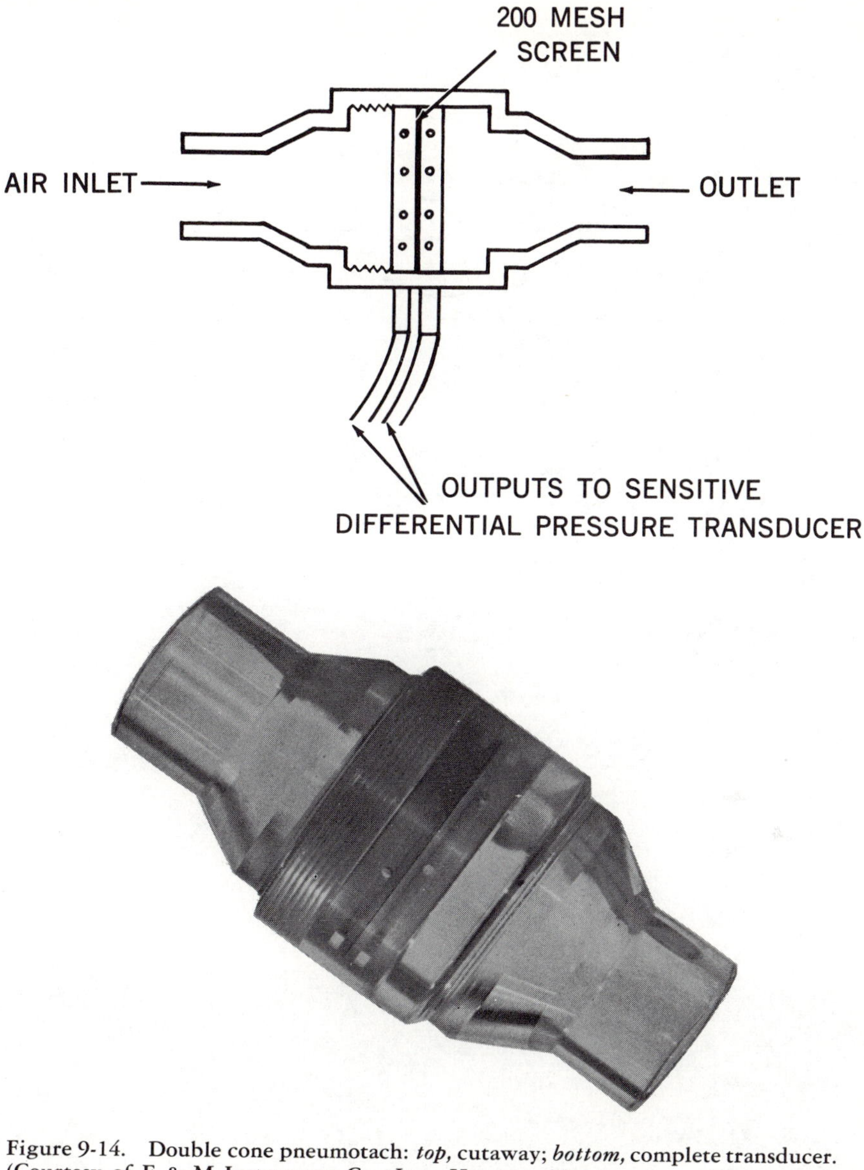

Figure 9-14. Double cone pneumotach: *top,* cutaway; *bottom,* complete transducer. (Courtesy of E & M Instrument Co., Inc., Houston, Tex.)

nating voltage induced across the secondary, and the output is thus reduced to DC, which is a more usable signal. The form of a typical calibration curve is shown in Figure 9-17(bottom). Notice that the DC output voltage in this differential transformer also deviates from linearity when extended beyond its usable range.

The differential transformer can also be used as a force transducer in the same way that the strain gauge can be used to measure displacement as well as force. Such a transducer, made by the Sanborn Company, is shown in Figure 9-18. In this transducer, the core and drive assembly are suspended by balanced stiff springs,

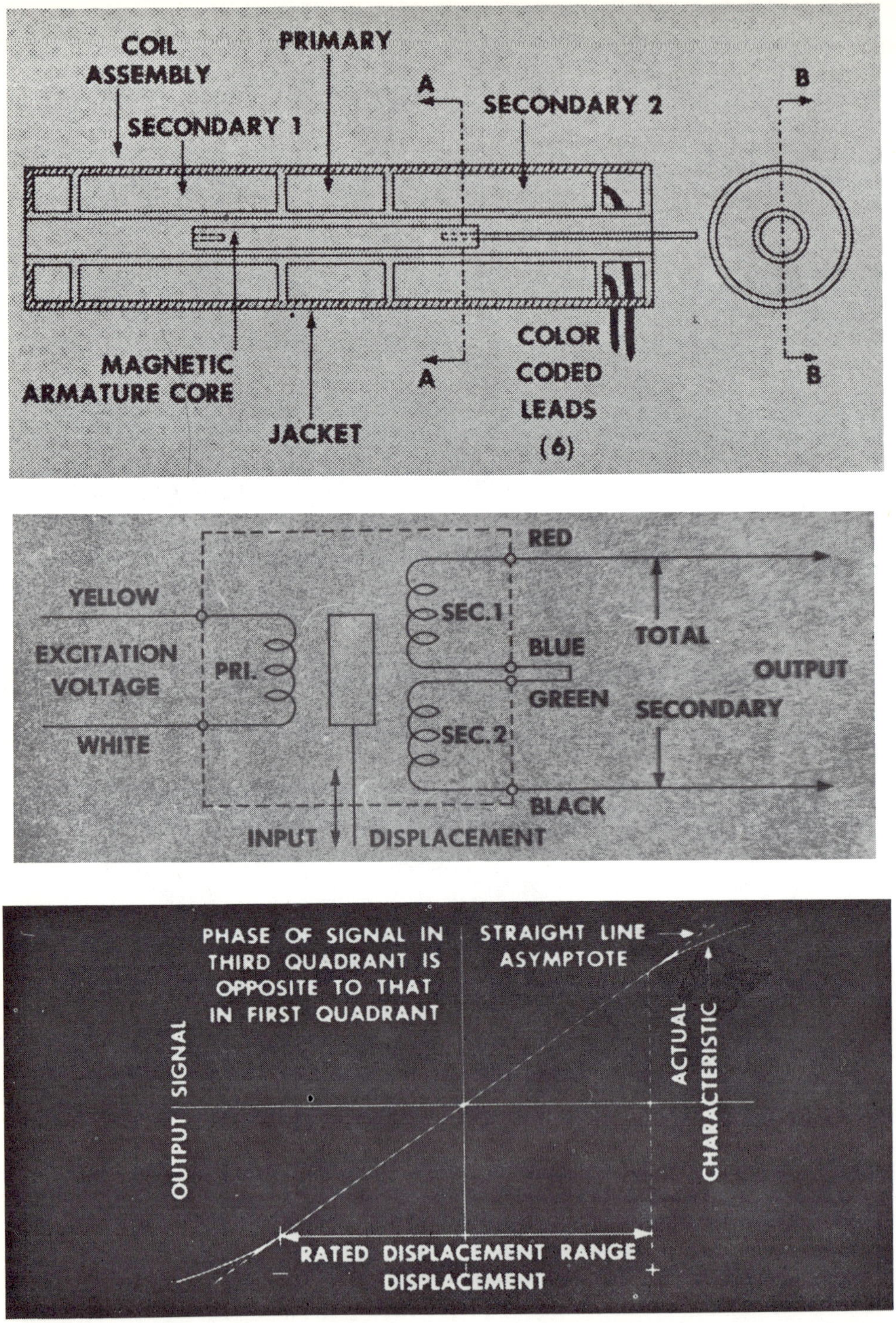

Figure 9-15. The differential transformer: *top,* cutaway diagram; *center,* schematic circuit; *bottom,* calibration curve. (Courtesy of Sanborn Co., Waltham, Mass.)

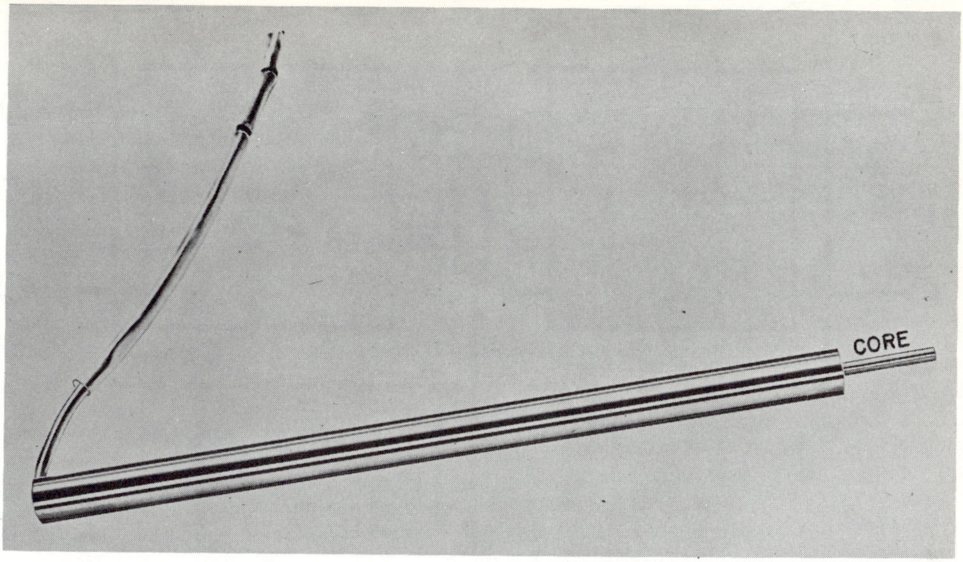

Figure 9-16. The Sanborn differential transformer displacement transducer. (Courtesy of Sanborn Co., Waltham, Mass.)

which are designed to allow the core to be displaced linearly with respect to the force applied to the spring and drive linkage. The application of an axial force causes proportional displacement of the core, which, in turn, induces a voltage across the secondary that is also proportional to the applied force.

The Sanborn FTA Myographic Force Transducer may be used as shown in Figure 9-19. In this application, a muscle is suspended in a nutrient solution in a temperature-controlled bath. The force of the muscle contraction is measured when the muscle is stimulated. Such an experimental setup may be used to study the physiology of the muscle or to use the amount of force developed by the muscle to determine the concentration of an ion in the nutrient solution.

A transducer which consists of a parallel plate capacitor can be used to measure displacement, force, or pressure. The capacitance of two parallel plates is

$$C = k\epsilon \frac{A}{D} \tag{9.4}$$

where C is the capacitance in farads, k is 8.85×10^{-10} when A is the area of the plate in square centimeters, and D is the distance separating the plates expressed in centimeters. When the dimensions are expressed in inches, the constant, k, is 22.4×10^{10}; ϵ is the dielectric constant which is equal to 1.0 when air is the dielectric medium.

There are several basic ways in which this system can operate as a transducer: (1) the distance between the plates can be varied as shown in Figure 9-20(top); (2) one plate can be moved with respect to the other so as to change the effective area between the plates as shown in Figure 9-20(bottom); and (3) the dielectric constant of the material between the plates could vary. This third method is almost never used. There are many disadvantages to the use of this system as a displacement transducer. Some of these include non-linearity or a very limited range over

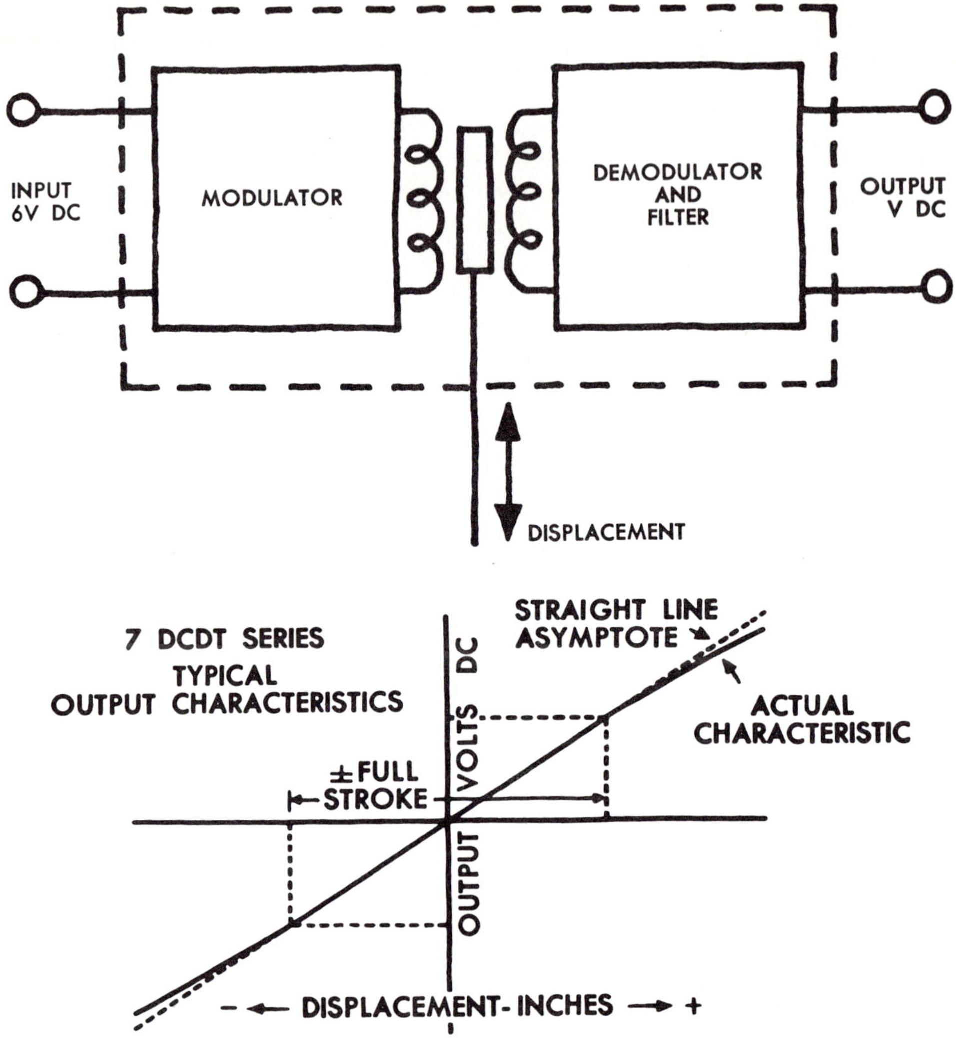

Figure 9-17. Another Sanborn differential transformer displacement transducer: *top,* Schematic diagram; *bottom,* calibration curve. (Courtesy of Sanborn Co., Waltham, Mass.)

which linearity can be maintained. The environmental temperature has a tendency to vary the physical dimensions which, in turn, have an effect on the output. As a rule, the change in capacitance which occurs is small enough that changes in the capacitance of the connecting cable which result from bending or movement of the coil are as great as or greater than the change in capacitance being measured.

It was pointed out at the beginning of this chapter that a transducer could be made from almost any electronic component. The triode is no exception. Figure 9-21 is a schematic diagram of a triode developed by the Radio Corporation of America to detect very small displacements. The plate of the triode is arranged so that as it is displaced it moves closer to the grid. The device itself, known as the RCA 5734, is shown in Figure 9-22. This *mechanoelectronic* transducer is usually

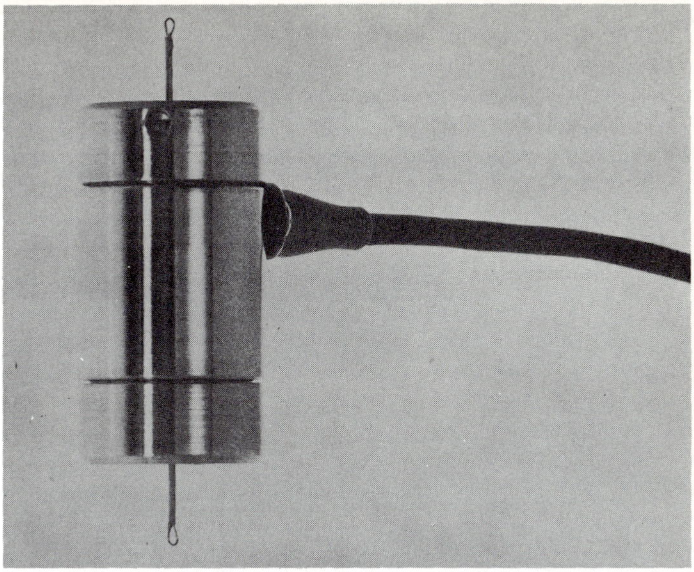

Figure 9-18. The Sanborn differential transformer force transducer. (Courtesy of Sanborn Co., Waltham, Mass.)

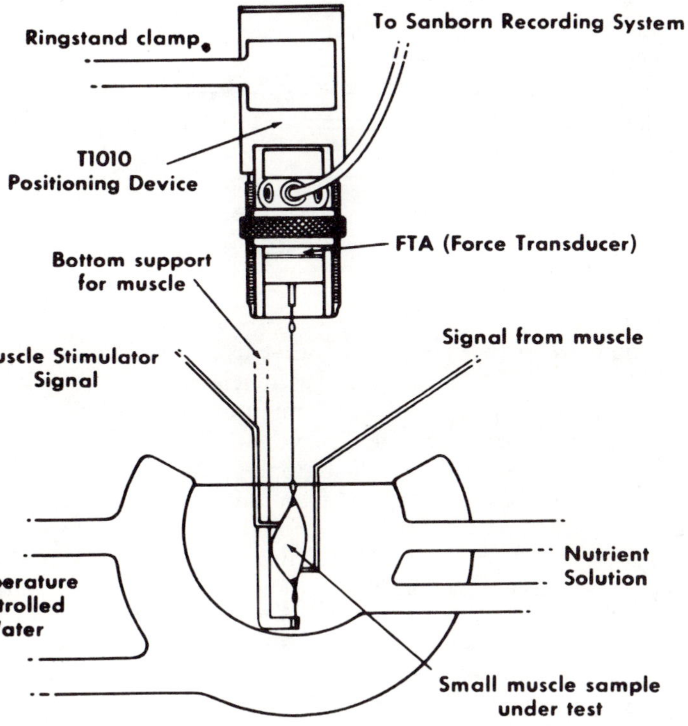

Figure 9-19. The Sanborn FTA myographic force transducer. (Courtesy of Sanborn Co., Waltham, Mass.)

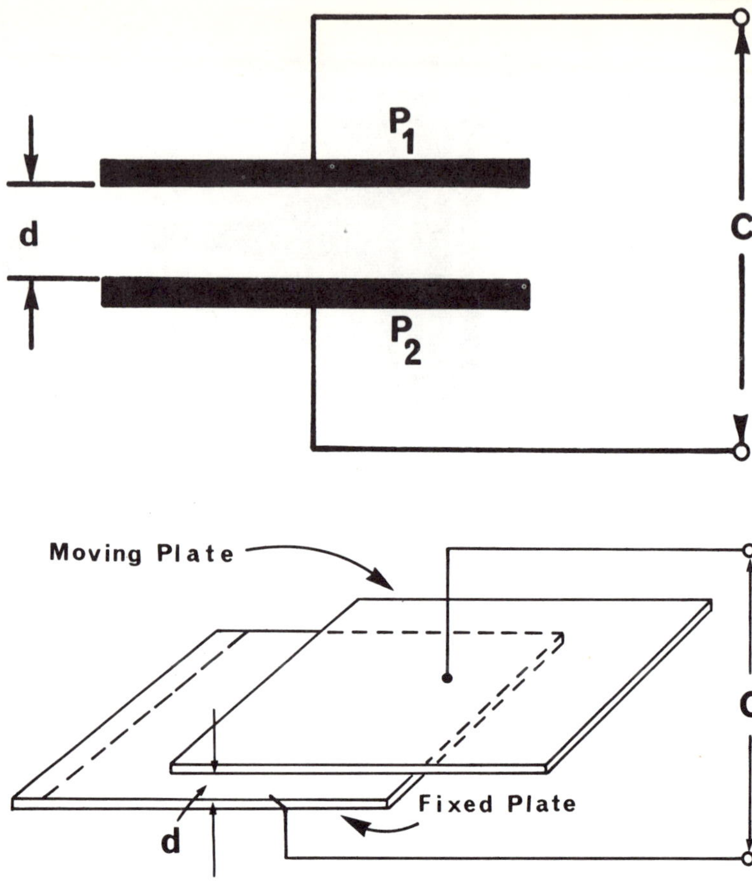

Figure 9-20. The capacitive displacement transducer.

used in a bridge circuit that contains another triode with characteristics similar to the RCA 5734. The 5734 can generate as much as 20 to 40 volts, if the proper plate resistor is used, with as little as 30 minutes of arc of plate displacement on either side of center. The calibration curve, however, is not linear. The linearity can be improved considerably by utilizing only a fraction of the usable range. This transducer is extremely fragile, and the plate must not be displaced too far. It also tends to get hot, and provision must be made to dissipate the heat.

Some materials, when in crystalline form, become polarized when distorted and develop a voltage across the crystal. This is called the *piezoelectric effect*, and transducers utilizing this principle are called *piezoelectric transducers*. When a crystal with this property is deformed some of the positive and negative charges are removed from their position within the crystal lattice. The result of such displacement is that equal and opposite charges develop at opposite surfaces of the crystal. A high-impedance device, such as an oscilloscope, a vacuum-tube voltmeter, or an amplifier with very high-input impedance, may be used to sense the potential difference or voltage signal.

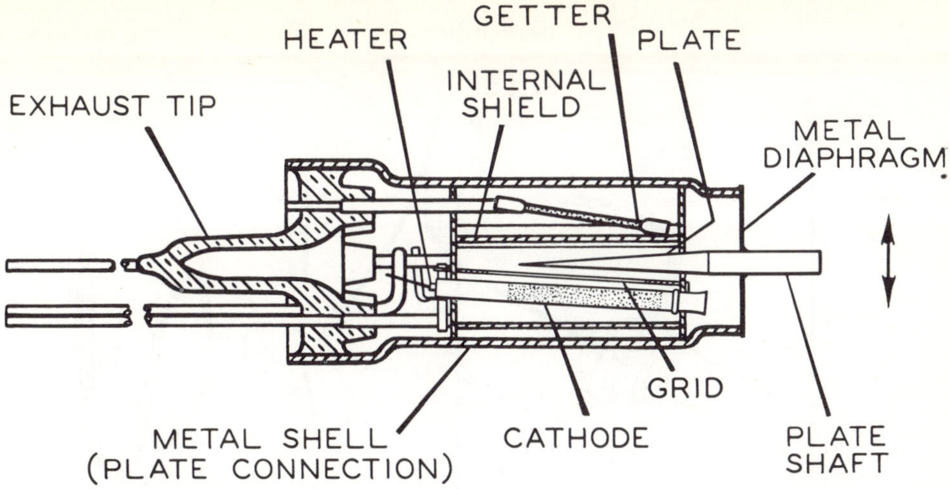

Figure 9-21. A schematic diagram of the transducer, RCA 5734. (Courtesy of Radio Corporation of America, Harrison, N.J.)

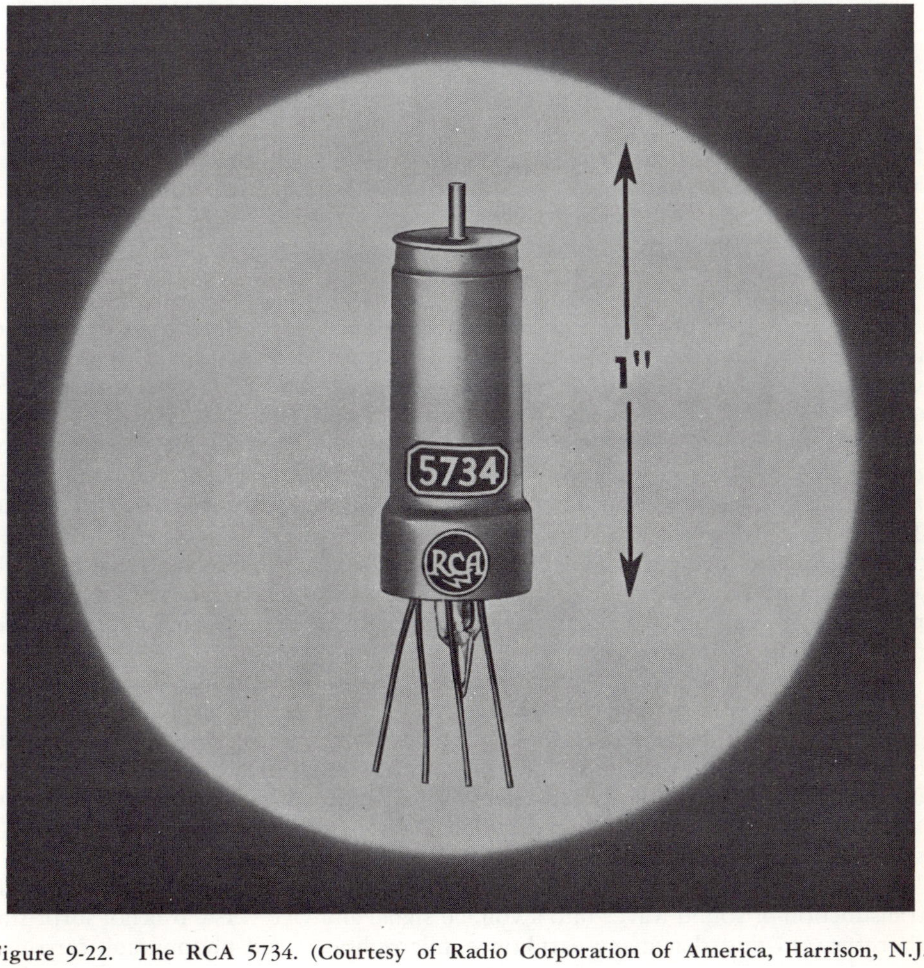

Figure 9-22. The RCA 5734. (Courtesy of Radio Corporation of America, Harrison, N.J.)

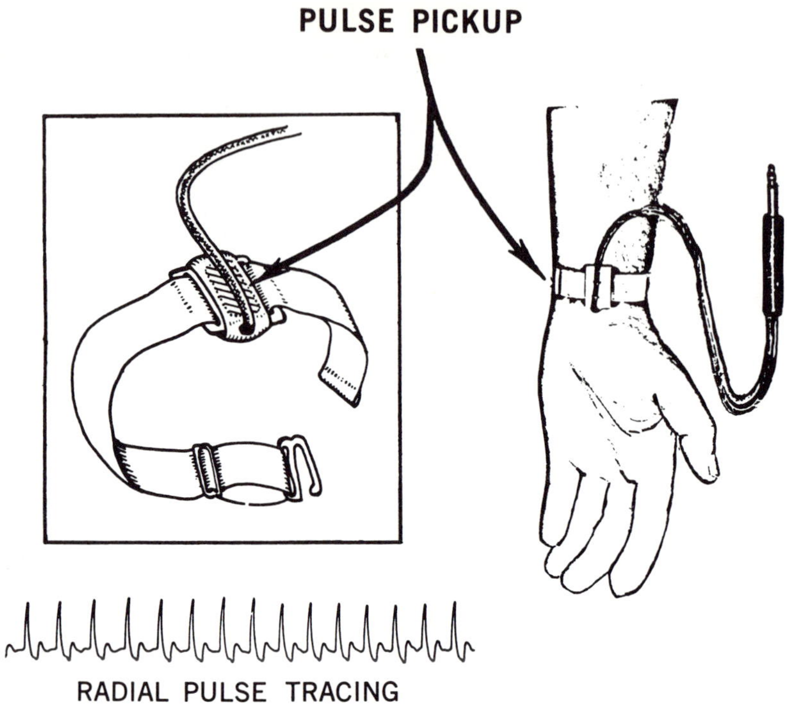

Figure 9-23. A piezoelectric pulse pickup (Courtesy of E & M Instrument Co., Houston, Tex.)

The charge displacement and resulting voltage depend on the force applied to the crystal and the relative direction of the force with respect to the crystal structure. Piezoelectric transducers are therefore *cut* from crystals so that the force can be directed along a particular axis. Crystals of materials such as quartz, Rochelle salt, and ammonium dihydrogen phosphate are used to make piezoelectric transducers. Many other crystals, however, exhibit the piezoelectric phenomenon. A man-made ceramic material of barium titanate also exhibits the piezoelectric effect. This material, however, can be manufactured in any shape or size and is not dependent on crystal structure. The piezoelectric effect may be reversed, that is, the application of an electric field causes a change in thickness of the crystal. This principle has been used to generate very small displacements which were then used to dynamically calibrate a pressure transducer.

The piezoelectric crystal may be used to measure very high frequencies, but it does not respond to very slow changes in pressure or displacement. The voltage developed across the crystal is not linear throughout its range, but a region may be found which is linear or almost linear. The most serious drawback of the piezoelectric crystal as a biomedical transducer is its temperature instability. The crystal structure is altered when the environmental temperature is changed, giving rise to spurious changes in the voltage output.

The frequency response of the piezoelectric crystal makes it ideal for the transduction of sound waves into a voltage signal and vice versa. It is not surprising, therefore, to find that many earphones, microphones, and phonograph cartridges utilize the piezoelectric crystal as a transducer. The biomedical instrumentalist can utilize the crystal from an inexpensive earphone or microphone to make a piezoelectric transducer for use as a cardiomicrophone or pulse pressure pickup. It is important to use an amplifier or recorder with an input impedance of at least 1 megohm. A commercial piezoelectric crystal transducer manufactured by the E & M Instrument Company is shown in Figure 9-23.

LIGHT SENSORS

In 1864, James C. Maxwell, a Scotsman, proposed the idea that light travels as an electromagnetic wave. Then, some years later, in 1887, about 10 years prior to the discovery of the electron by J. J. Thomson, a German physicist named Heinrich Hertz tried to demonstrate the accuracy of Maxwell's idea. He did so by generating an electric spark which was allowed to pass through the air from one electrode to another. Hertz discovered that if he illuminated the air path between the electrodes with ultraviolet light, the path became a better conductor. In the next year, Wilhelm Hallwachs measured the current which was generated when the light from a carbon arc lamp was allowed to fall on a zinc plate which was connected to an electroscope (see Chapter 1). Hallwachs showed that when ultraviolet light is allowed to strike a zinc plate, the plate is left with a positive charge. This phenomenon was called the *photoelectric effect.* In 1905, Albert Einstein applied the concept of conservation of energy to the photoelectric process and explained that the light striking the plate has a certain amount of energy depending on its wavelength (c/ν). This energy may be imparted to an electron under certain conditions, and, when it is, the electron moves toward the surface of the metal. The amount of energy necessary to cause the electron to escape from the surface is called the work function of the metal. This is the minimum amount of energy that the photon must give to the electron in order to observe this effect. The energy which the photon imparts to the electron in excess of the work function appears as the kinetic energy of the released electron. Mathematically, this is stated as follows:

$$E_{\text{electron}} = \frac{mv^2}{2} = hv - p \qquad 9.5$$

where E is the kinetic energy of the electron, m is the mass of the electron, v is the velocity of the released electron, h is Planck's constant, that is, 6.625×10^{-27} erg seconds, ν is the frequency of the light, and p is the work function of the metal.

There are several varieties of devices which are currently used as transducers which convert light into usable voltage and current signals. These include the vacuum phototube, the gas phototube, the photocell, the photodiode, and the phototransistor. These devices may be classed as *photoemissive, photoconductive,* or *photovoltaic* in nature.

The vacuum phototube (Fig. 9-24) is the closest to Hallwachs' zinc plate and is considered to be a photoemissive transducer. Light striking the cathode causes electrons to be emitted which may then travel to the anode and become available as current in the external circuit. The cathode is shaped to collect the incident light. Likewise, the anode is made as small as possible so that it does not shadow the cathode. It has to be large enough, however, to permit large numbers of electrons to be absorbed (this is important only when the current is high).

The response of the vacuum phototube to various wavelengths of light is called its *spectral response.* The over-all spectral response of the phototube is a combination

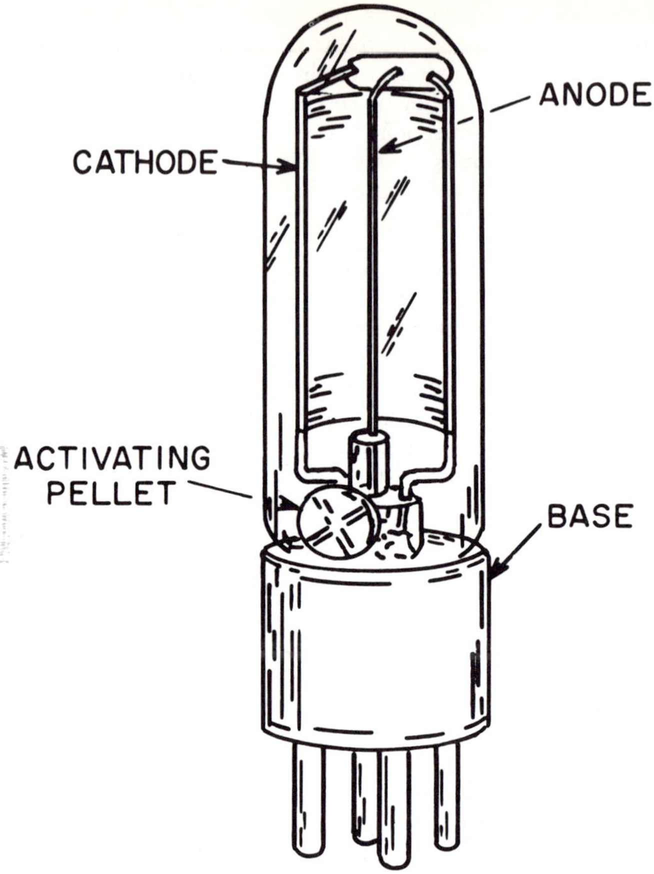

Figure 9-24. The vacuum phototube. (Courtesy of Radio Corporation of America, Harrison, N.J.)

of effects. An important factor is the material from which the cathode is made. Figure 9-25(A) illustrates the relative spectral responses of a variety of alkali metals which are commonly used as cathode material. Most vacuum phototubes today have cathodes which are made from more than one element. The spectral response of the phototube is referred to as one of several S-numbers, *e.g.*, S-1, S-4, S-14. The most commonly used spectral response is the S-1 response shown in Figure 9-25(B). The cathode used in tubes having an S-1 response is a silver-oxygen-cesium alloy. The glass envelope also influences the spectral response of the phototube.

Figure 9-26 is the current-voltage characteristic curve for the RCA 6570 vacuum photodiode as a function of the incident light flux. This particular vacuum phototube has an S-1 response. The phototube is used in the region of the curve where the current is almost constant for a large range of applied anode voltage. When used properly, avoiding anode shadow and using the correct spectrum, the vacuum phototube is extremely linear, that is, the current signal is linear with respect to the incident light flux. The response time of the phototube is relatively fast, and hence the frequency response of the transducer is fairly high.

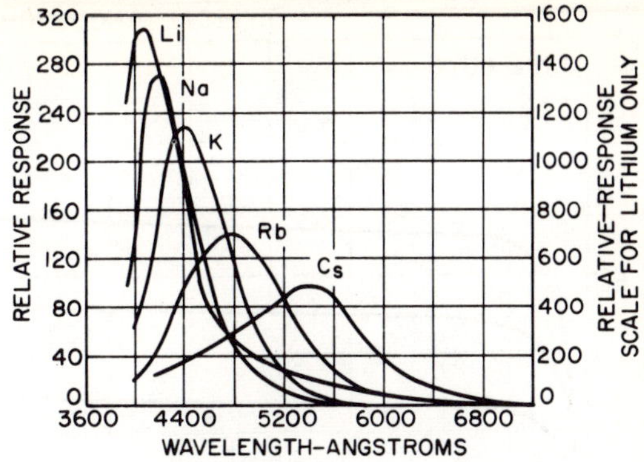

Fig. 9-25A

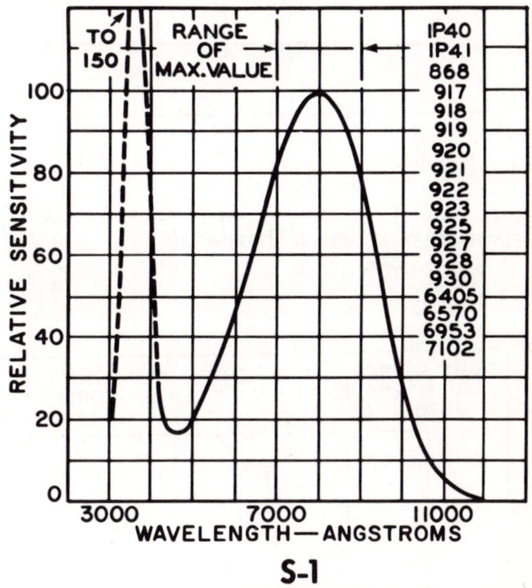

Fig. 9-25B

Figure 9-25. A, The relative spectral response of several alkali metals; B, the S-1 response. (Courtesy of Radio Corporation of America, Harrison, N.J.)

Another type of phototube is like the one shown in Figure 9-24 except that it is filled with a gas that ionizes under the proper conditions. This light-sensitive detector operates at a higher current than does the vacuum phototube. Most of these tubes are designed to operate at 90 volts. The operating point is defined as the intersection of the load line, which is the locus of points on the characteristic curve all having the same load resistor, usually 1 megohm, with the anode characteristic when the light flux is 0.1 lumen. This type of phototube has a slower response than its vacuum-tube equivalent. The output in terms of current is also not linear with respect to the light flux, as may be seen in Figure 9-27, which is the characteristic curve for the RCA type 930 gas-filled phototube. These gas-filled phototubes are best suited for triggering a relay or similar device.

A useful modification of the evacuated phototube is the *multiplier phototube.* The "transducer" itself is identical, in principle of operation, to the photosensitive devices already described. The photosensitive cathode is arranged so that the emitted electrons are directed to a positive plate which is called a *dynode.* The

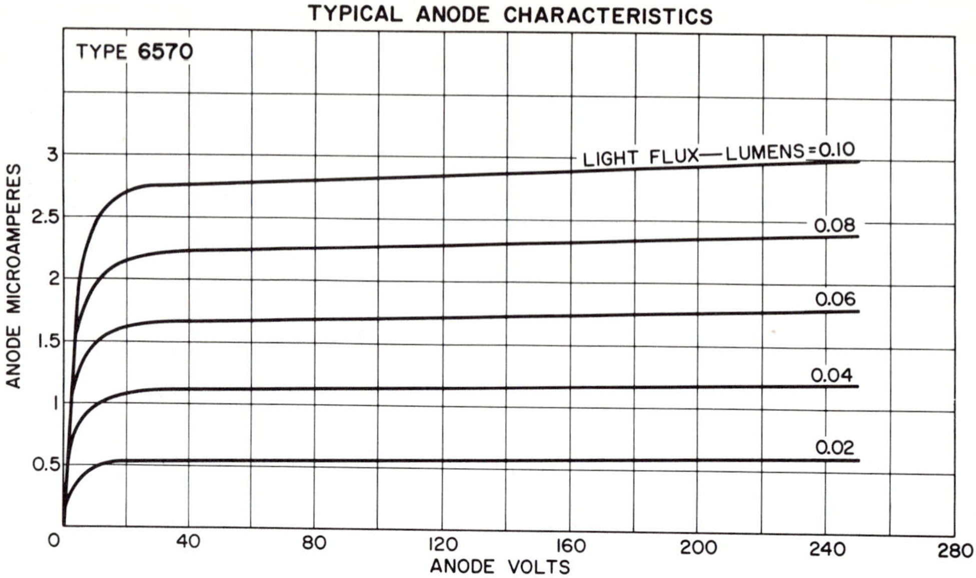

Figure 9-26. Characteristic curve for the RCA 6570 vacuum photodiode. (Courtesy of Radio Corporation of America, Harrison, N.J.)

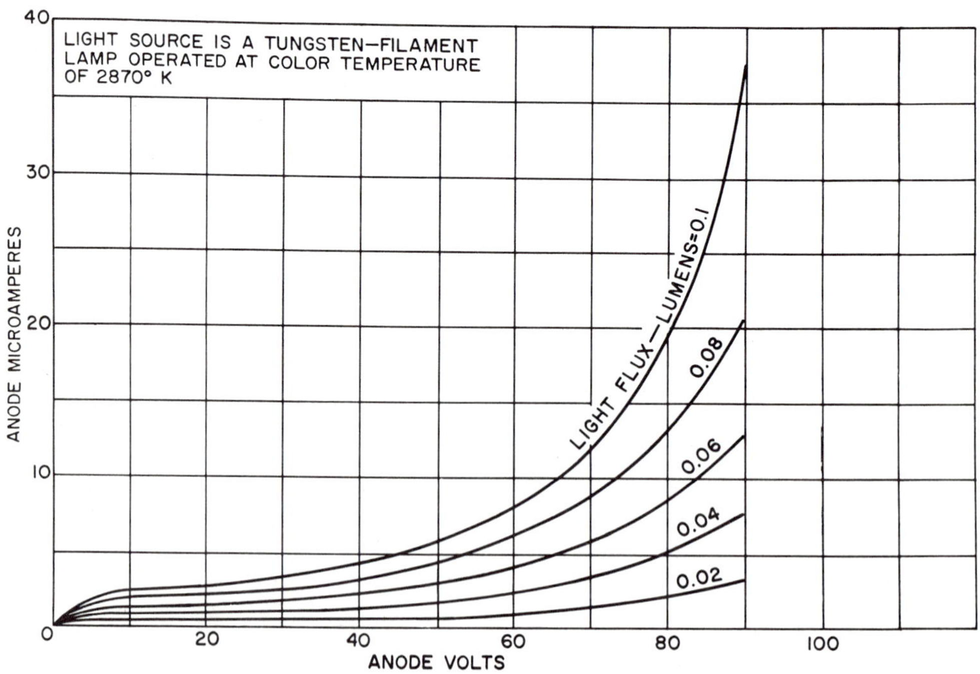

Figure 9-27. Characteristic curve for the 930 gas-filled phototube. (Courtesy of Radio Corporation of America, Harrison, N.J.)

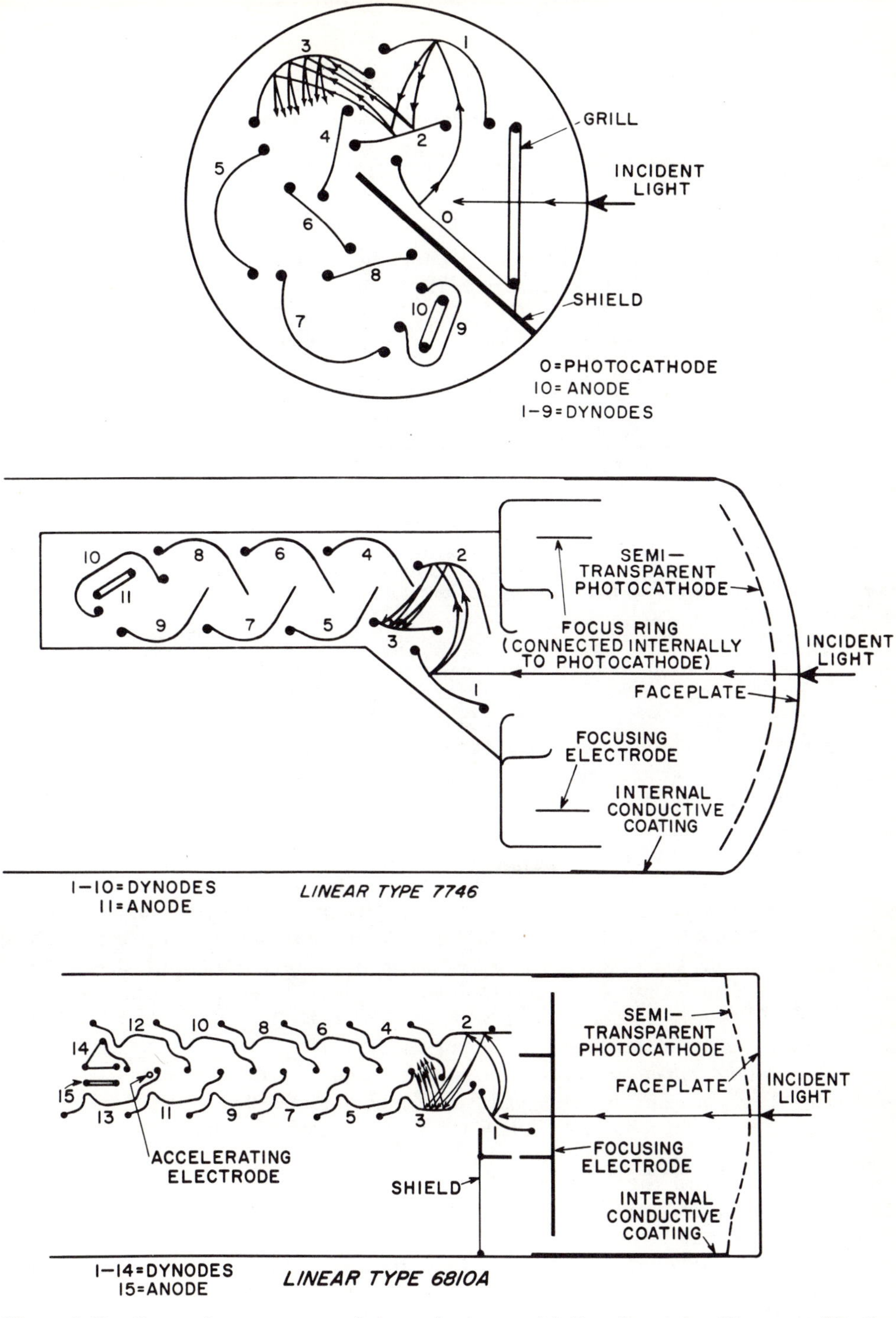

Figure 9-28. Internal arrangement of electrodes in a multiplier phototube. (Courtesy of Radio Corporation of America, Harrison, N.J.)

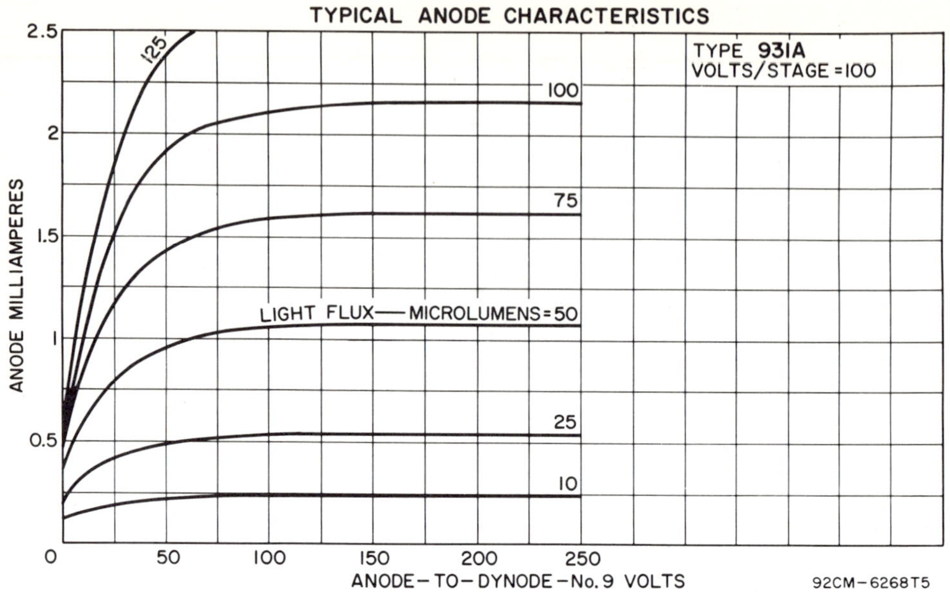

Figure 9-29. Characteristic curve for the RCA 931A multiplier phototube. (Courtesy of Radio Corporation of America, Harrison, N.J.)

dynode is coated with a material which also has a low-work function, and many secondary electrons are released for each electron that strikes its surface. The dynode is shaped to focus the electrons toward another dynode which is at a still higher potential. At each dynode the number of electrons is multiplied and the result is a rather high current from a relatively small amount of light. The arrangement of the cathode, dynodes, and anode of several types of multiplier phototubes is shown in Figure 9-28. The characteristic curve for the RCA 931A multiplier phototube is shown in Figure 9-29. This photosensitive device is often used in conjunction with a scintillation crystal to detect gamma-ray emission. The gamma ray causes a photon to be released (a scintillation) within the crystal. This photon activates the multiplier phototube, which when used with a counting rate meter is an indication of the number of scintillations per second caused by gamma emission from the isotope. These multiplier phototubes must be used with caution as they require about 100 volts per dynode for a total of about 1000 volts.

A photosensitive device in which all of the current flow occurs in a solid material is called a *photocell.* All of the light sensors described thus far have been *phototubes.* The photocell is made of a semiconductor material which is said to be *photoconductive.* When a suitable light strikes photoconductive material, it releases the electron from the crystal lattice and it is free to be conducted within the material. The loss of the electron, of course, leaves a *hole.* If the semiconductor material is of the N type it is the electrons which are conducted, whereas in the P type it is the holes which are the majority carriers. There are four substances which are being used commercially at the present time: cadmium sulfide, germanium, silicon, and gallium arsenide.

Figure 9-30 is a typical RCA cadmium sulfide photocell. It consists of a ceramic base, a layer of photoconducting cadmium sulfide, a very thin metallic electrode of

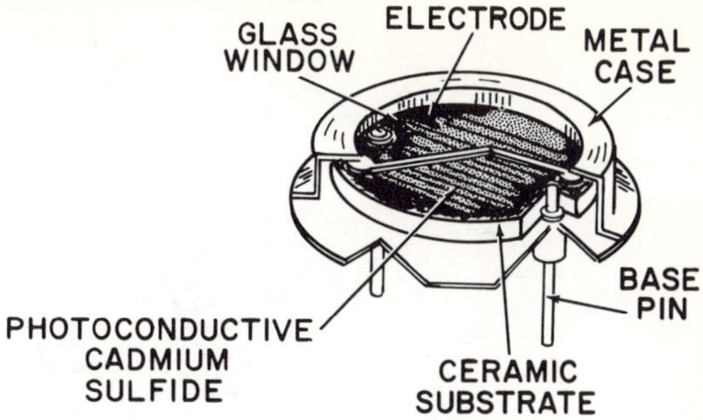

Figure 9-30. A typical RCA cadmium sulfide photocell (Courtesy of Radio Corporation of America, Harrison, N.J.)

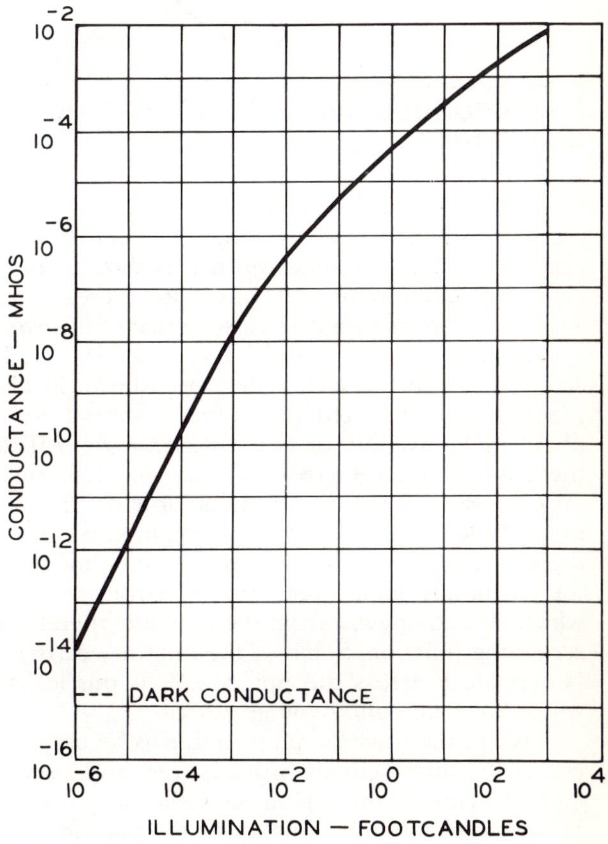

Figure 9-31. Calibration curve for a cadmium sulfide photocell. (Courtesy of Radio Corporation of America, Harrison, N.J.)

tin, indium, or gold, and a protective case with a glass cover. This transducer is incorporated into a simple circuit as if it were a resistor. When light, within spectral limitations, shines on the photocell the resistance changes. If the current through the circuit is kept constant, the voltage across the photocell changes as a function of the amount of light flux. The change is often not perfectly linear as may be seen in Figure 9-31. The response time for this type of photodetector is not rapid but is adequate for many biomedical applications. The time constant for a dark-adapted

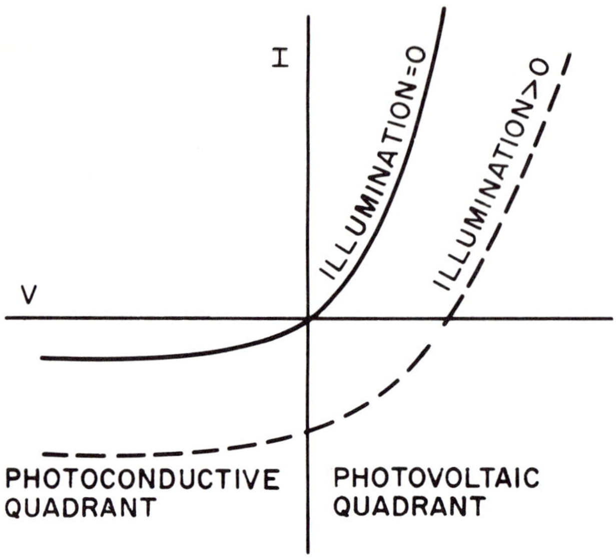

Figure 9-32. Characteristic curve for a junction photodiode. (Courtesy of Radio Corporation of America, Harrison, N.J.)

cadmium sulfide sensor is about 70 milliseconds when illuminated with 10 foot-candles of light. The time constant is shorter for higher levels of illumination. The cadmium sulfide photocell sensitivity decreases with increasing sensitivity and has a spectral response which is close to that of the eye.

A photocell which has a P-N junction is called a *junction photocell.* The characteristic curve for such a device is shown in Figure 9-32. Notice that when the junction is illuminated the curve is shifted downward. It may be operated as a photoconductive unit or as a *photovoltaic device.* The photovoltaic cell is one in which there is a junction of two different materials such as N-type and P-type semiconductors or an N-type semiconductor and a metal or a P-type semiconductor and a metal. When two materials are brought into close contact, there is an adjustment of the energy which results in a potential being developed across the junction. When the junction is illuminated, more electrons are liberated and more holes are formed, which flow to or away from the junction, thereby adding to the potential difference across the junction. Some of the minority carriers, electrons in P type and holes in N type, leak across the junction. It is this leakage which constitutes the current which flows in a photovoltaic device.

When the junction photocell is to be used in the photoconductive mode, it is connected into a circuit with a battery and a series load resistor. The cell is *reverse biased* to deplete the region between the P material and the N material of carriers. When the junction is illuminated the photons striking the diode cause a photon to be injected across the junction. This electron then flows in the external circuit and causes the voltage drop across the series resistor. The latter is used as the primary signal. The number of photons necessary to cause an electron to be emitted determines the *quantum efficiency* of the photodiode.

The germanium P-N junction, when used as a photoconductive detector, has a sensitivity which is greater than the vacuum phototube but less than the cadmium sulfide cell. It is, however, very rapid in response and can be used when the optical excitation exceeds audio frequencies. The germanium photocell contributes very little noise to the circuit, and its response to illumination is linear.

Silicon photovoltaic cells convert incident illumination directly into electrical power. These devices are often called solar batteries. N-type semiconductor is diffused into a large single crystal slice of P-type silicon, by treating the surface of the P-type material with phosphorus. The electrode contact on the bottom is often solder, while a series of metallic grid lines are evaporated onto the top to form the other electrode. The top is also coated with silicon oxide to minimize reflection from the surface. Two photojunction diodes may be incorporated into a *phototransistor* or into a *silicon-controlled rectifier.*

Photodiodes employing gallium arsenide are presently being manufactured by the Hewlett-Packard Associates (an affiliate of the Hewlett-Packard Company) in Palo Alto, California. These photodiodes have a *quantum efficiency* of 0.5 (2 photons cause 1 electron to be ejected). The hpa device (hpa 4201) is extremely fast (DC to 10^8 cps) and responds to light of 6000 Å to 10,000 Å.

The Hewlett-Packard Associates also manufacture a gallium arsenide injection luminescent diode (hpa 4104). This device is a source of infrared radiation over a very narrow spectral band. For each 1000 electrons of *forward bias* current there results approximately 1 photon of external radiation at a wavelength of 9000 Å. The device is intended for use at room temperature without external cooling. In the present device (hpa 4104) the effective emission diameter is 0.020 inch, through which 0.1 milliwatt of infrared energy can be radiated. This source of infrared light can be modulated at nanosecond rates (1 nanosecond $= 10^{-9}$ sec).

An entire new field of electronics, called *optoelectronics,* has emerged using these devices. Wunderman* has described such applications of these gallium arsenide devices as a wide-band photon-coupled isolator (hpa 4301), photochoppers, computer card reader, an input stage with extremely high input impedance, and a variety of other uses in digital logic circuitry.

There is a tremendous potential for the use of these devices in biomedical instrumentation. The gallium arsenide type of device may be used to sense the flow of blood in an unopened blood vessel or to measure "skin blood flow." (I have been working on development of this device at the time of writing this book.)

Other applications will include a variety of oximeter devices made possible by the unique property of being able to alter the chemistry of the semiconductor so it will emit a predetermined unique or at least a narrow band of light of any desired wavelength or band of wavelengths. The device may also be used to detect very small displacements without making physical contact with the object being displaced. Photodiodes of this type will also be used in the near future as radiation detectors.

All of the photosensitive devices generate a certain amount of current in the dark. This is called the *dark current,* and it is obvious that this dark current contributes to the noise in any application. The dark current can become a serious problem in the multiplier phototube. Figure 9-33 is a composite curve supplied by the Radio Corporation of America showing the range of luminous flux for which the various photosensitive detectors are useful. Figure 9-34 was also supplied by RCA; it illustrates the usable frequency response of the various detectors, except for the newest types of materials which have much higher frequency response characteristics.

Biomedical transducers incorporating phototubes and photocells have already been introduced: the photoelectric myograph (Fig. 9-11), the Bourdon-tube pres-

---

*Wunderman, I.: Electronics. p. 58, July 27, 1964. McGraw-Hill Book Co., New York.

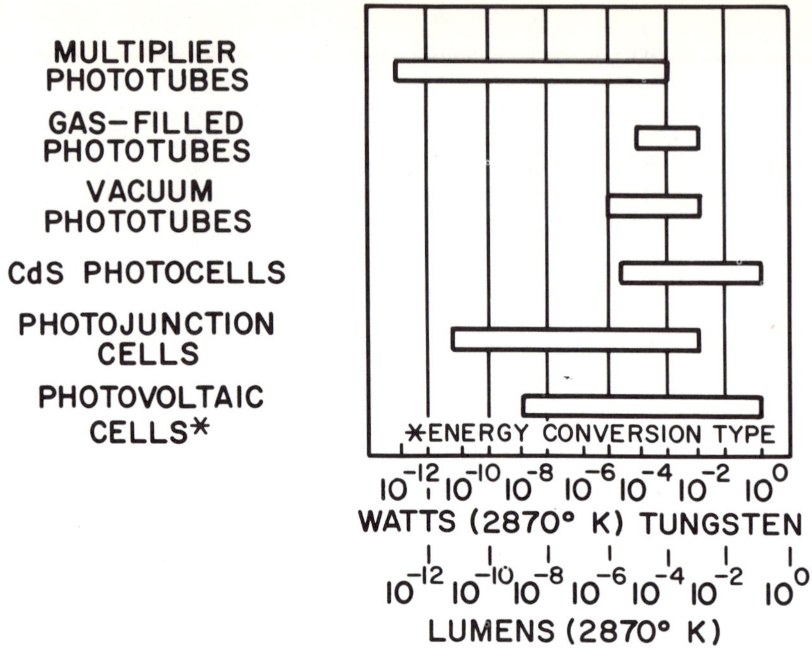

Figure 9-33. The spectral sensitivity of a variety of photosensitive devices. (Courtesy of Radio Corporation of America, Harrison, N.J.)

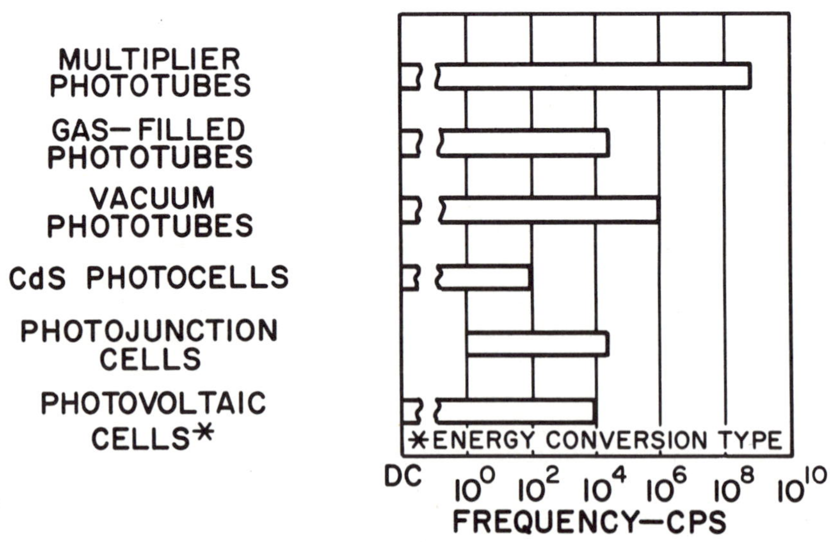

Figure 9-34. Frequency response of a variety of photosensitive transducers. (Courtesy of Radio Corporation of America, Harrison, N.J.)

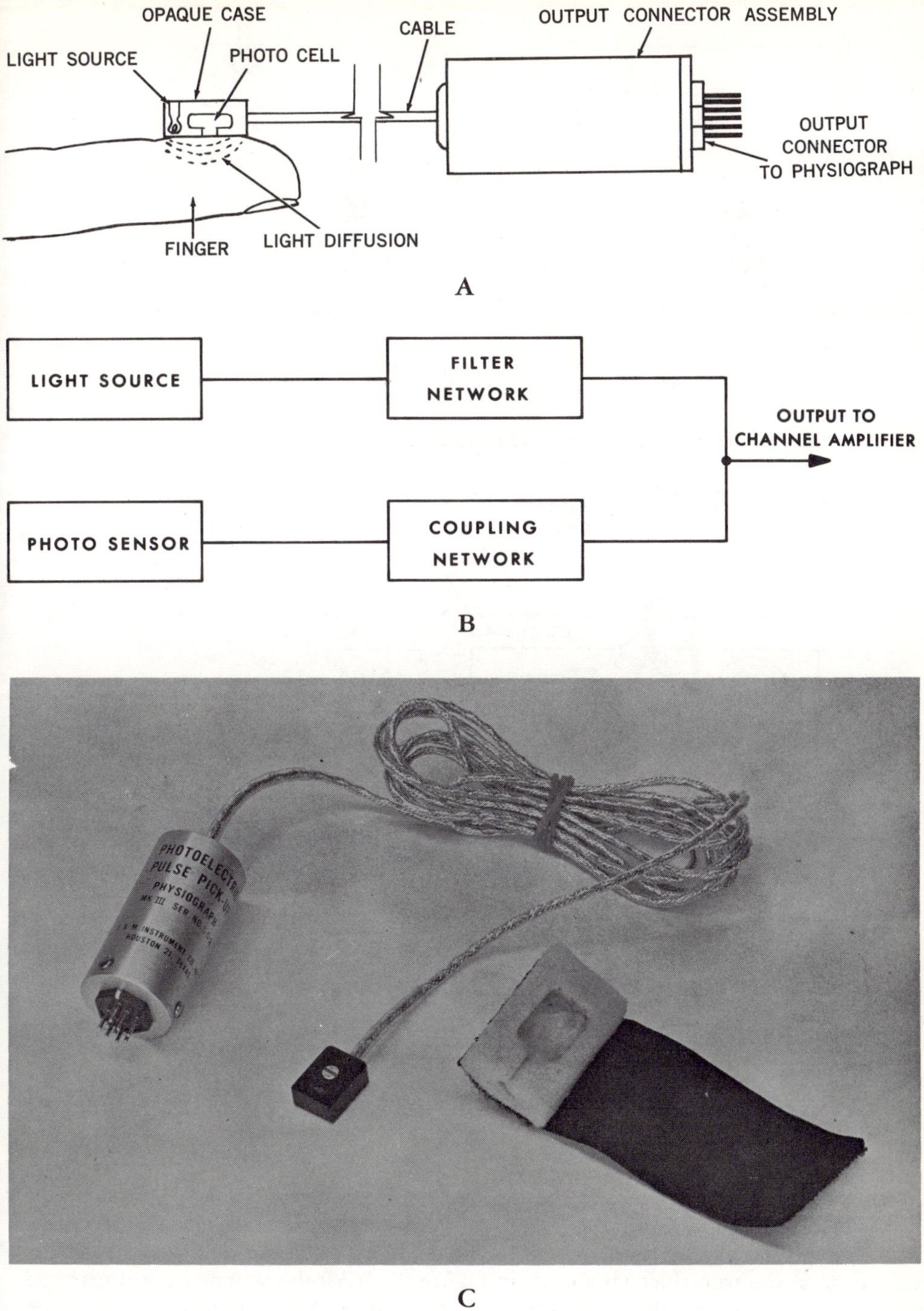

Figure 9-35. The photoelectric pulse pickup: A, The cutaway; B, the schematic; C, the complete transducer. (Courtesy of E & M Instrument Co., Inc., Houston, Tex.)

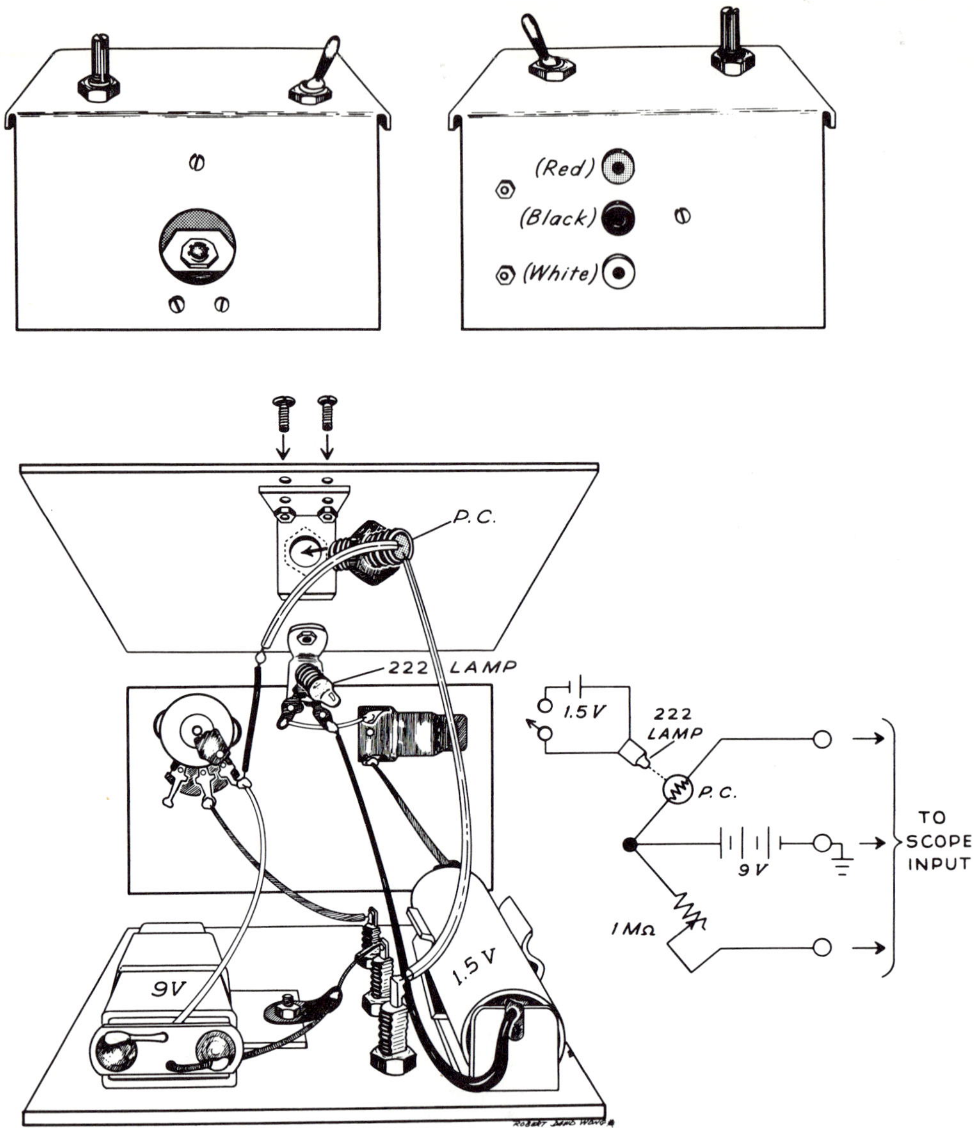

Figure 9-36. A photoelectric plethysmograph. (Courtesy of Norman N. Goldstein, University of California Extension, Berkeley, Calif.)

sure manometer (Fig. 9-12), and the bellows pneumograph (Fig. 9-13). Figure 9-35 is an illustration of a commercial device available from the E & M Instrument Company which may be used to sense changes in blood volume pulses in the skin. A cutaway of the transducer is shown in Figure 9-35(A); the schematic is illustrated in Figure 9-35(B); and the complete instrument is shown in Figure 9-35(C). The spectral emission of the lamp and the spectral sensitivity of the photocell are chosen so that the photocell will sense the change in the light absorbed by the blood pigments. Figure 9-36 is an example of a similar device designed by Norman N.

Goldstein, Jr.* This transducer was designed for use with the IMPScope (manufactured by the Heath Company) but will work as shown with any high-input impedance device which has a push-pull input, such as an oscilloscope or a recorder with a high enough input impedance.

PHYSIOLOGICAL GAS ANALYSIS

The measurement of hydrogen ion concentration as pH (the negative of the log to the base 10); the partial pressure of oxygen or oxygen tension, P_{O_2}, in mm mercury; and the partial pressure of carbon dioxide or carbon dioxide tension, P_{CO_2}, in mm mercury in blood, cells, tissues, and body fluid provides valuable physiological data as to the respiratory and metabolic state of the animal. The Beckman Model 160 physiological gas analyzer, shown in Figure 9-37, is an instrument which illustrates the principles used to measure the above parameters.

The transducer for measuring the partial pressure of oxygen in solution is shown diagrammatically in Figure 9-38. It consists of a 0.0005-inch platinum wire cathode and a silver anode. The platinum wire is cemented into a fine silver tube. The cement insulates the platinum from the silver. The platinum wire extends into an electrolyte bath (KCl) which is kept in contact with both the silver anode and the platinum cathode with a membrane. A battery maintains the platinum wire at a polarizing potential of −0.68 volts with respect to the silver anode. This thin membrane, which can be of either polypropylene 0.001 inch or Teflon 6 microns thick, allows only oxygen to pass from the sample to the platinum wire inside the captive electrolyte. When the oxygen gets to the platinum wire it is reduced, and electrons are released which then flow in the external circuit as current. The current is proportional to the partial pressure of oxygen.

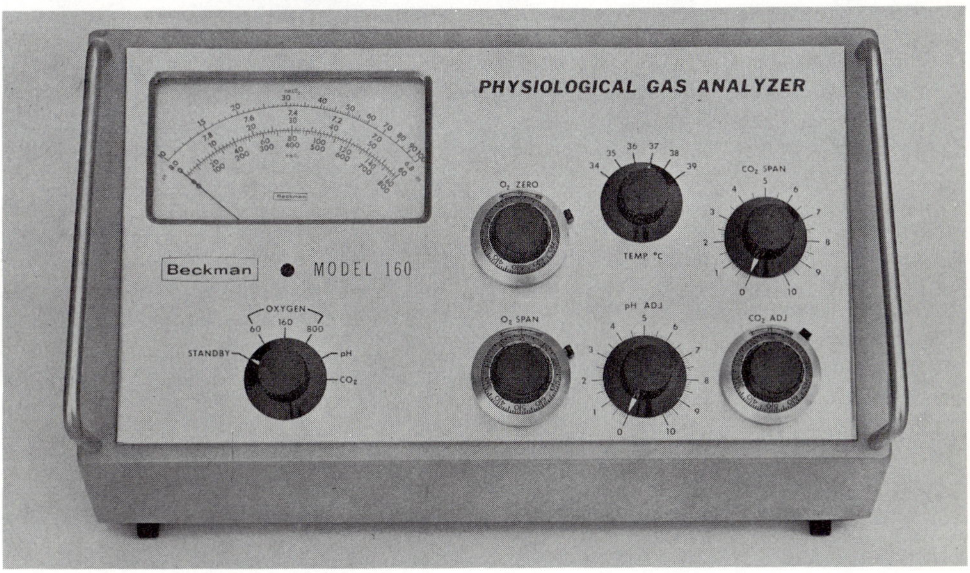

Figure 9-37. The Beckman model 160 physiological gas analyzer. (Courtesy of Beckman Instruments, Inc., Palo Alto, Calif.)

*Director, Project IMPS, University Extension, University of California, Berkeley, California.

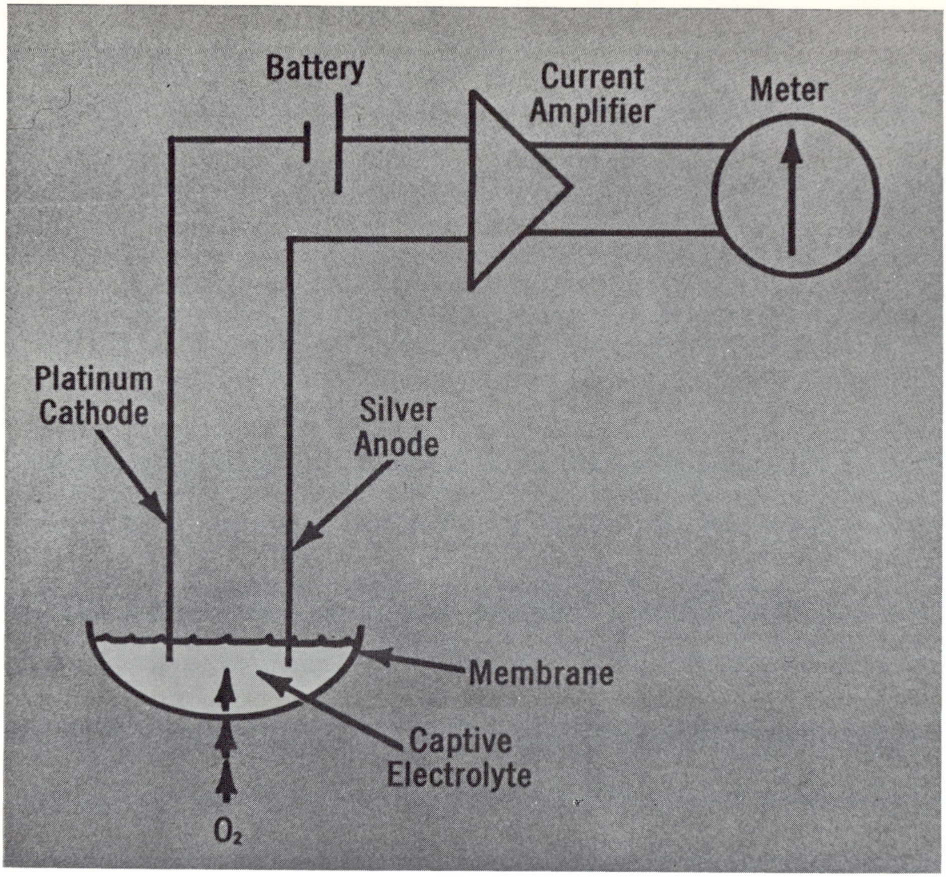

Figure 9-38. A diagram showing the principle of the operation of the Beckman oxygen electrode. (Courtesy of Beckman Instruments, Inc., Palo Alto, Calif.)

A very small oxygen transducer built into an 18-gauge Riley arterial needle for *in vivo* measurement is shown in Figure 9-39(top). The tip of the needle assembly is enlarged in Figure 9-39(bottom). Figure 9-40 is a larger oxygen electrode assembly. This transducer consists of a 0.001-inch platinum wire which is sealed in glass. The anode in this macrotransducer is silver-silver chloride. The cathode and anode are dipped in electrolyte and shielded from interfering artifacts such as solutions with a high or low pH and other reducing chemicals. It does, however, permit the entrance of oxygen molecules at the tip.

The oxygen transducers are calibrated in a chamber (see Fig. 9-43), using gas with known concentrations of oxygen. The output *current* is directly proportional to the amount of oxygen in the sample. It is important to bear in mind that both the amount of oxygen in solution and the measurement of oxygen tension are sensitive to changes in temperature. This transducer must be kept at a constant temperature during use and calibration.

The measurement of pH is based on the potential difference which exists between two solutions of different pH separated by a glass membrane. Figure 9-41(top) illustrates the basic arrangement. The *glass electrode* consists of a silver electrode, coated with silver chloride, surrounded by a solution of known pH, and

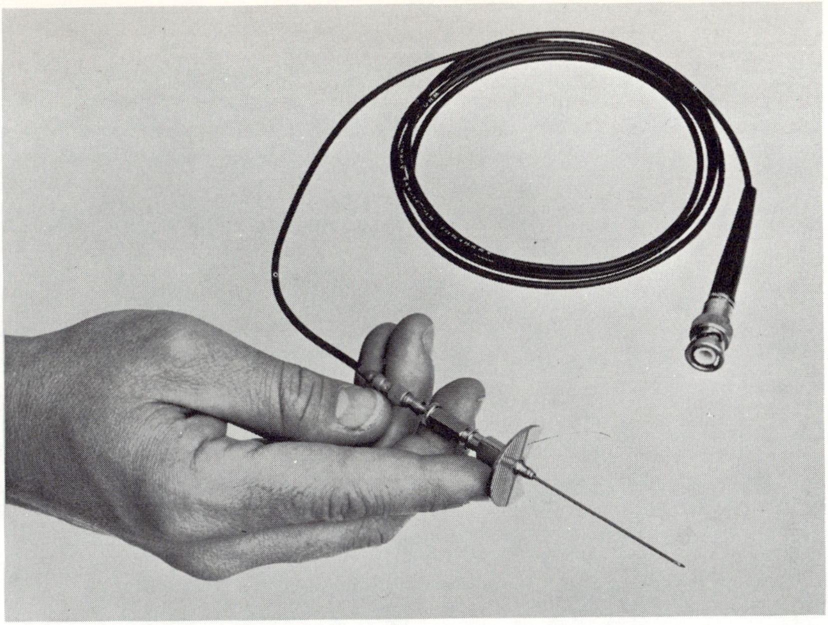

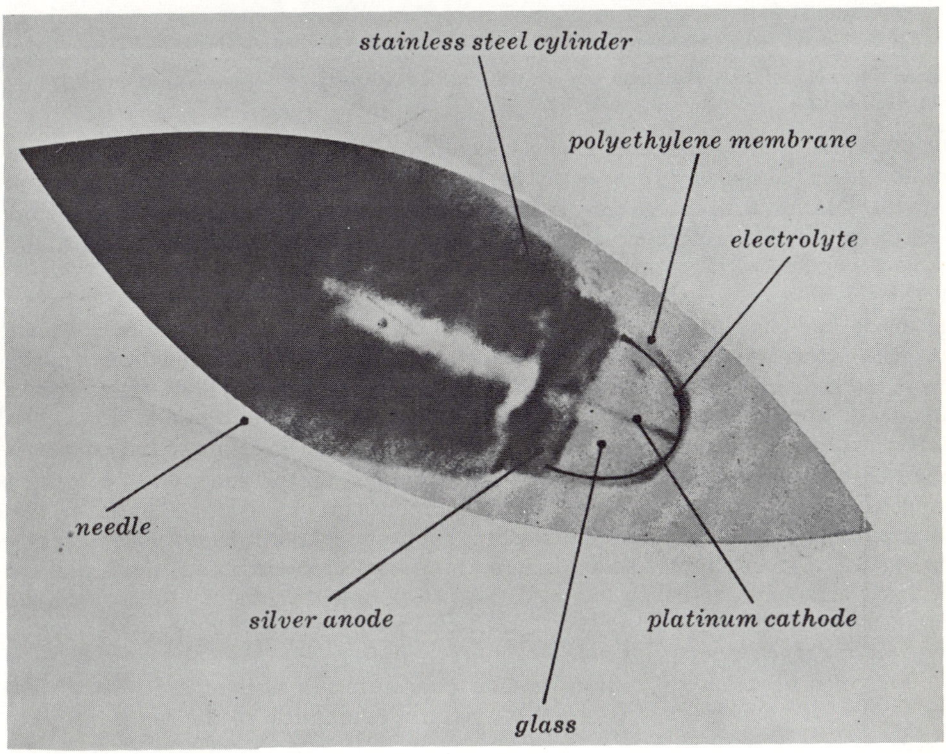

Figure 9-39. *Top,* The Beckman oxygen transducer; *bottom,* enlargement of the tip of above. (Courtesy of Beckman Instruments, Inc., Palo Alto, Calif.)

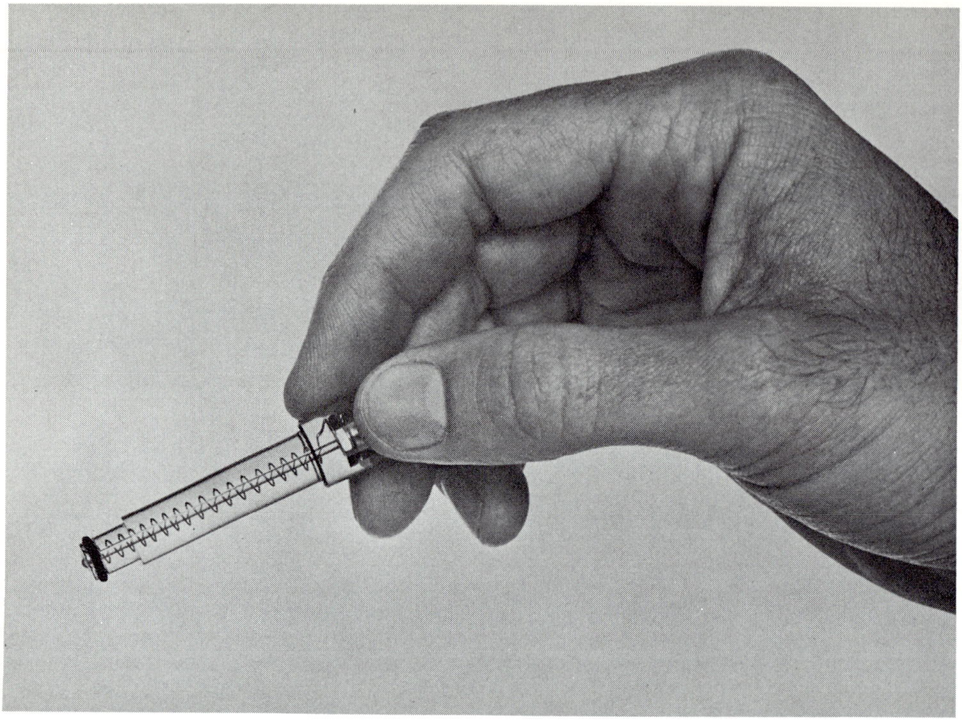

Figure 9-40. The larger Beckman oxygen electrode. (Courtesy of Beckman Instruments, Inc., Palo Alto, Calif.)

enclosed in a glass membrane. This glass electrode is suspended in the solution of unknown pH. The glass prevents the passage of ions (other than the hydrogen ion) from the solution with the unknown pH to the solution with the known pH inside the glass electrode. The reference electrode consists of a calomel half-cell (Hg_2Cl_2:Hg) and is connected to the solution of unknown pH by means of a potassium chloride salt bridge. The external circuit must measure the changes in potential between the inside and the outside of the glass membrane without drawing current. Since the glass offers an extremely high resistance to the flow of current, the voltage across the membrane can be measured only by means of an amplifier with an input impedance greater than 10^{11} ohms or a precision slide-wire potentiometer such as the one shown in Figure 3-8. An amplifier with such a high-input impedance is called an *electrometer.* It has not been possible, until the present time, to make a completely transistorized circuit with a high enough input impedance for this application. Electrometer amplifiers using solid state components have recently appeared. These amplifiers use field-effect transistors in the input stage or a positive feedback arrangement using standard transistors and thermistors to provide the required very high input impedance (Burr-Brown Research Corporation, Tucson, Ariz.; Zeltex, Inc., Concord, Calif.).

The measurement of carbon dioxide concentration utilizes a Severinghaus-type* electrode which consists of a glass electrode, identical to the pH transducer

*Severinghaus, J.W., and Bradley, A.F.: Electrodes for Blood pO_2 and pCO_2 Determination. J. Appl. Physiol. 13:515, 1958.

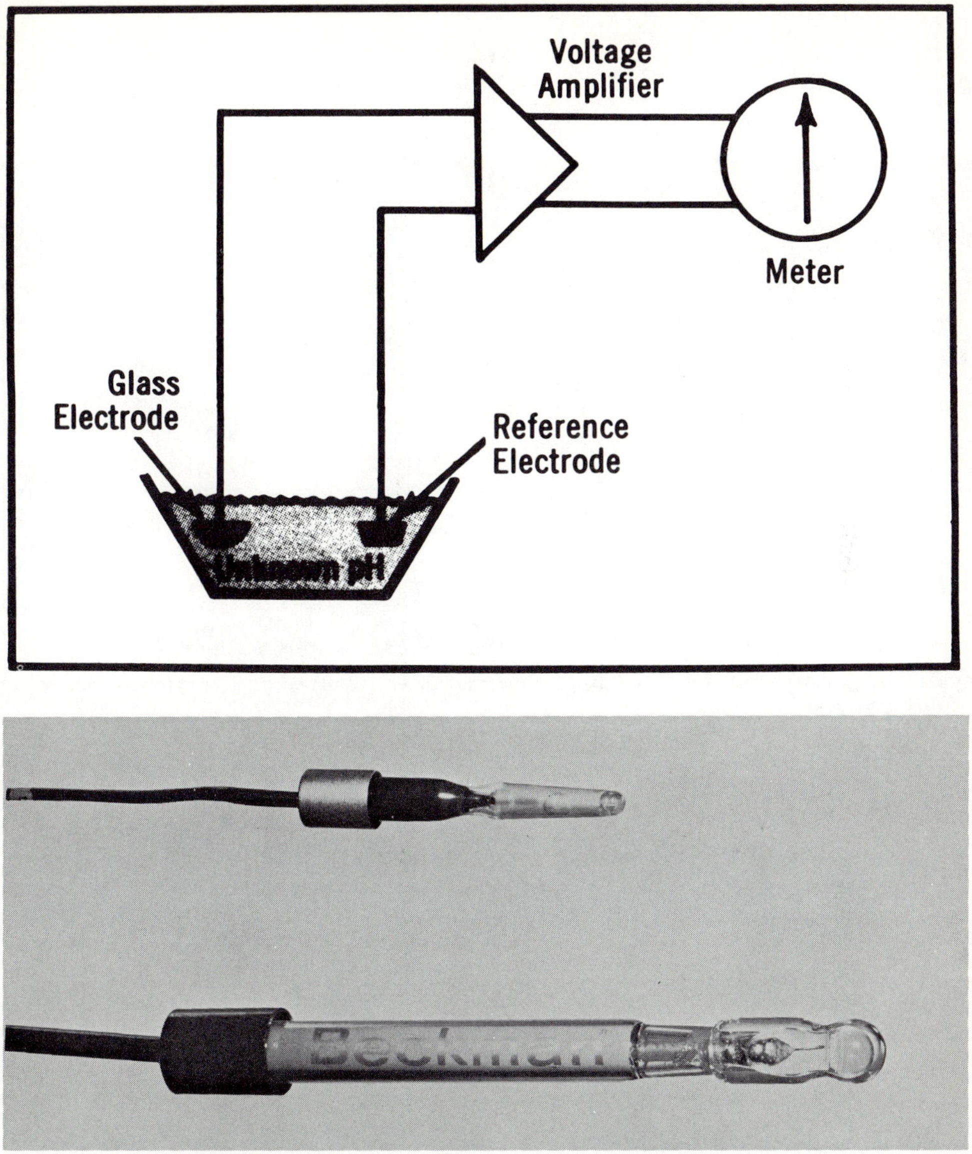

Figure 9-41. *Top,* The principle of operation of the Beckman pH electrode; *bottom,* The Beckman pH electrode. (Courtesy of the Beckman Instruments, Inc., Palo Alto, Calif.)

described above, except that a thin layer of sodium bicarbonate solution is held captive by a thin Teflon membrane (Fig. 9-42). When the entire unit is immersed in the blood sample the carbon dioxide ions diffuse through the Teflon membrane and react with the sodium bicarbonate solution to form carbonic acid which changes the pH of the captive sodium bicarbonate solution. Since the pH decreases one unit for a tenfold increase in carbon dioxide tension (P_{CO_2}), the pH is a linear function of the logarithm of the partial pressure of carbon dioxide.

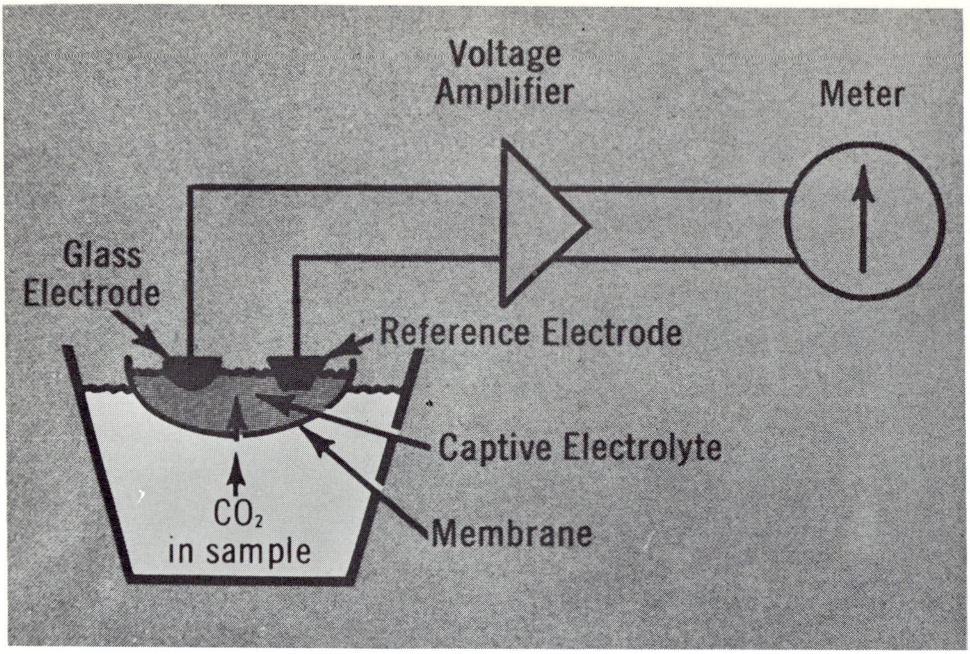

Figure 9-42. A Severinghaus-type CO_2 sensing transducer. (Courtesy of Beckman Instruments, Inc., Palo Alto, Calif.)

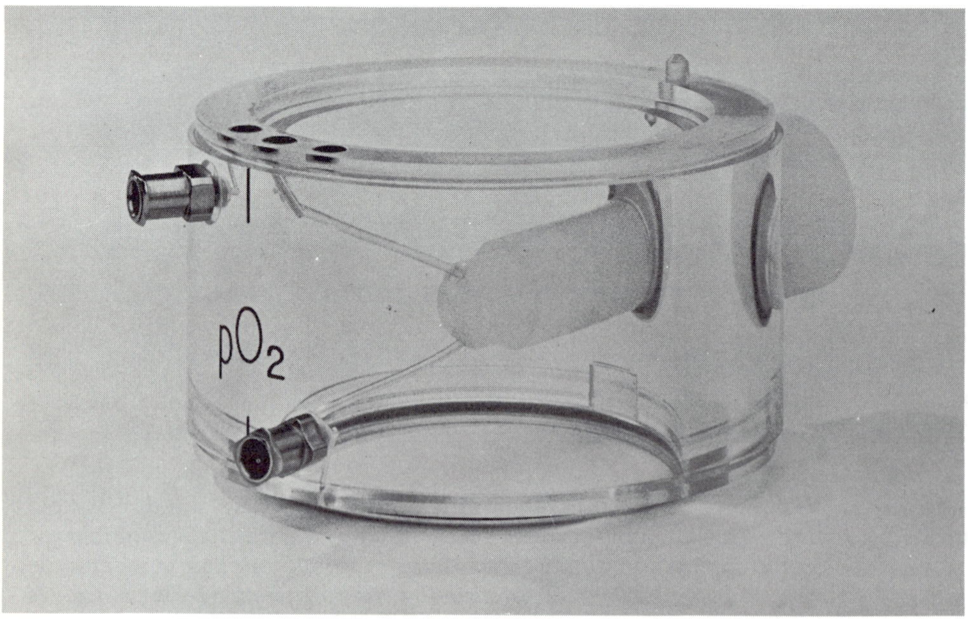

Figure 9-43. A cuvette for the measurement of oxygen concentration. (Courtesy of Beckman Instruments, Inc., Palo Alto, Calif.)

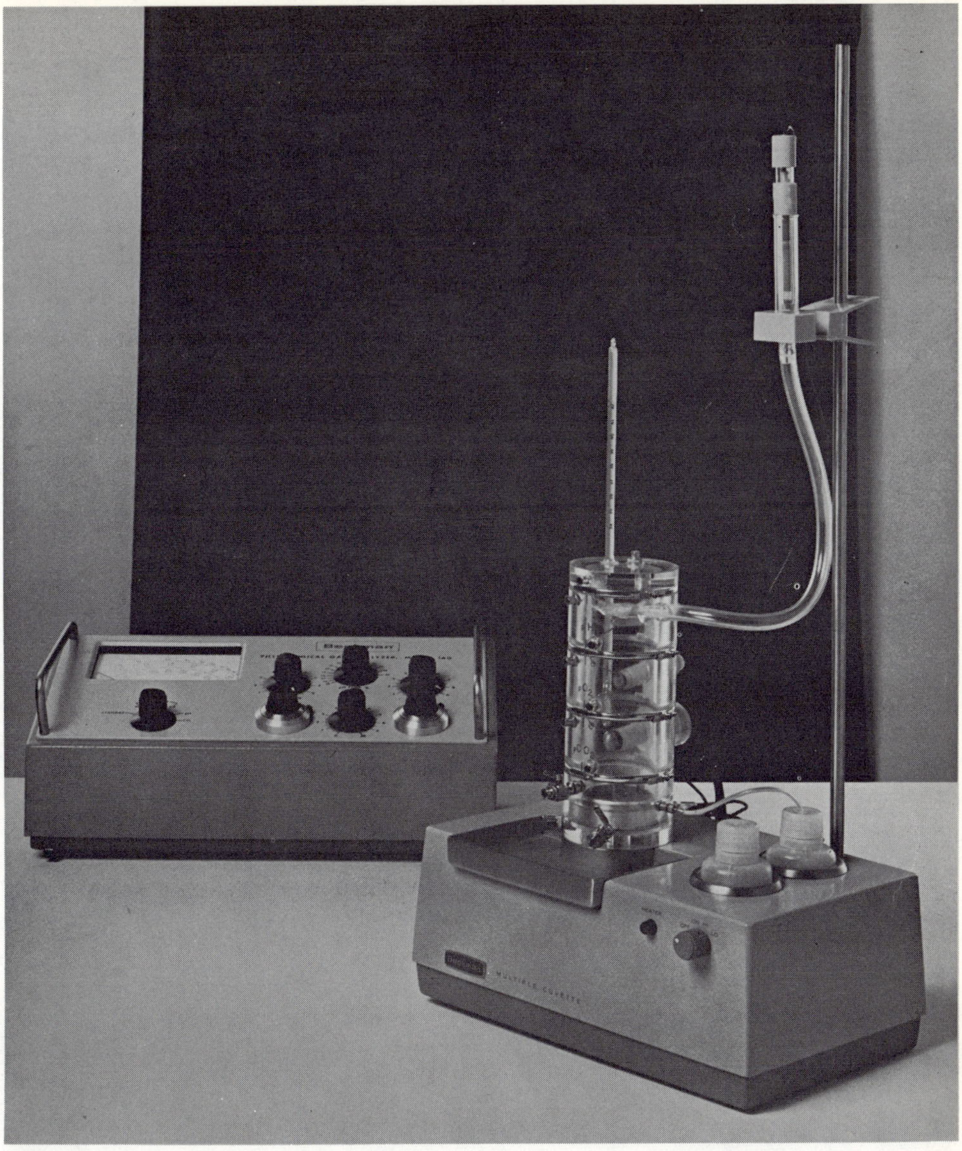

Figure 9-44. The Beckman multiple cuvette arrangement for measuring P_{O_2}, P_{CO_2} and pH at the same time. (Courtesy of Beckman Instruments, Inc., Palo Alto, Calif.)

The Beckman physiological gas analyzer may be used with a multiple cuvette arrangement. The cuvette for the measurement of oxygen tension is shown in Figure 9-43. The oxygen macroelectrode fits into the back of the cuvette. As little as 0.3 ml of sample may be used in this cuvette. Other cuvettes for the measurement of pH and P_{CO_2} may be added on top of the oxygen cuvette. The complete setup is shown in Figure 9-44. Notice that the calomel reference electrode is attached to the ring stand on the top and to the right of the cuvette column. The bottom module is a reservoir and temperature-controlled bath.

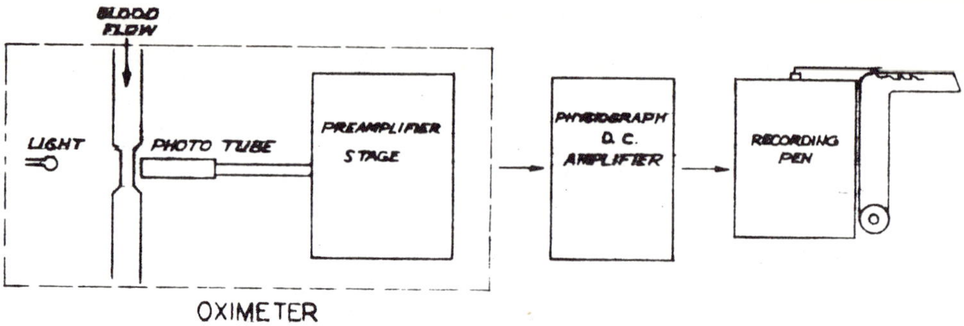

A

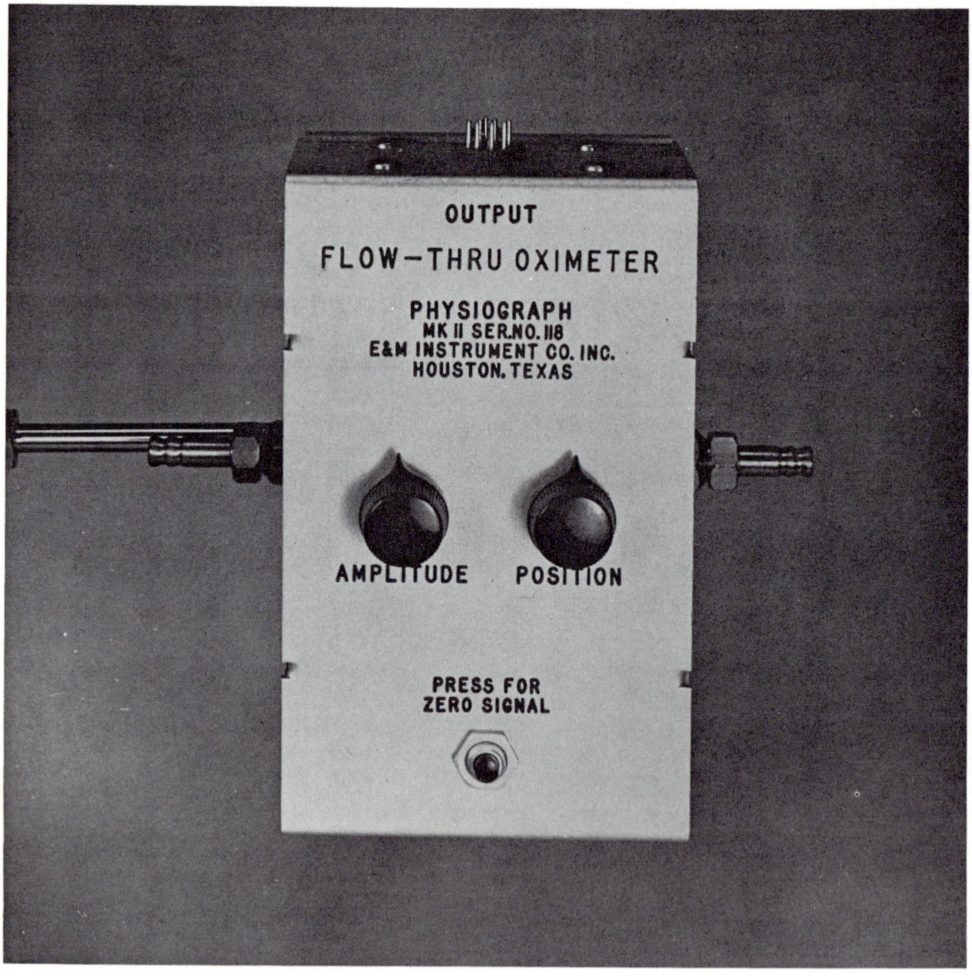

B

Figure 9-45. A flow-through oximeter based on the absorption of light: A, a schematic diagram; B, the complete transducer. (Courtesy of E & M Instrument Co., Inc., Houston, Tex.)

The block diagram of the instrument is shown in Figure 9-45A. Note that the signal is first chopped and then amplified with a high-input impedance "preamplifier." The signal is then further amplified by an AC amplifier, demodulated, and further amplified with a DC amplifier. The signal may then be read by means of a meter or a suitable recorder. The feedback network used depends on whether P_{O_2} or P_{CO_2} is being measured.

Another method of measuring the amount of oxygen in the blood is shown in Figure 9-45(A). Here, the blood flows through a flattened glass cell, and changes in the amount of light absorbed by the oxyhemoglobin of the blood are detected by a photocell. In this transducer the output is not only sensitive to oxygen concentration but also to flow. For this reason, there must be sufficient flow to cause complete mixing within the chamber. The complete transducer is shown in Figure 9-45(B). This same principle, but not this particular instrument, may be used to measure the blood flow in an artery if the proper wavelength, not appreciably absorbed by a form of oxyhemoglobin (over 8000 Å), is used.

This chapter has described a few of the basic varieties of biomedical transducers. Some of those described are easy to duplicate while others are almost impossible to fabricate in the laboratory of the scientist. When a transducer is made commercially available, large amounts of money are spent developing specialized production techniques. It is extremely difficult to duplicate this effort in the university or hospital shop or laboratory.

It is important, however, that the instrumentalist be familiar with a large variety of these transducer techniques. In some cases, a simpler model of the transducer may be made for use in the laboratory. In any case, it is important to understand as much as possible about an instrument in order to use it effectively. I have seen some otherwise extremely intelligent individuals attempting to measure high-fidelity pulsatile blood pressure with a large air bubble in the dome of the fluid-filled transducer. Not all errors in the use of transducers and instruments are as obvious as this, however, and defects will be apparent only after very careful study. The transducer is often the weakest link in a chain of instruments.

REFERENCES:

Biophysical Research Methods, Fred M. Uber (Interscience Publishers, New York, 1950).

Instrumentation in Scientific Research, Kurt S. Lion (McGraw-Hill Book Co., New York, 1959).

Instrumentation Methods for Physiological Studies, Norman N. Goldstein, Jr. (University of California Berkeley Extension, Berkeley, Calif., 1964).

Medical Physics, vol. 1, 2, 3, Otto Glasser, ed. (Yearbook Publishers, Chicago, 1944, 1950, 1960).

RCA Phototubes and Photocells, Technical Manual PT-60 (Radio Corporation of America, 1963).

10

The Output Signal and What to Do with It

general remarks

low-frequency recorders

medium-frequency recorders

high-frequency recorders

magnetic tape recorders

summary

GENERAL REMARKS

Each biomedical instrument provides some form of *output signal.* In some cases, this signal may be a high-voltage and low-current signal, and in other cases it may have relatively low voltages and high currents. Occasionally the output signal is the direct output of the transducer. In order to use this output signal it is necessary to display it in some way. The device displaying the output signal is called a *read-out* device.

There are many forms of read-out devices: The meter, the oscilloscope, the direct-writing recorder, the optical recorder, and the magnetic tape recorder are among the most common. The output usually varies as a function of time, and it is usually important to follow the changes in the parameter being measured. Some of the changes occur rapidly and others slowly. Although there are a variety of read-out devices that would serve any particular need, it is, nevertheless, possible to choose the wrong read-out system.

Let us discuss, in general, the characteristics which an ideal read-out instrument should have. The ideal read-out instrument for a given application will, of course, depend on the parameter being measured and how the data are to be used. The instrument designer or purchaser should ask himself the following questions:

1. How fast is the voltage (or current output, representing the parameter being measured) changing with respect to time? Is the user interested in rapid changes or is the average response all that is required?
2. What is the range of the voltage and current amplitude to be displayed?
3. What is the output impedance of the instrument preceding the read-out device?

4. What is the input impedance of the read-out device?
5. Must the information required be recorded or is visual display all that is necessary?
6. How portable must the device be?
7. Is the cost of the recorder justified by its performance?

The first question is perhaps the most important one, although all factors must eventually be considered. In almost every chapter of this text, frequency response has been considered. The read-out device must not be an exception. Table 8.1 on page 166 lists the frequency characteristics of a variety of important biomedical parameters. It is obvious that if a parameter such as the measurement of blood pressure requires an instrument to respond to changes occurring from 0 to 50 cps, then the entire instrumentation from the transducer to the read-out device must also have the same frequency-response range. On the other hand, it is not only unnecessary but unwise to use a read-out device that is capable of responding, for example, to 2000 cps. Most electronic noise (unwanted signal) is 60 cps, which is the fundamental frequency of alternating house current. When there is no signal representing the parameter to be measured above 50 cps, all of the signal above 50 cps that is seen in the display must be noise. The read-out device must not be capacitively coupled to the output of the measuring instrument if it is to respond to changes in DC voltage (0 cps).

Other parameters such as the measurement of oxygen concentration in the blood do not require the use of read-out instruments which respond rapidly because the oxygen electrode itself, which is the transducer, cannot respond very fast. The measurement of some parameters, however, such as the *action potential* of a nerve requires fidelity to over 2000 cps.

Every display device, whether a recorder or an oscilloscope, has an upper and a lower frequency limit. The upper limit is defined as the frequency at which the amplitude drops to 95 per cent of its value at very low frequencies. If a sinusoidal signal with an amplitude of 1 millivolt (or some other suitable value) is applied to a read-out device such as a recorder, the deviation of the pen is adjusted to some value close to that which will be used under operating conditions, let us say 10 mm. The frequency of the test input signal is increased, and recordings are taken at regular intervals. That frequency at which the amplitude is only 9.5 mm is the *upper frequency limit* of the recorder. This recorder is said to be *flat* to the upper frequency limit. As the frequency is increased, the response may rise above 100 per cent, that is, the amplitude may be higher at some high frequencies than it is at very low frequencies. This is usually the result of a resonance phenomenon. The addition of an appropriate *damping resistor* may increase the usable frequency response of the recorder.

It is important to use a precision oscillator to supply the test input signal and to be certain that the amplitude of the oscillator output is remaining constant. This can be done by monitoring the output of the oscillator with an oscilloscope. It is also important to include all parts of the recorder such as input couplers and filters. The average user of a recorder might be amazed at the frequency response of the instrument he is presently using.

The second important characteristic of the read-out device is the voltage or current range of the instrument. This is usually clearly stated by the manufacturer and should present no difficulty. A knowledge of the amplitude range expected from the preceding instrument is, of course, required.

Almost all recording devices except the oscilloscope and the magnetic tape recorder require some variety of galvanometer. A galvanometer is a *current device,* that is, it is the flow of current that generates the magnetic field which causes the suspended mirror or writing pen to move. The input impedance of a current-sensing device is usually very low as is the impedance of the galvanometer coil itself.

In some recorders the input signal is allowed to flow through the galvanometer coil directly. In others, however, the input signal is modified by one or more amplifiers, and it is only the last amplifier which *drives* the galvanometer coil. Devices of this latter type usually have very high input impedance. The higher the input impedance the less current the amplifier draws from the preceding instrument. Such an instrument is basically a *voltage-sensing* device. It would be a true voltage-sensing device only if it were to have infinite input impedance. This is, of course, impossible, but an input impedance of from 1 to 10 megohms is not unusual.

The *output impedance* of the instrument whose signal is to be recorded and the *input impedance* of the display instrument must both be considered in the selection of a proper read-out device. In general, it is better for an instrument to have low-impedance output. This prevents minute currents induced from fields set up by electromagnetic devices from generating large noise signals. Maximum power (voltage times current) is transferred to the recording instrument when the output impedance of the instrument and the input impedance of the recorder are matched (equal).

Most read-out devices, however, sense voltage rather than current or power. Because of this fact, it is important that the input impedance be as high as possible. Maximum power transfer does not occur, but the voltage is more efficiently sensed.

As previously mentioned, there are all sorts of read-out devices. It is sometimes necessary to make a permanent record of the output of the instrument. This may be done by using a recorder which operates an ink pen. The recording may also be made by the movement of a pen which melts the wax coating on special paper, or the paper may be sensitive to a spark which passes through it. The recorder may also be designed so that the galvanometer causes a mirror to move, which, in turn, causes a light beam to move across a strip of moving light-sensitive paper.

There are two basic varieties of recording devices. In all recorders the current which passes through the galvanometer coil causes it to rotate. Thus, if a pen is attached to the rotating galvanometer coil it scribes an arc. Actually it is desirable for the pen to move back and forth perpendicular to the movement of the strip-chart. When the pen moves in an arc, the recording is said to be *curvilinear,* and when it moves perpendicular to the movement of the paper, the recording is called *rectilinear.* The rectilinear tracing is compared to the curvilinear tracing in Figure 10-1.

There are many advantages to using a rectilinear recorder:

1. A ruler can be used to measure the amplitude.
2. It is easier to mark time lines and to correlate the time lines with the recorded waveform.
3. It is easier to compare simultaneous recordings on more than one channel.
4. It is somewhat easier to interpret a rectilinear waveform than its curvilinear equivalent.

In general, most physiologists prefer to use the rectilinear tracing whenever possible. It must be pointed out, however, that the treated paper used for rectilinear heat-tracing is considerably more expensive than paper used for the curvilinear ink

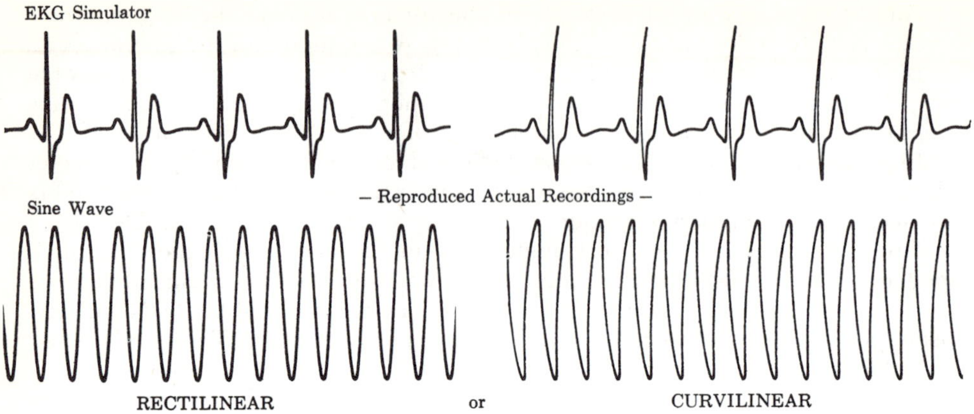

Figure 10-1. The rectilinear tracing compared to the curvilinear tracing. (Courtesy of E & M Instrument Co., Inc., Houston, Tex.)

pen recorder. It must also be pointed out that there is really no such thing as rectilinear recording when a galvanometer device is used.* Let us examine Figure 10-2 in which *R* represents the length of the recording pen or, in the case of the optical recorder, the distance from the galvanometer mirror to the paper.

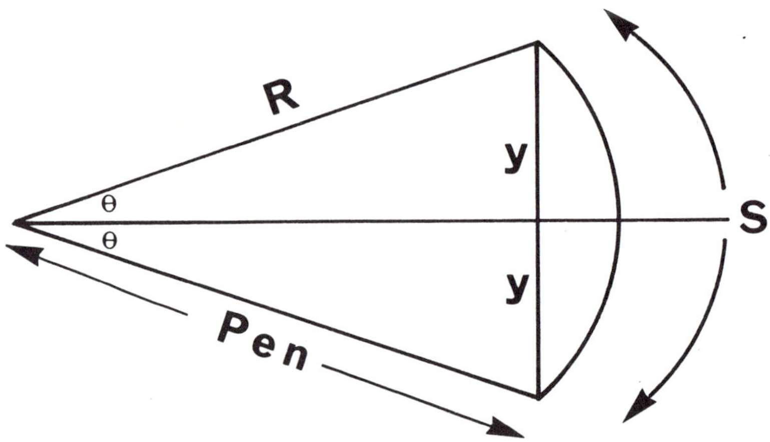

Figure 10-2. Calculating the error in a curvilinear tracing.

The arc scribed by the movement of the pen is called *S*. The shortest distance between the points of measurement on the arc *S* is the straight line, $2y$. This distance, $2y$, is the measurement of the arc length with a straight ruler. The difference in length between the arc length *S* and the measured distance $2y$ is the error associated with the use of a curvilinear tracing. The central angle which intersects the circumference to form the arc *S* is 2θ. The measured deflection is given by:

$$2y = 2R \text{ Sin } \theta \qquad 10.1$$

*By this I mean that some form of linkage must be used to convert the rotational motion of the galvanometer to the rectilinear motion of the pen.

The true arc length S is approximately equal to:

$$S \cong R(2\theta) \cong 2R\theta \qquad (\theta \text{ is in radians}) \qquad 10.2$$

The per cent error is thus:

$$\% \text{ error} = \frac{S - 2y}{S} \times 100 = \frac{2R\theta - 2y \operatorname{Sin} \theta}{2R\theta} \times 100 = \left(1 - \frac{\operatorname{Sin} \theta}{\theta}\right) \times 100 \quad 10.3$$

It may be seen from equation 10.3 that if the pen were 10 cm long and were deflected an arc length which when measured with a ruler was 2 cm, then the error would be only 0.2 per cent. The error would be about 0.7 per cent if the measured deflection were 4 cm, and only 2.75 per cent for a deflection of 8 cm. The greater the pen length the less is the error. Therefore, in an optical system with a distance of 1 meter between the galvanometer mirror and the photographic paper, the error would be indistinguishable. It is not possible to make an ink-writing pen 1 meter long because it would be too heavy and would load the galvanometer coil. There is a recorder manufactured in Europe which squirts the ink from a nozzle located in the center of the galvanometer coil. This system, in addition to being almost rectilinear, has a very high frequency response.

Since the user would prefer, if possible, a rectilinear tracing, several systems for converting the direct-writing pen into a more rectilinear system have evolved. At present there are three ways of doing this: by using a mechanical linkage, a servo-driven system, or a heat stylus. Each of these systems will be described later in this chapter when the particular recorder using that system is described. It may be said at this point, however, that the errors involved in using these systems must be carefully analyzed. The use of the heat stylus, as will be seen later, has an error associated with $\theta/\tan\theta$ which is over two times the error discussed in equations 10.1 to 10.3.

The read-out device may also be an oscilloscope, as shown in Figure 3-4, or it may be an oscilloscope which has a memory built in to automatically keep the tracing on the scope face until the user erases it by pushing a button (*e.g.,* Tektronix Memoscope). The display may also be an ammeter or a voltmeter. In the last few years the multichannel tape recorder has been used more frequently to record a variety of signals.

In each case, the instrumentalist must decide which of the multitude of commercially available recorders and visual displays is best suited to the needs of the instrument producing the signal to be displayed and to the requirements of the research project. Not the least important factor is the cost of the equipment. An expensive piece of equipment that does not do the job properly is worse than a less expensive device that is better suited.

The oscilloscope and a variety of meter movements have already been discussed in Chapter 3. This chapter will be devoted, for the most part, to a discussion of recording read-out devices. *The commercial instruments chosen for inclusion in this chapter do not represent the only available instruments nor do they necessarily represent the best choice.*

LOW-FREQUENCY RECORDERS

There are a variety of parameters such as the measurement of environmental temperature, oxygen tension, carbon dioxide tension, hydrogen ion concentration, mean blood pressure, and many others which do not require high-frequency

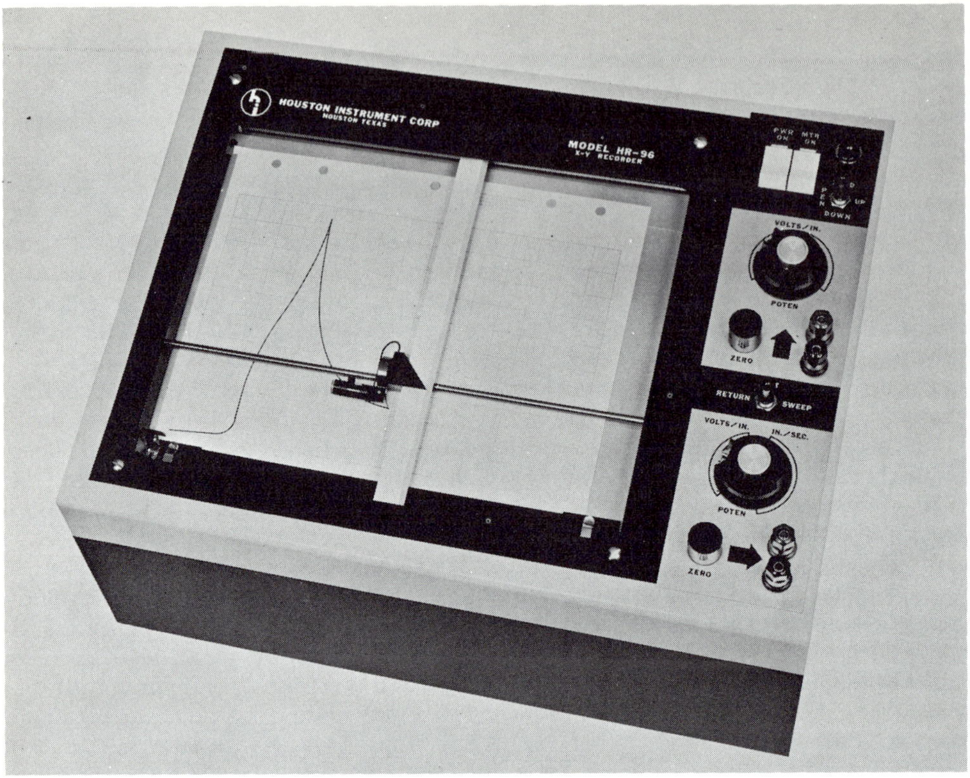

Figure 10-3. The X-Y recorder (Courtesy of Houston Instrument Corp., Houston, Tex.)

response recorders. In some of these cases, the data could be read from a meter and written down. This would, of course require a certain number of man-hours, and the man may miss important transients.

Recorders which respond faithfully from DC to about 5 cps may be classed as low-frequency recorders. These recorders are also often referred to as *strip-chart* recorders because they write on continuous strips of paper. Almost all low-frequency recorders register the waveform as a function of time. An exception to this is the X-Y recorder such as the one shown in Figure 10-3 which plots any two mutually independent variables on a sheet of graph paper.

Figure 10-4 is an example of a low-frequency recorder which can be obtained as a current-sensing or a voltage-sensing device and with a variety of voltage and current ranges. When used as a current-sensing device, the input signal activates the coil directly. As has been pointed out previously, this requires a low-output impedance from the previous instrument. The manufacturer of this particular instrument, Texas Instruments, Inc., also provides a similar recorder which employs an amplifier with a high-input impedance to convert the input signal, which is usually a low-current millivolt signal, to a signal with sufficient voltage and current to drive the galvanometer of the recorder. One model of this particular recorder will record faithfully up to 5 cps and another to 2 cps.

Figure 10-4. The Texas Instrument recorder. (Courtesy of Texas Instruments, Inc., Houston, Tex.)

The Texas Instrument recorder utilizes a unique mechanical linkage to translate the normally curvilinear motion. The galvanometer (Fig. 10-5) is located at *G*, the line *MPT* is the writing arm, and the pen point is located at *T*. *G* is the fixed pivot of the meter movement, and *PG* is the swinging galvanometer arm. *PM*, *PT*, and *PG* are made equal; therefore, if *M* is limited to movement back and forth along line *MG*, the pen point at *T* will always have linear motion perpendicular to line *MG*. The mechanical linkage does, however, limit the frequency response.

Figure 10-5. The mechanical linkage use by Texas Instruments to convert curvilinear recording to rectilinear recording. (Courtesy of Texas Instruments, Inc., Houston, Tex.)

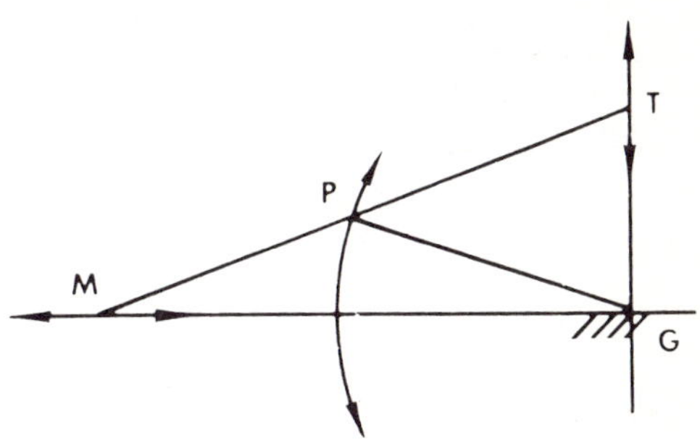

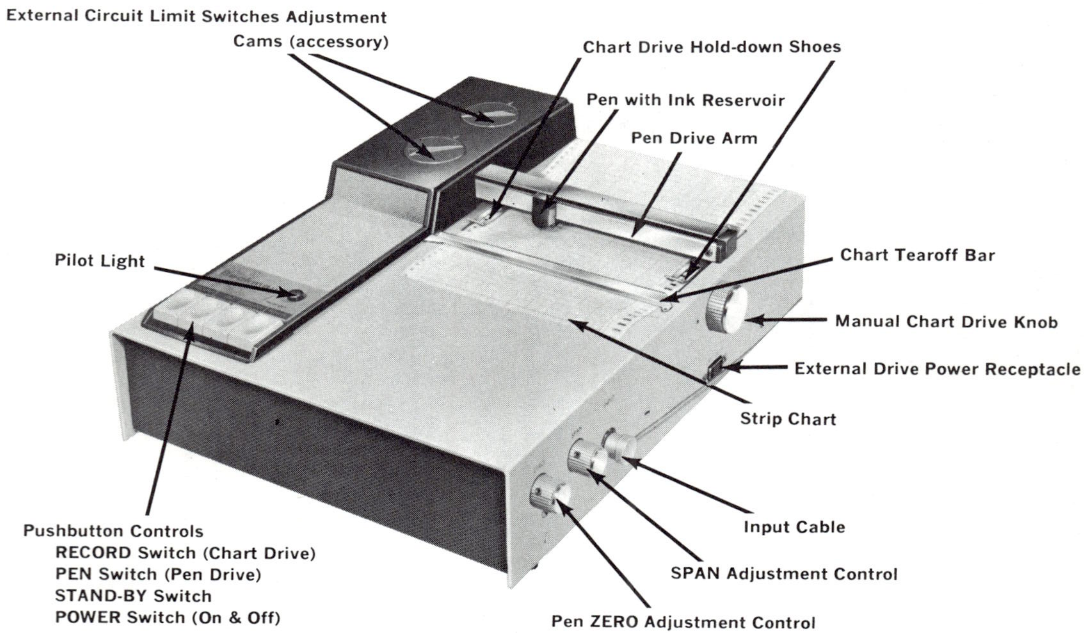

Figure 10-6. The Beckman strip-chart recorder. (Courtesy of Beckman Instruments, Inc., Palo Alto, Calif.)

Recorders manufactured by both the E & M Instrument Company (the Physiograph) and the Offner Division of Beckman Instruments also are available with a similar mechanical linkage.

Another type of low-frequency response recorder is the *potentiometric strip-chart recorder* which utilizes a null-balancing measuring system. In operation, the circuit continuously compares the input signal voltage with a stable reference voltage supplied by a mercury cell and balances out any difference between the two voltages by means of a servo-motor-driven slider which moves along a linear potentiometer (the slide wire). The pen is attached to this slider and records its displacement in canceling out discrepancies between the signal and the reference voltage. This type of system is one of the previously mentioned methods of generating a rectilinear tracing. Its accuracy is limited to the care with which the servo-system is designed.

In the Beckman *potentiometric* strip-chart recorder (Fig. 10-6) the input impedance is very high (40KΩ to 500KΩ)) since essentially no current is drawn with the pen in the null position. In some instruments of this type, the input impedance can be over 10 megohms. This particular recorder has a continuously adjustable range from 10 to 100 millivolts full-scale. Its frequency response, however, is limited to 1 cps.

MEDIUM-FREQUENCY RECORDERS

Recorders in this class have frequency response characteristics from DC to 200 cps. Most of these recorders are direct writing, that is, they use some sort of pen which writes on a strip of moving paper. Almost all of the recorders in this class are designed to accept a wide range of voltage signals. The majority of these recorders will measure a wide variety of specialized biomedical parameters. Various input amplifiers are supplied to serve as an integral part of the recorder system. The medium-frequency recorders are probably the most important read-out devices for use with biomedical instrumentation. Almost all biomedical parameters fall in this frequency range as can be seen by examining Table 8.1. Four widely used commercial instruments will be described in some detail.

The first of these is the completely transistorized recorder designed by Franklin Offner, Ph.D., and now manufactured by the Offner Division of Beckman Instruments. Figure 10-7(A) is a photograph of the type R Offner recorder, and Figure 10-7(B) shows type RS, a two-channel table model which is electronically equivalent to the larger type R unit.

There are four basic modules which make up the recorder:

1. A power supply (Fig. 10-8[A]) which provides all of the necessary power for the entire unit.

2. A type 481* "preamplifier" module (Fig. 10-8[B]) which has a sensitivity range of 10 microvolts to 50 volts per cm of pen deflection. It may be used as a push-pull or differential amplifier in which neither of the two input terminals is connected to ground, or it may be used as a single-ended amplifier in which one side or the other is grounded. The 481 module exhibits exceptional gain stability, low noise, and excellent linearity. It is a chopper type of amplifier, that is, it uses a

---

*The numbers 481 and 482 are numbers assigned by the manufacturer. When the unit is modified or improved an alphabetical letter is added, e.g., 481B.

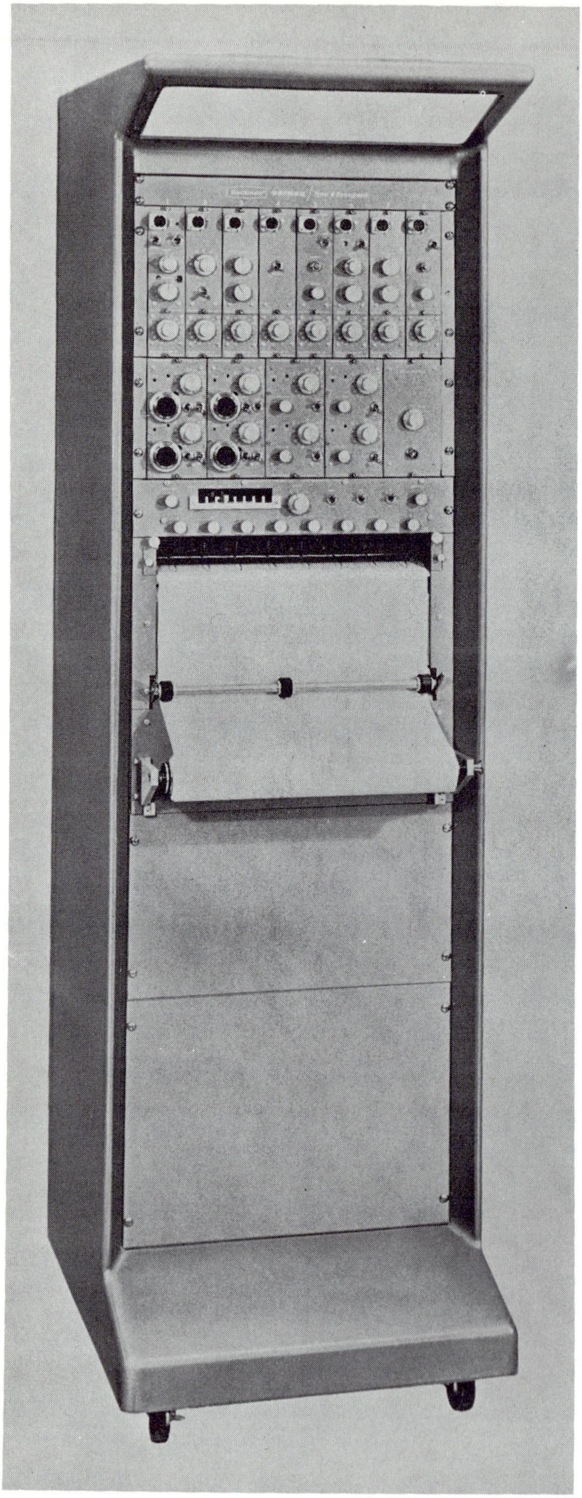

Figure 10-7A. The Beckman/ Offner type R recorder. (Courtesy of Beckman Instruments, Inc., Chicago.)

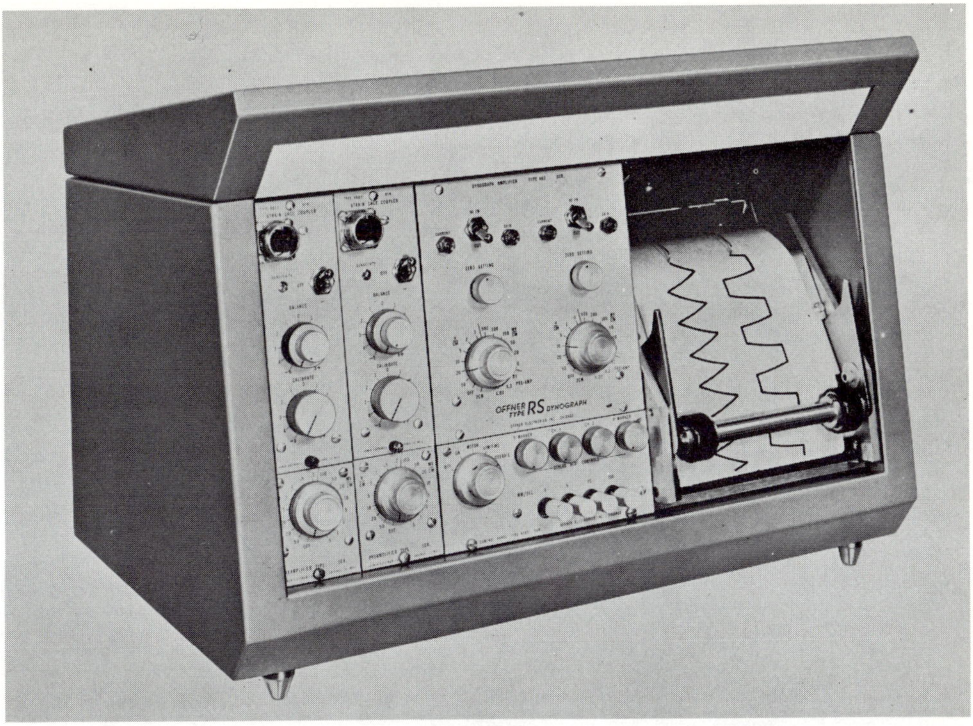

Figure 10-7B. The Beckman/Offner type RS recorder (Courtesy of Beckman Instruments, Inc., Chicago.)

chopper, operating at 400 cps, to chop incoming DC signals. These chopped signals are then amplified by a transistorized AC amplifier. Next, the resulting amplified AC signal is demodulated by another set of contacts on the same chopper, leaving the 481 preamplifier output as an amplified DC signal.

The frequency response of this module, without the chopper, is over 2000 cps. The 400-cps chopper will, of course, lower the frequency response. The entire recorder has an over-all frequency response which is flat to 125 cps, within 10 per cent to 150 cps, and within 20 per cent to 200 cps. These figures are valid only when ink pens are used (or the electric-spark arrangement). When a heat stylus is used the frequency response must be lower because the pen is heavier and because the specially treated paper is limited as to how fast it can melt. The high-frequency response is accomplished by using a special bit of circuitry in the type 482* power amplifier which amplifies the high frequencies more than the low frequencies. This compensating circuit may be switched on or off. Thus, in some applications in which it is undesirable to allow frequencies over 50 cps, these frequencies can be eliminated. The input impedance of the entire unit is 2.0 to 12.5 megohms, depending on the coupler used.

---

*The numbers 481 and 482 are numbers assigned by the manufacturer. When the unit is modified or improved an alphabetical letter is added, e.g., 481B.

Figure 10-8A. Power supply module for Beckman/Offner type R recorder. (Courtesy of Beckman Instruments, Inc., Chicago.)

As may be seen in Figure 10-8(B), the 481 preamplifier has provision for plug-in couplers. These couplers are compact modules which provide complete circuitry to adapt the preamplifier for use with various transducers and electrode systems. These couplers, *together with the appropriate transducer or electrode,* allow for measurement of voltage, current, power, pressure, strain, velocity, temperature, EEG, and ECG, as well as many other physical and biophysical parameters. A few of these coupler modules are shown in Figure 10-9.

3. The third part of the recorder is the type 482A power amplifier, shown in Figure 10-10. The first two stages consist of two push-pull DC amplifier stages used in the common-emitter configuration. The remainder of the 482A power amplifier consists of several stages which are designed to drive the galvanometer coil of the pen motor. This circuitry employs a unique design, the end result of which is a power amplifier which does not require a center-tapped galvanometer coil, and which has very low idling current. This means that when no signal is applied

Figure 10-8B. Type 481 preamplifier module for the recorder shown in Figure 10-8A. (Courtesy of Beckman Instruments, Inc., Chicago.)

to the amplifier, the current through the coil is very low. The galvanometer coil utilizes about 1 ampere at about 8 volts. A simplified block diagram of the recorder is shown in Figure 10-11. The majority of the circuitry detail is not included in this illustration.

4. The remainder of the recorder is shown in the middle section of Figure 10-7(A). This part of the assembly houses the chart paper, the pen motors, which include the writing pens, and the motor and associated gear assembly for moving the chart paper. This recorder has an exceptional mechanical system for driving the chart paper uniformly and smoothly. This is an important consideration when deciding which recorder to purchase for laboratory use. A side view of the arrangement for holding the paper and the pen is shown in Figure 10-12 (A) and (B). Notice that the pen makes contact with the paper on what is called a *forming bar.* This recorder has made provision for using both the heat stylus for rectilinear recording and the ink pen for curvilinear tracing. As may be seen in Figure 10-12(A),

Figure 10-9. A few of the plug-in couplers for the Beckman/Offner Recorders. (Courtesy of Beckman Instruments, Inc., Chicago)

Figure 10-10. A dual type 482A power amplifier module for the Beckman/Offner Recorder. (Courtesy of Beckman Instruments, Inc., Chicago.)

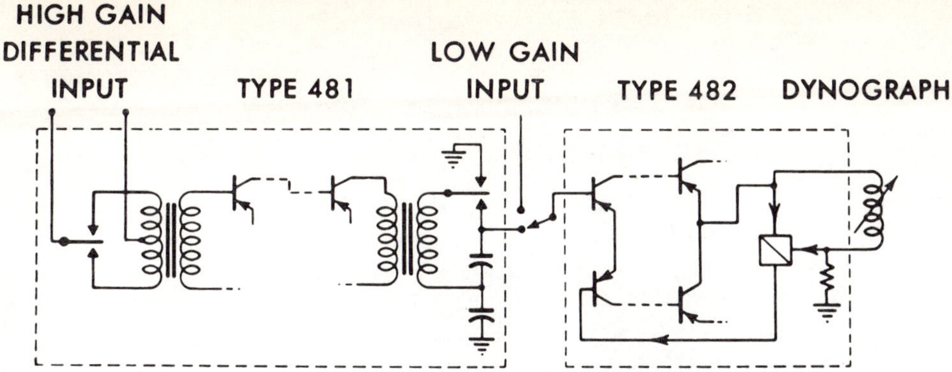

Figure 10-11. The block diagram of the Beckman/Offner type R recorder. (Courtesy of Beckman Instruments, Inc., Chicago.)

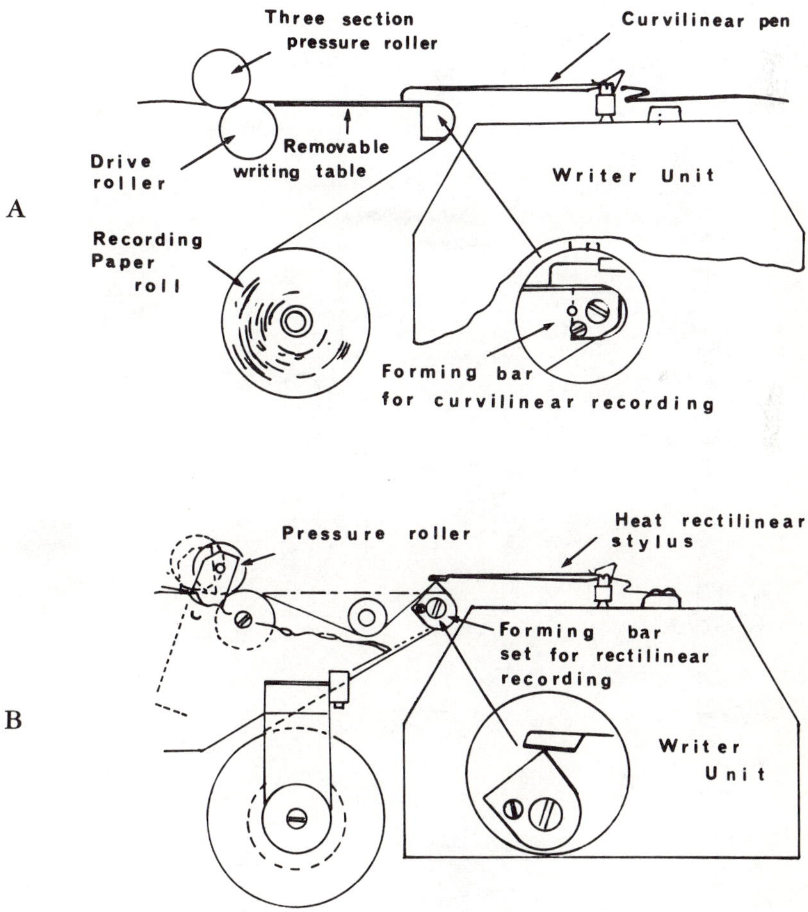

Figure 10-12. A, The mechanical assembly for holding the chart paper for curvilinear recording; B, for rectilinear recording. (Courtesy of Beckman Instruments, Inc., Chicago.)

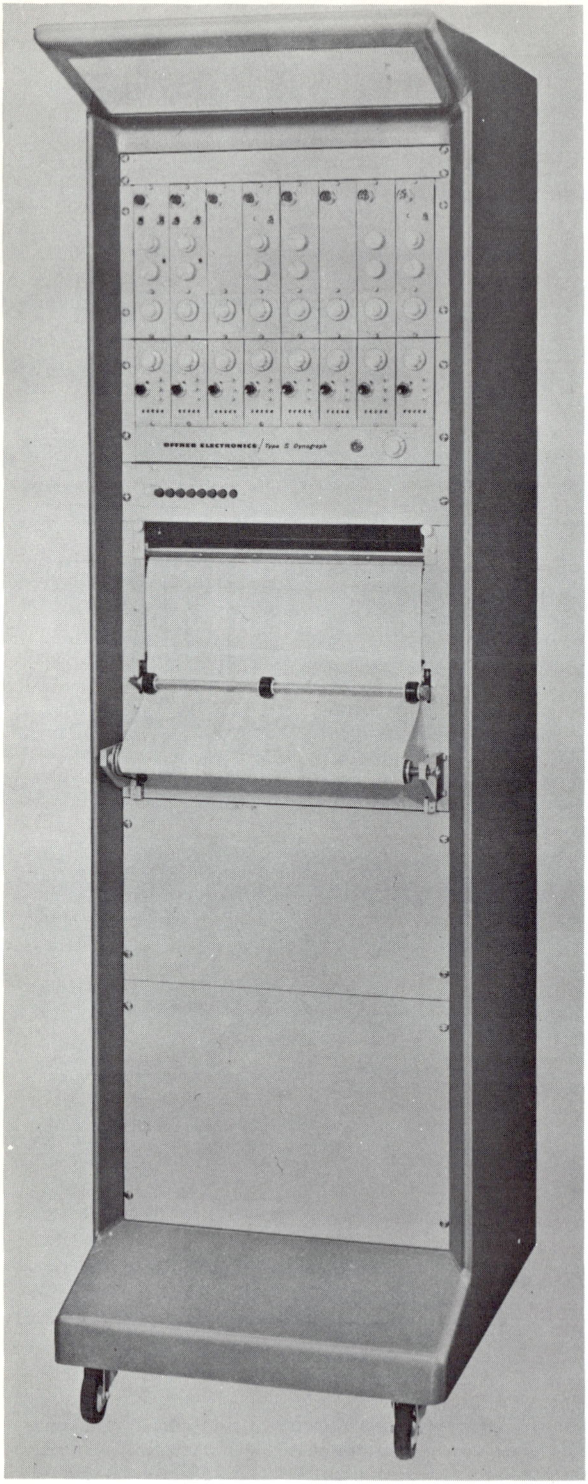

Figure 10-13. The Beckman/Offner type S ink rectilinear recorder. (Courtesy of Beckman Instruments, Inc., Chicago.)

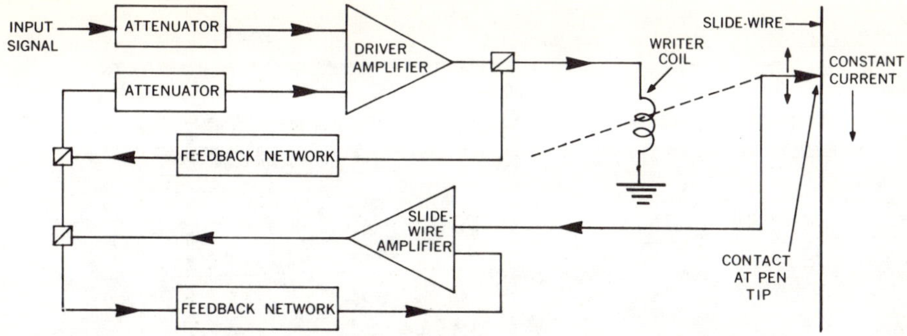

Figure 10-14. The block diagram for type S recorder. (Courtesy of Beckman Instruments, Inc., Chicago.)

the paper is merely moved past the ink pen, providing a flat table for the pen to scribe an arc. In Figure 10-12(B), however, the forming bar has been rotated so that the paper is pulled tight over the sharp edge. The heat stylus makes contact with the paper only on this edge. As the stylus moves in an arc, a straight line is written until the stylus has moved far enough so that the hot wire is no longer touching the forming-bar edge. It will be recalled that there is some error in measuring the curvilinear arc with a straight ruler. When the error associated with measuring the rectilinear tracing of the heat stylus, shown above, is more than twice as great as that when the curvilinear arc is measured with a straightedge, the error with the heat pen is related to $(\tan \theta)/\theta$.

Since it is desirable to use rectilinear recording whenever possible, Beckman Instrument Company designed a recorder assembly which has essentially the same characteristics as the recorder described above except that the power amplifier and the pen motor have been altered to provide an ink rectilinear tracing. Figure 10-13 is a photograph of the Offner Type S ink-rectilinear dynograph. By avoiding the use of a heat stylus, this recorder obviates the relatively costly special chart paper. The basic principle on which the ink-rectilinear system works is shown in a block diagram in Figure 10-14. A signal causes the driver amplifier to rotate the driver coil. This causes the pen to move along a slide wire; the pen also writes while it moves. The position of the pen on the slide wire causes a *positional feedback signal* to activate the special *slide-wire amplifier*, which then applies either positive or negative feedback to the driver amplifier to correct the position of the pen. In this way, the pen is continually corrected and forced to the proper position, thereby avoiding the errors due to curvilinear or heat rectilinear recording. The ink in the above system must be forced through the pen under pressure.

Another medium-frequency recorder is shown in Figure 10-15. This recorder, manufactured by the Grass Instrument Company, employs vacuum-tube circuitry. The entire preamplifier unit may be changed to allow the recorder to be used for recording various parameters such as ECG, EEG, and pressure. The low-level DC preamplifier has a sensitivity range from 0.01 millivolts to 20 volts. The instrument shown in Figure 10-15 has a switch which allows the highest frequency response, down 50 per cent at 60 cps, to be lowered to a one-half amplitude frequency response of 15, 3, 0.5, or 0.1 cps. This recorder uses an ink-fed curvilinear recording system.

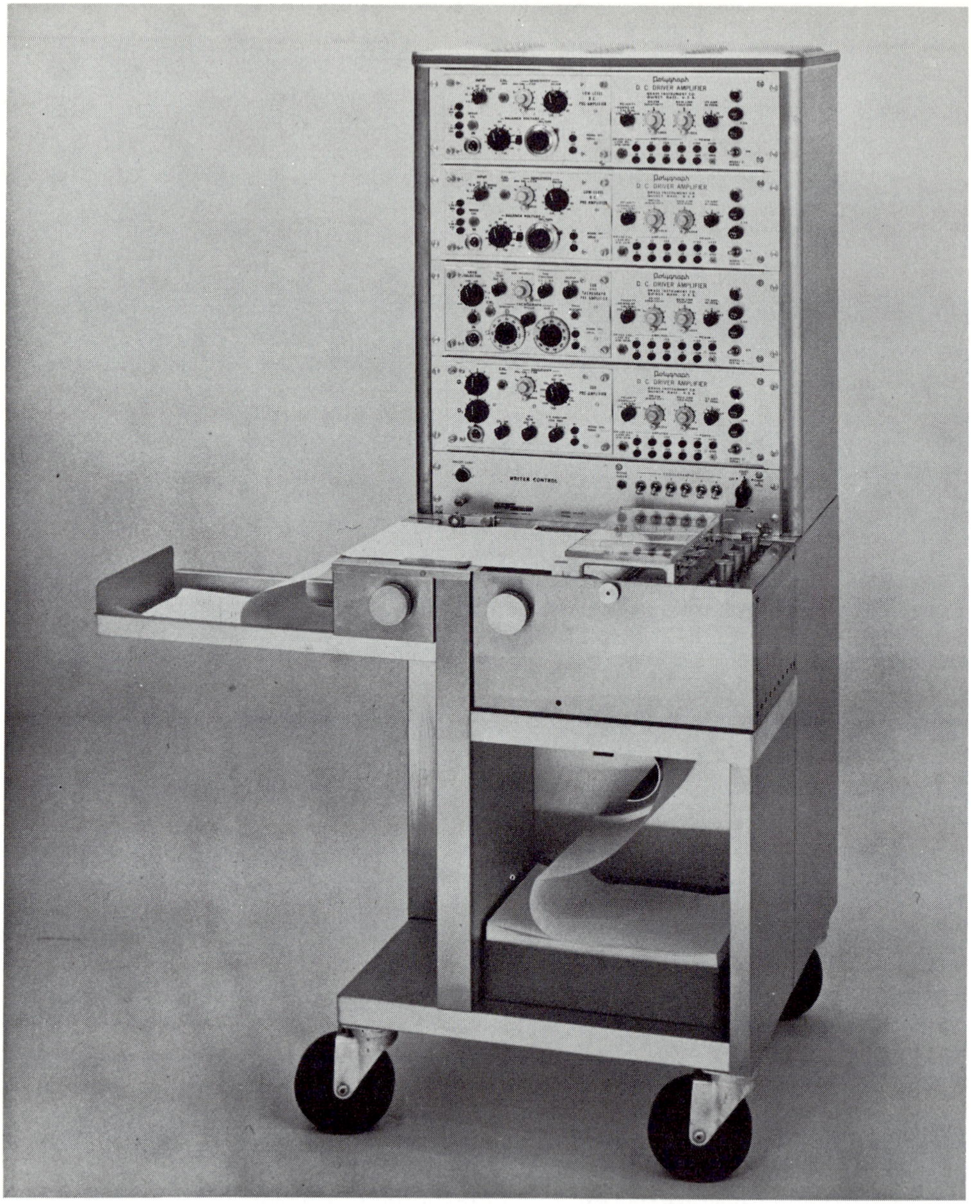

Figure 10-15. The Grass polygraph. (Courtesy of Grass Instrument Co., Quincy, Mass.)

One of the most commonly used biomedical recording systems is manufactured by the Sanborn Company. Model 964 shown in Figure 10-16 is an example of the Sanborn recorder. This particular assembly provides four simultaneous channels of information. The input amplifiers or preamplifiers may be chosen from a wide selection. Some of these input amplifiers are complete units and others utilize a plug-in module. Two Sanborn preamplifiers are shown in Figure 10-17.

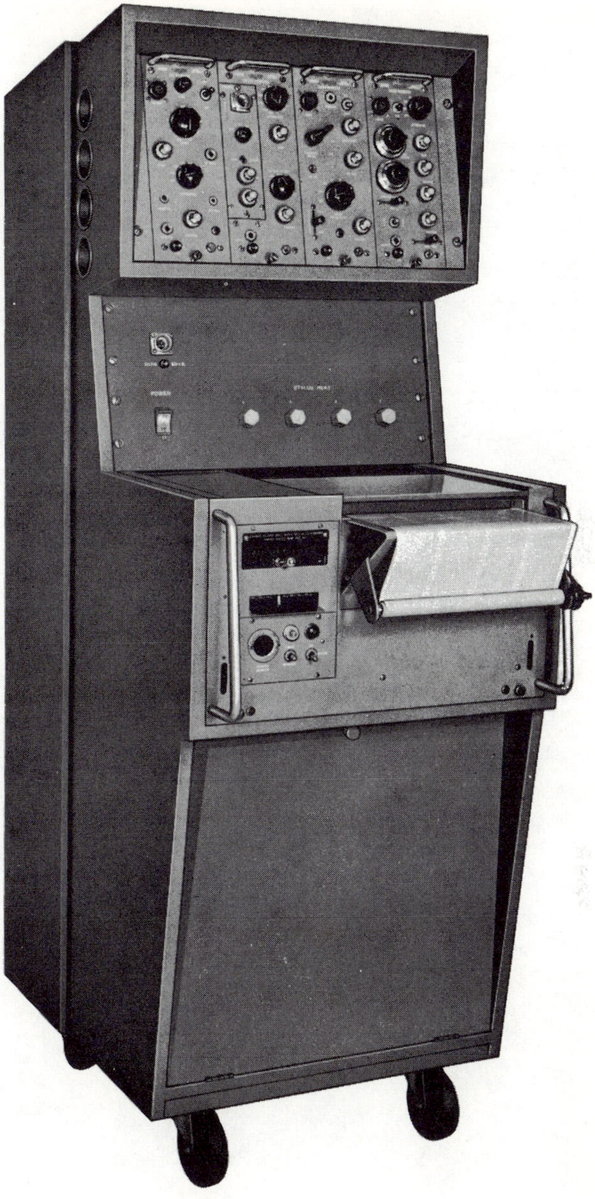

Figure 10-16. The Sanborn Oscillographic Recording System, model 964. (Courtesy of Sanborn Co., Waltham, Mass.)

Model 964 utilizes almost all vacuum-tube circuitry. The manufacturer's specifications indicate that it has a frequency response of DC to 125 cps; I found that it was less than that on the unit tested. The galvanometers use low-resistance coils (about 15 ohms) which are enclosed for protection. They have a natural frequency of 45 cps with high torque to insure accuracy over the entire DC to 125 cps range. Current feedback circuits within the power amplifier virtually eliminate thermal drift, which is less than 0.25 divisions of pen deflection per 10°C temperature change from 20° to 40°C.

Figure 10-17. Two preamplifiers for the Sanborn model 964 recorder. (Courtesy of the Sanborn Co., Waltham, Mass.)

A close-up of the writing unit is shown in Figure 10-18. This recorder uses the heat stylus arrangement as shown in Figure 10-19. Notice that the heat stylus uses a sharp forming bar as did the Offner recorder shown in Figure 10-12(B). The error in using this system is the difference between the mean arc S_m and the length of the recorded line y (Fig. 10-20).

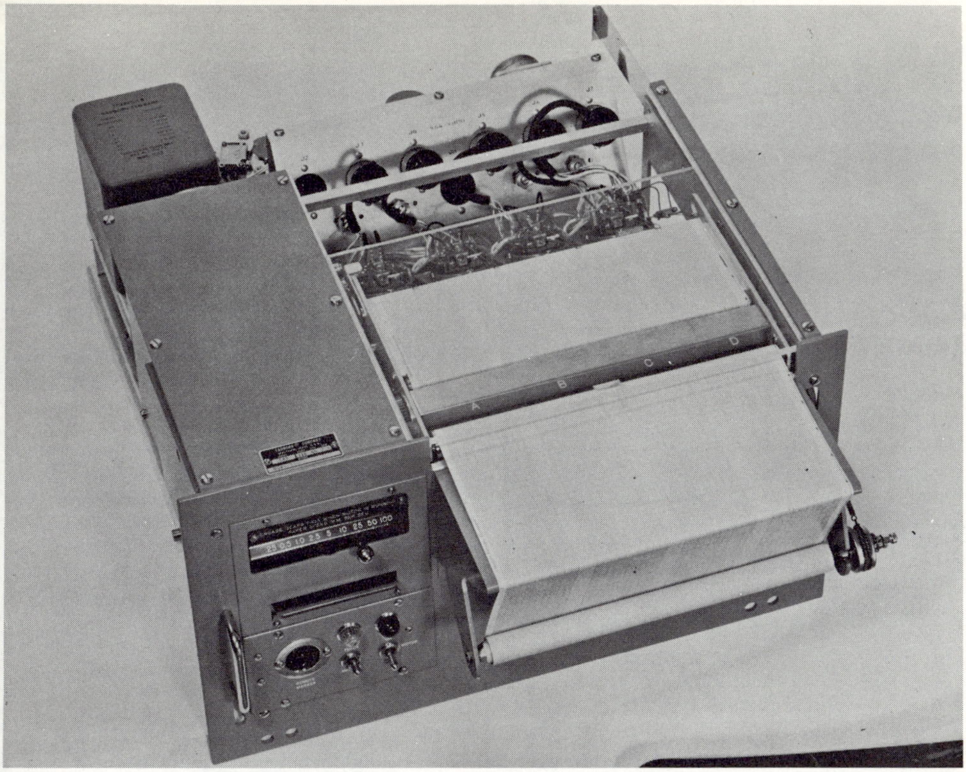

Figure 10-18. The mechanical assembly of the Sanborn model 964 recorder showing the heat styli and chart paper. (Courtesy of Sanborn Co., Waltham, Mass.)

$$\tan \theta = \frac{y}{R - E} \qquad 10.4$$

where y is the length of the recorded line and R is the over-all length of the pen from the axis of the galvanometer to the farthest point on the heating element which intersects the forming bar. E is the length of the heating element used to scribe line y. The length of the pen deflection y is given by:

$$y = (R - E) \tan \theta \qquad 10.5$$

Therefore the central angle is:

$$\theta = \tan^{-1} \frac{y}{R - E} \qquad 10.6$$

and the arc length at the tip of the pen is:

$$S_2 = R\theta \qquad 10.7$$

The arc length at the inside edge of the heating element is:

$$S_1 = (R - E)\theta \qquad 10.8$$

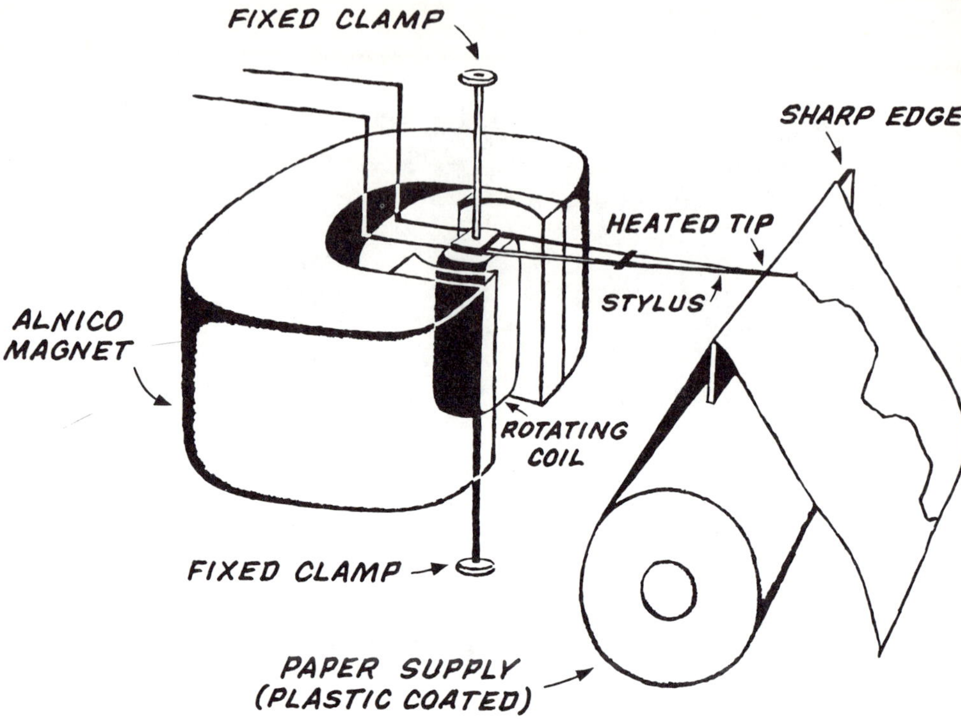

Figure 10-19. The Sanborn pen motor showing the arrangement for rectilinear recording. (Courtesy of Sanborn Co., Waltham, Mass.)

The mean arc length is thus:

$$S_m = \frac{R\theta + (R - E)\theta}{2} \qquad 10.9$$

and the per cent error is:

$$\% \text{ error} = 1 - \frac{(R - E)\tan\theta}{\dfrac{R + (R - E)\,\theta}{2}} \qquad 10.10$$

Therefore, if the pen (R) is 11 cm long and the heating element (E) is 1 cm long, the per cent error for a deflection of 2 cm is 1.69 per cent. In this analysis the mean arc length was used; the error would be even greater if the arc length S_2 were used. The heat stylus is usually heavier than an ink writing pen would be; this tends to decrease the frequency response.

The E & M Instrument Company also manufactures a complete recording system, the Physiograph, which is ideal for student physiology or pharmacology laboratories. The complete four-channel unit, which has one timer and event marker channel and three channels for data, is shown in Figure 10-21. Another Physiograph recorder is available with six channels. The recorder is supplied with twelve fixed chart speeds from 0.01 to 5 cm per sec, or continuously variable speeds from 30 cm per hr to 6 cm per sec. The chart speed control and a timer which

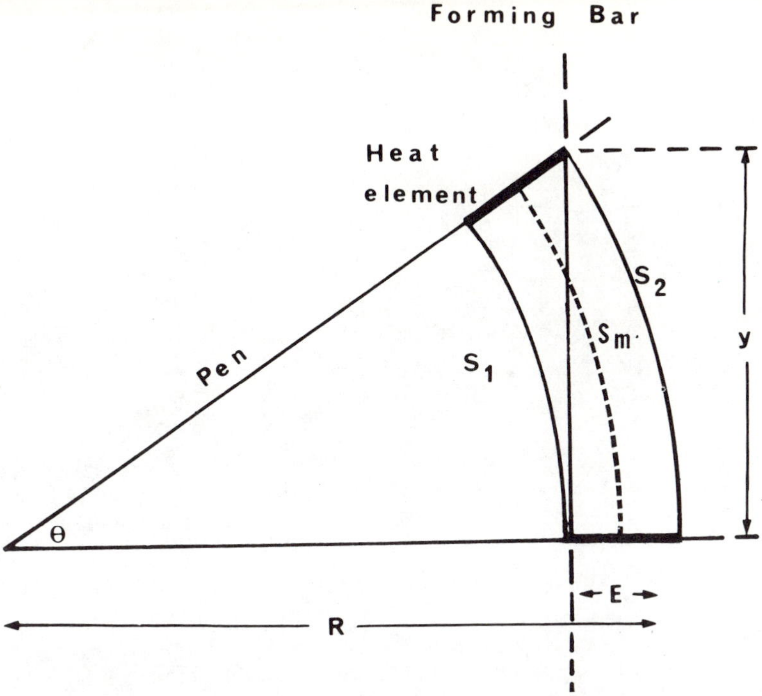

Figure 10-20. The calculation of error using a heat rectilinear pen.

delivers a 1-, 5-, 30-, or 60-second timing pulse are both electronically controlled. The circuit is designed so that the timer marks are deflected downward and the event marks are deflected upward.

The pen motor may be obtained for either rectilinear or curvilinear ink writing. The pen for rectilinear recording has a total excursion of 6 cm while that for curvilinear tracing has an 8-cm excursion. A gravity-feed ink supply is used. The ink for each pen is stored in a small plastic vial which can be squeezed to facilitate flushing of the system. The over-all frequency response for both the rectilinear and curvilinear systems is 50 per cent down at 40 cps. Although this frequency response is not adequate for studies requiring high-frequency detail and certain sophisticated biophysical studies, it is entirely adequate for any student laboratory and may be used in many research projects. Increasing the frequency response would add considerable expense to the cost of the system. It is advantageous not to use an extremely high frequency response recorder in the student laboratory as the entire system is more delicate and may be more susceptible to student breakage.

The Physiograph recorder uses a rather unique DC driver amplifier. The module itself contains its own power supply and is therefore not dependent on the other channels for operation. This particular driver amplifier keeps the grid voltage at −290 volts with respect to ground, and the plate output is therefore near ground potential. This facilitates the use of monitors such as the DC oscilloscope. It also simplifies the circuit as it eliminates cathode-follower stages and voltage-divider networks necessary to keep the input and the output at the same potential. *The only disadvantage is that the user must be careful to avoid touching the pins of the input connector so as not to get an electric shock.*

Figure 10-21. The Physiograph Four. (Courtesy of E & M Instrument Co., Inc., Houston, Tex.)

The console (Fig. 10-21) is designed only for use with the various transducers manufactured by the E & M Instrument Company. It supplies all necessary accessory voltages for the transducers and the preamplifiers needed for the remainder of the system. The E & M Instrument Company provides a number of accessory preamplifiers, so that the system may be used to monitor almost any biomedical parameter. Many of the E & M transducers such as the myograph, the bellows pneumograph, and the Bourdon pressure transducer are designed to plug directly into the pen-driver amplifier. An example of an experimental arrangement used to correlate events in a turtle heart is shown in Figure 10-22. Notice the effect of stimulation of the right vagus nerve on the various cardiac events.

It is necessary to use the preamplifier, shown in Figure 10-23, for recording the electrocardiogram (ECG or EKG), the electroencephalogram (EEG), the electromyogram (EMG), and any transducer employing a microphonic element. This amplifier is condenser-coupled with a maximum sensitivity of no less than 40 microvolts per cm of pen deflection. It is necessary to establish a good symmetrical ground when using an amplifier of this sort. Some techniques for doing this are discussed in Chapter 12.

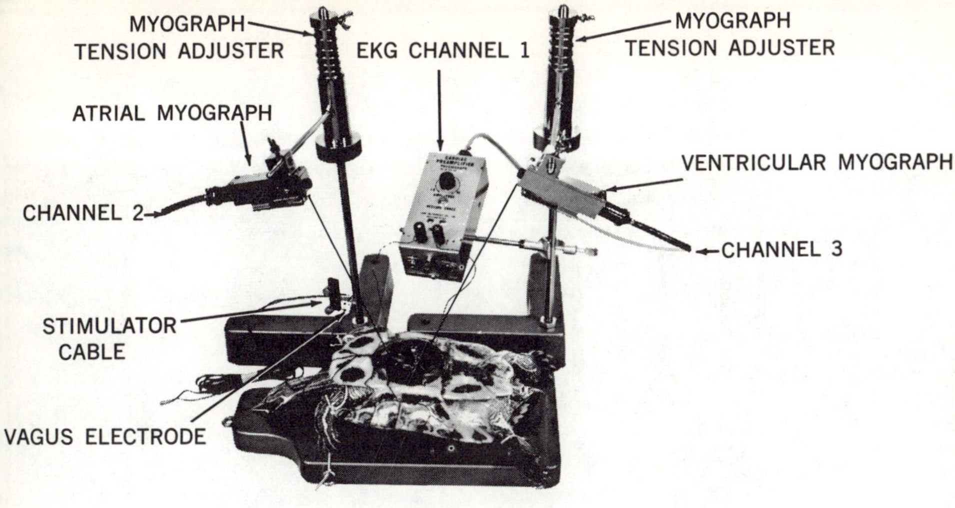

Fig. 10-22A

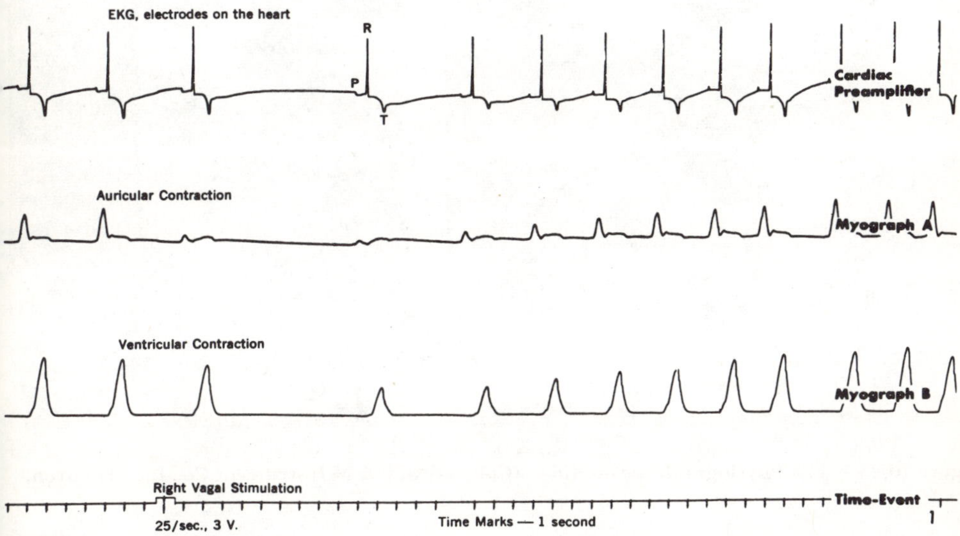

Fig. 10-22B

Figure 10-22. Events of the cardiac cycle: A, experimental arrangement; B, sample data. (Courtesy of E & M Instrument Co., Inc., Houston, Tex.)

The front panel (Fig. 10-23) has a selector switch labeled EKG, EEG, EMG, and MIC. This switch changes the high-pass filter of the amplifier. The EKG setting has a time constant of 2 seconds or a low frequency cut-off reaching half-amplitude at 0.03 cps. The time constant in the EEG position is 0.3 seconds (half-amplitude cut-off at 0.2 cps); the EMG has a time constant of 0.03 seconds (half-amplitude cut-off at 2 cps); and the microphone (MIC) setting has a time constant of

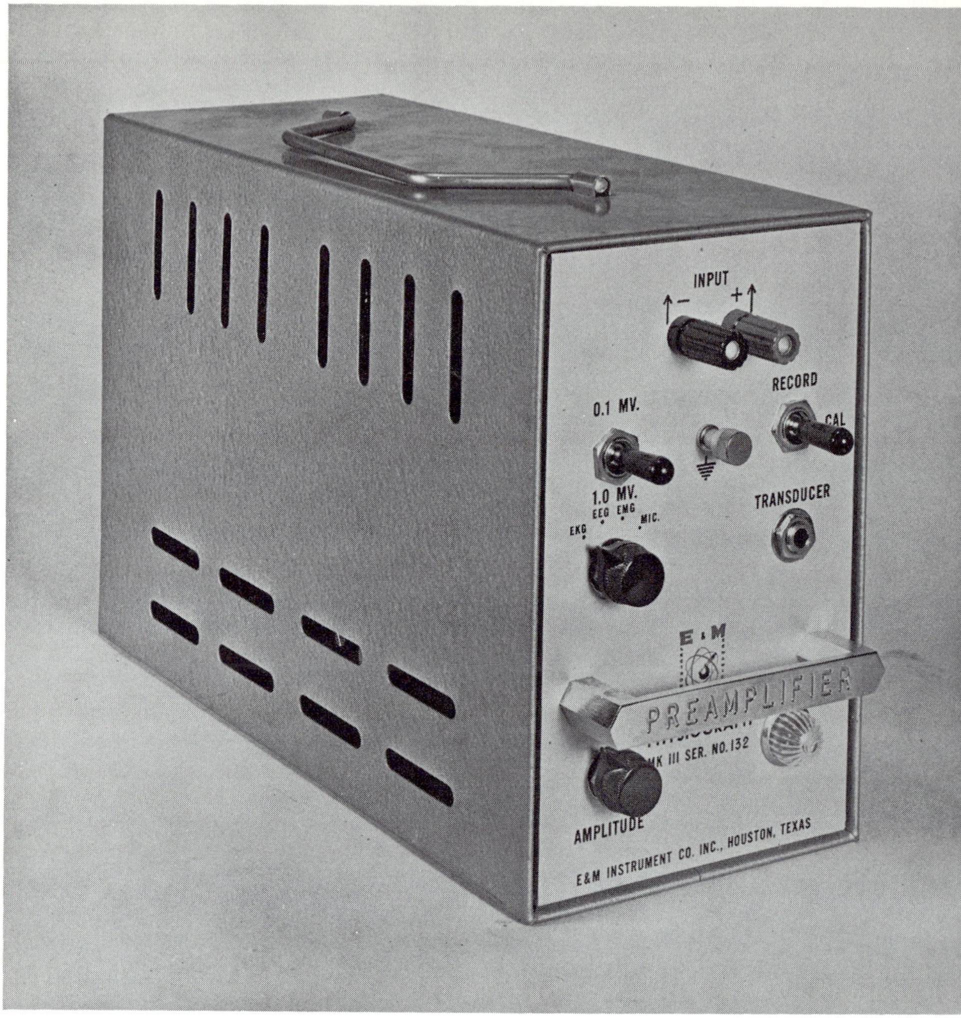

Figure 10-23. The Physiograph preamplifier. (Courtesy of E & M Instrument Co., Inc., Houston, Tex.)

0.3 seconds, but the signal is attenuated, and may be used for heart sounds or pulse recordings. In cases in which the heart rate is faster than in the human, it will be necessary to use a setting with a higher frequency response. In other words, it may be necessary to use the EEG or EMG setting for recording a rapid EKG pulse.

In summary, the Physiograph recording system is a well-organized, relatively inexpensive recording system which is extremely well suited for use in a physiology teaching laboratory. Similar student recording systems are available from the Beckman Instrument Company, the Gilson Instrument Company, and others. The Physiograph and the other recording systems mentioned above in this section have been chosen for inclusion as examples of existing medium-frequency recording systems.

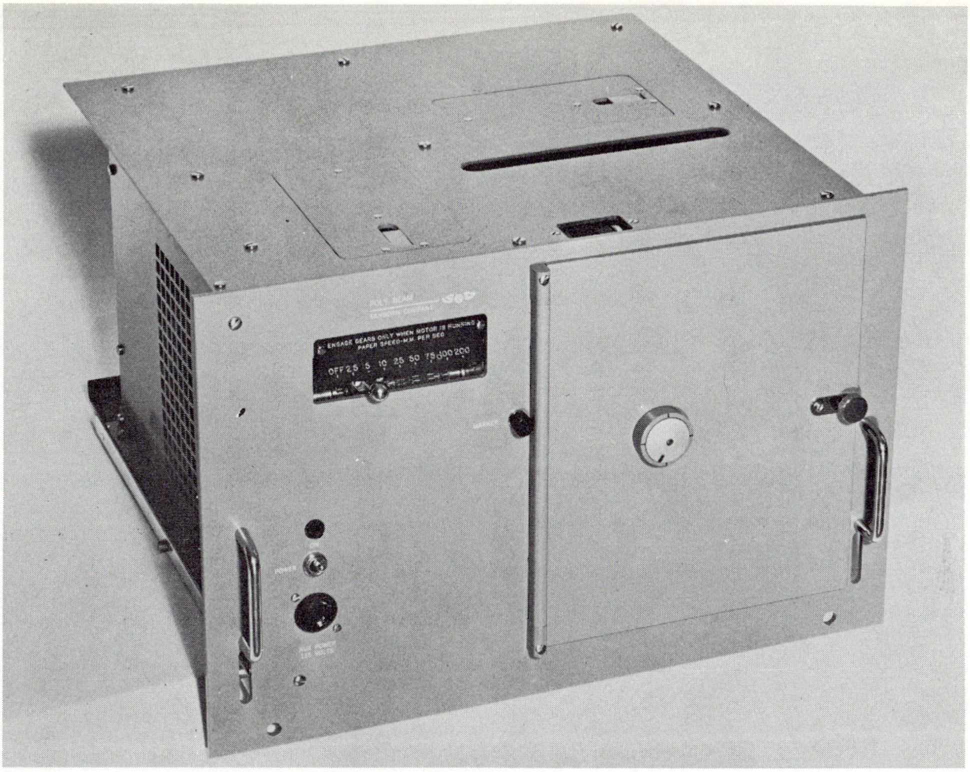

Fig. 10-24. The Poly Beam recorder model 568-100A. (Courtesy of Sanborn Co., Waltham, Mass.)

HIGH-FREQUENCY RECORDERS

Recording assemblies with a frequency response over 200 cps are generally optical systems. Until recently all such systems used visible light and photographic chart paper. An example of a commercial recorder of this type is the Poly Beam recording system manufactured by the Sanborn Company. The cartridge holding the photographic paper can be detached from the recorder and taken to a dark room for development. An attachment (model 563 rapid developer) may be used for automatic development with bromide paper in a one-solution system. It would appear that dark-room development is necessary for good quality records suitable for publication.

There are several advantages and disadvantages in using such an optical system. The obvious disadvantage is the requirement of dark-room development, which includes the need for using expensive photographic paper and the bother of using solutions and drying the chart. Advantages include very nearly rectilinear writing, because of the extremely long beam length (when compared to the length of an ink-writing pen), adjustment of line width, overlapping records, and time lines which go all the way across the chart paper.

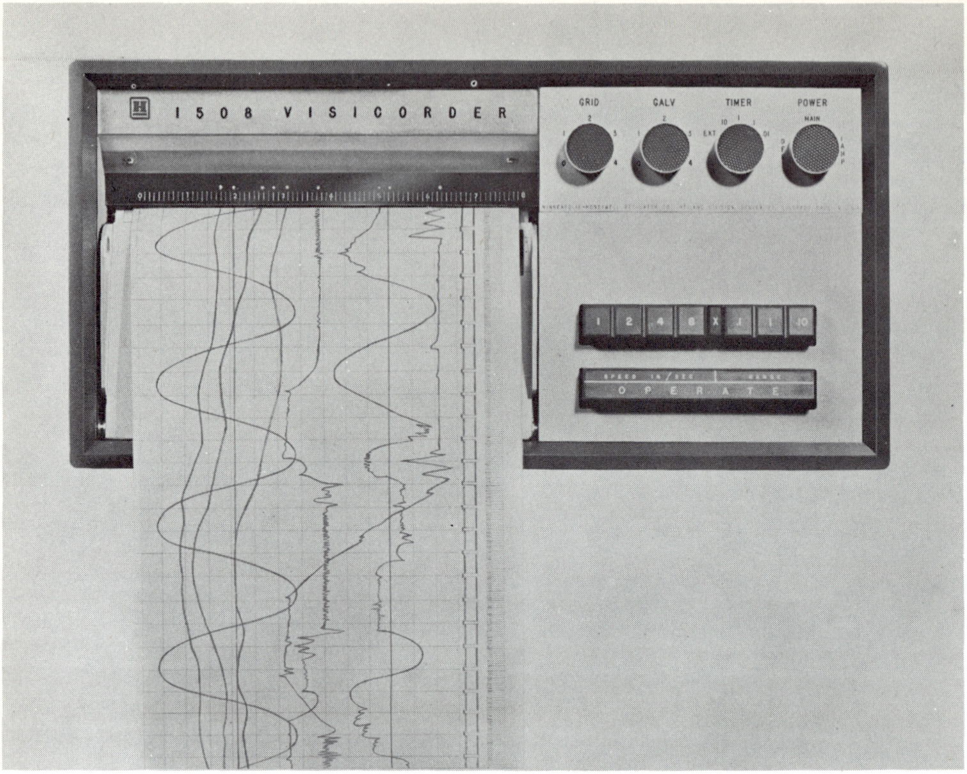

Figure 10-25. The Visicorder. (Courtesy of Minneapolis-Honeywell Regulator Co., Denver, Colo.)

The galvanometers used in the Sanborn Poly Beam system are similar to the Honeywell subminiature galvanometers shown in Figure 10-26. They have an undamped natural frequency of 500 cps which is flat, within $\pm$ 5 per cent, to 300 cps. This unit will accommodate up to eight channels. The Sanborn optical recorder may be used in any of the various Sanborn consoles in place of the unit shown in Figure 10-18 and is compatible with all their preamplifiers.

The Honeywell Visicorder, one model of which is shown in Figure 10-25, is an optical recorder which uses a mercury arc lamp as a source of ultraviolet light. The photographic paper used with this recorder is sensitive to ultraviolet light and when exposed will develop without chemicals in room light. Chemicals must be used, however, to stop the development or the record will not remain usable indefinitely. It will, nevertheless, last for a considerable time if kept out of the light. Development in room light is not instantaneous if the chart is moving past the lamp rapidly, although it may be speeded somewhat by exposure of the chart paper to fluorescent light as it comes out of the recorder. This recorder is provided with time lines which appear all the way across the record; a trace identification system may also be obtained as an accessory.

The galvanometer used with the Honeywell recorder is shown schematically in Figure 10-26(left). These galvanometers are completely sealed and sufficiently

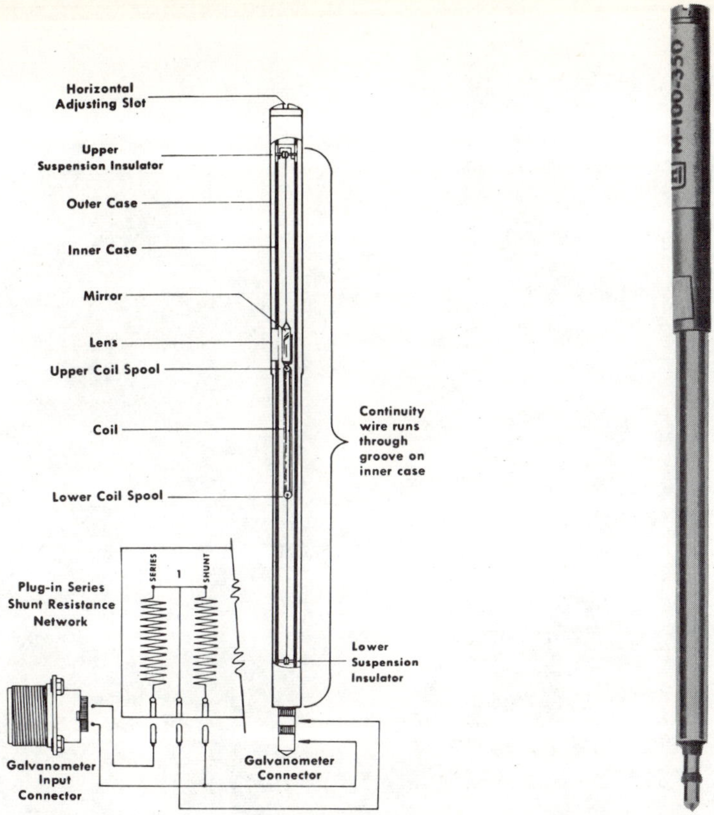

Figure 10-26. *Left,* A schematic diagram of the electromagnetic type of subminiature galvanometer; *right,* the Honeywell subminiature galvanominiature. (Courtesy of Minneapolis-Honeywell Regulator Co., Denver, Colo.)

sensitive to allow the primary signal from many transducers to be recorded directly. The magnet assembly used in the recorder is made to accommodate the tiny plug-in galvanometers without additional pole pieces. It is not necessary to have galvanometers on each side of the galvanometer being used as it is in some magnet assemblies. Small, easily removed, printed circuit boards are built into the magnet frame to allow an attenuating resistor, which is in series with the coil, and a shunting resistor, which is in parallel with the coil, to be selected. These resistors should be high quality metal-film resistors with $\pm$ 1 per cent accuracy. They are used for impedance matching and to adjust the sensitivity of the galvanometer to the job at hand. The galvanometers may be obtained with electromagnetic or fluid damping so that the frequency response is flat to as low as 24 cps or as high as 4800 cps. There are, however, very few biomedical applications which need extremely high frequency response. Although this recorder is very well constructed and excellent in design, it is much more difficult to use than a direct-writing instrument.

There are some applications for which it is convenient to display the output

Figure 10-27. Grass kymograph camera. (Courtesy of Grass Instrument Co., Quincy, Mass.)

of one or more instruments on an oscilloscope. In order to make a permanent record of the display, when it varies over a period greater than one sweep cycle, it is necessary to photograph the face of the oscilloscope. The kymograph camera manufactured by the Grass Instrument Company, shown in Figure 10-27, may be used for this purpose. In use the horizontal sweep of the scope is turned off. The result is a dot on the oscilloscope face which moves up and down with the applied signal. The Grass kymograph camera contains a 100-foot roll of 35-mm film. The optical system of the camera focuses the oscilloscope display on the film as the film is moved past a slit arrangement. The result, after dark-room development, is a continuous 35-mm film strip. The Grass kymograph camera will also take single pictures of the oscilloscope face.

Almost every oscilloscope manufacturer will supply a still camera for taking single pictures of the oscilloscope. These cameras use Polaroid film, and hence do not need a dark room for development. Several oscilloscopes are also available which will "remember" the picture displayed on the face of the scope. This is done by means of electronic circuitry similar to that employed in a computer memory. The oscilloscope display may then be erased at any time by pushing a button.

MAGNETIC TAPE RECORDERS

The use of magnetic tape to record biomedical data is perhaps the most recent recording innovation. There are many advantages which speak in favor of its ultimate acceptance as the recording medium of choice. Data may be stored on a tape in less space than can a graphic recording. Once it is recorded it can be played back any number of times without loss of the original information. It can be used for a number of different purposes. For example, it can be played back through an oscilloscope for quick observation. It can be used to prepare classroom demonstrations, or it can be fed directly into a digital computer for analysis. The tape can be played back through filters or modifying circuits such as integrators or differentiators which may then be used as the input for a computer. The tape can also be played into an analog computer, and if the equation is not correct, it can be changed and the data remain on the tape to be reanalyzed.

Figure 10-28. Magnetic tape recorder. (Courtesy of Precision Instrument Co., Palo Alto, Calif.)

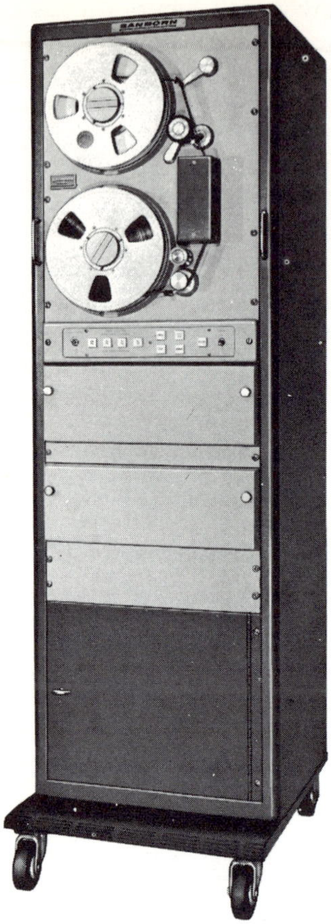

Figure 10-29. The Sanborn Magnetic Data Recorder. (Courtesy of Sanborn Co., Waltham, Mass.)

The magnetic tape recorder, such as the Precision Instrument Company's Model PI-6100 shown in Figure 10-28, can be used in conjunction with a high quality direct-writing recorder to increase the usable frequency response of the direct-writing recorder. The PI-6100 instrument records and plays back at three speeds, 37.5, 3.75, and 0.375 inches per second (ips). A 10,000-cps signal, for instance, which is far beyond the means of any medium-frequency and most high-frequency direct-writing recorders, can be recorded on magnetic tape at 37.5 ips and subsequently played back at 0.375 ips, thus producing a 100-cps signal which is within the capabilities of most pen recorders. Similarly a 1000-cps signal can be reduced to a 10-cps signal by recording at 37.5 ips and playing back at 0.375 ips. The tape recorder, in both cases, literally extends 1 second of data to 1 minute and 40 seconds. The Ampex SP-300 is a similar instrument. This instrument has the additional advantage that it can be obtained with either 1/2 inch or 1/4 inch magnetic tape.*

Another advantage of the magnetic tape recorder is that it can be used to record every detail of an experiment. Let us say that a complicated experiment is being run in which eight or ten instruments are being used to measure as many different

*One-half inch magnetic tape is more likely to be compatible with tape decks found on computers at the time of writing this book.

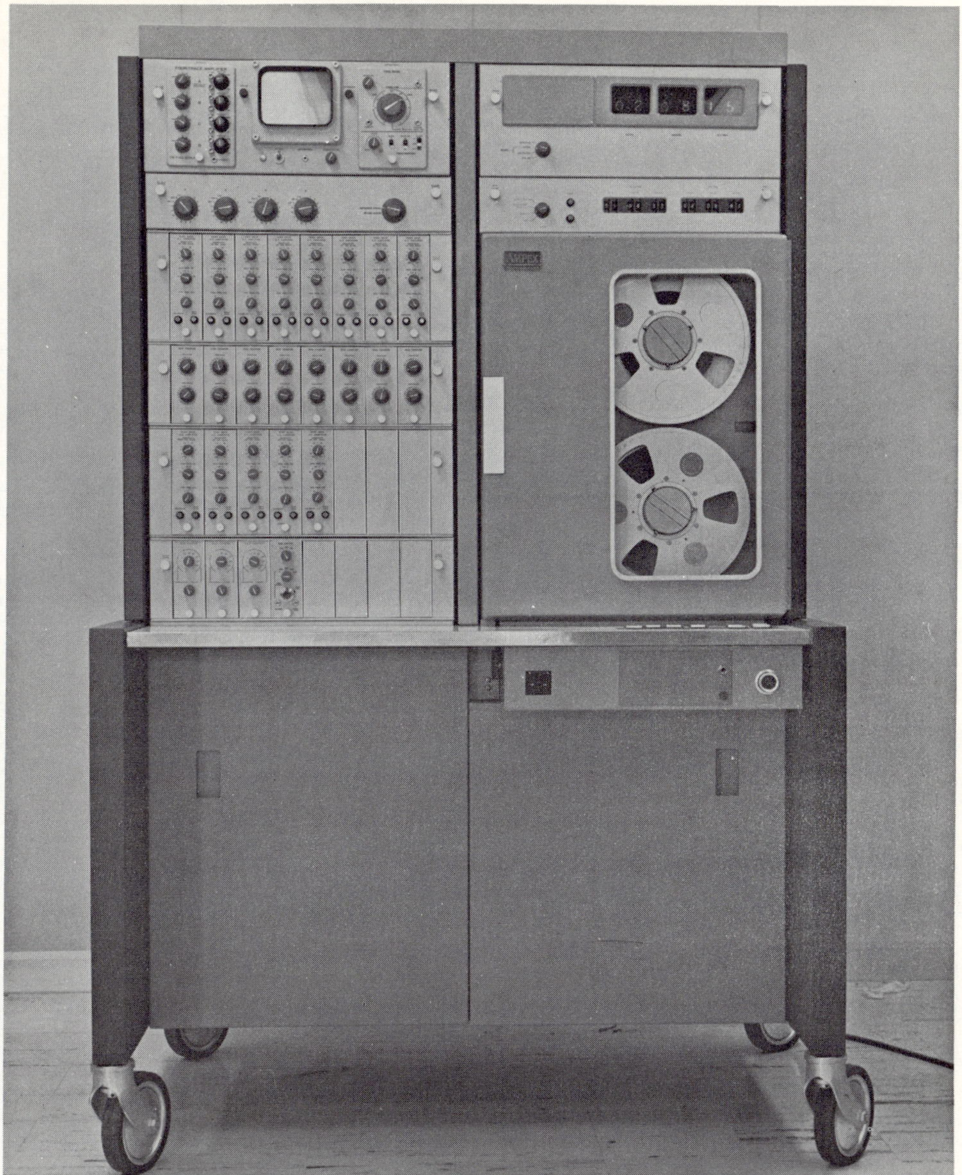

Figure 10-30A. The Ampex Data Acquisition System. (Courtesy of Ampex Corp., Redwood City, Calif.) See also Figure 10-30B on page 248.

parameters. An experiment of this sort requires a variety of investigators, assistants, and technicians (or physicians, surgeons, nurses, and technicians) each with his or her own job. The magnetic tape recorder, in addition to recording all of the measured information, can also be used as a voice monitor. In this way all of the conversation is recorded. This may include instrument setting, procedural changes, or any

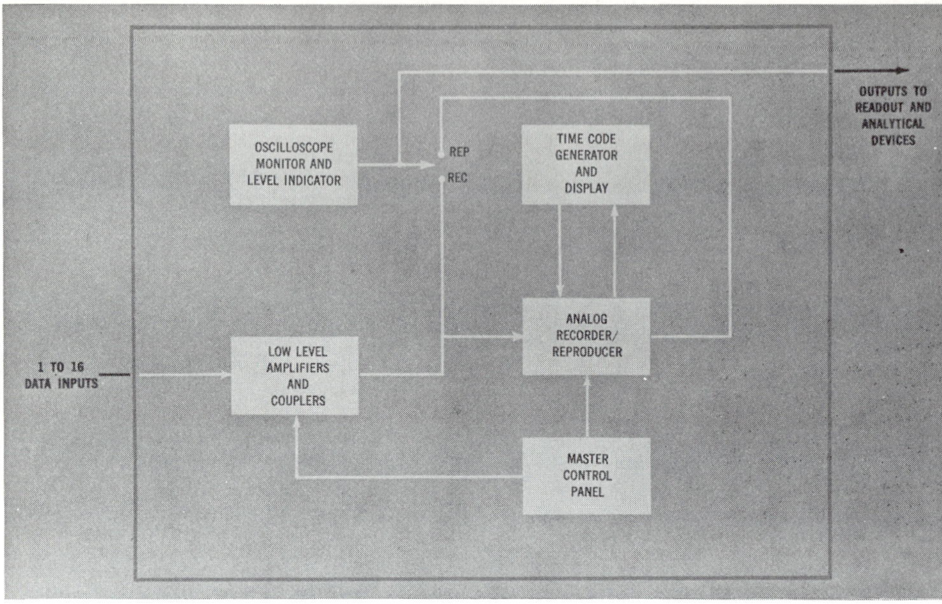

Figure 10-30B. DAS-100 data acquisition system functional diagram. (Courtesy of Ampex Corp., Redwood City, Calif.)

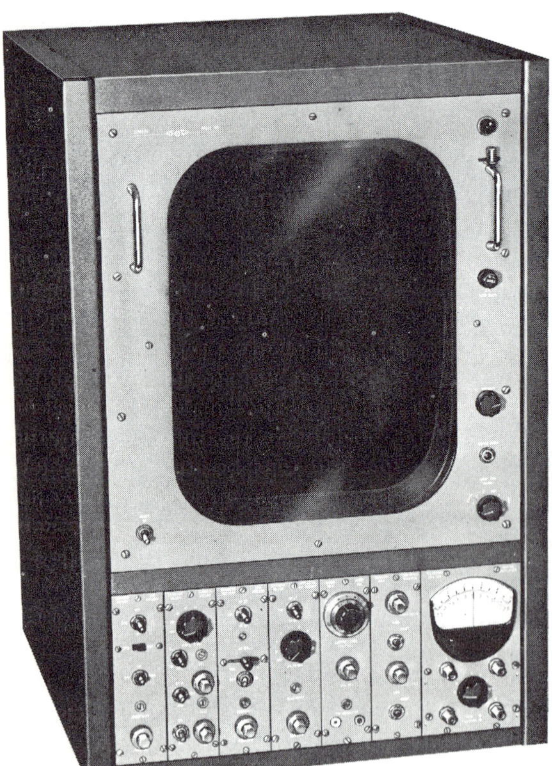

Figure 10-31. Monitoring system.

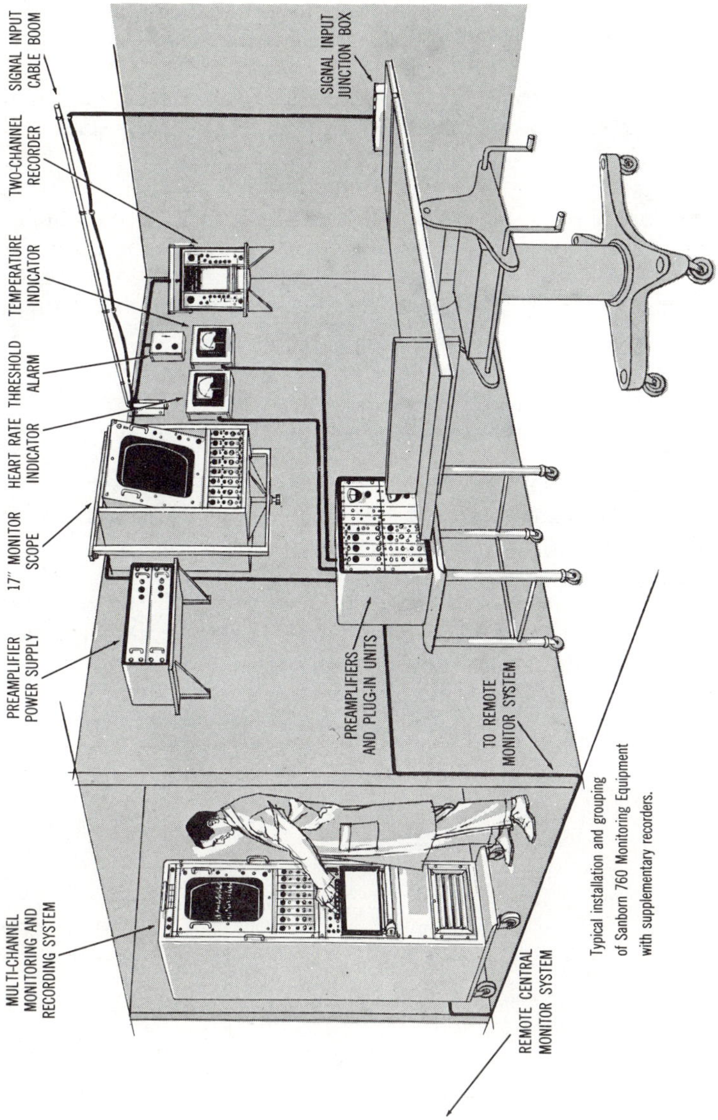

Figure 10-32. An operating room with electronic instrumentation. (Courtesy of Sanborn Co., Waltham Mass.)

number of small failures which may otherwise be missed. Later, when all of the excitement is past, the tape can be stored or erased, depending on the investigators' desire. It is, indeed, an odd sensation to hear an experiment being replayed and watch the data roll off a direct-writing recorder when there is no experiment in progress.

There are many magnetic data recorders suitable for use as biomedical instruments. Figure 10-29 is one of the instruments designed by the Sanborn Company. At the time of this writing the price of the tape recorder is the same as or slightly higher than data recorders of other types. Data recorders are available which are compact and portable, and others are available which are larger, such as the Data Acquisition System (DAS) manufactured by the Ampex Company (Fig. 10-30).

SUMMARY

This chapter has considered some of the read-out devices which may be obtained to enable the biomedical instrumentalist to utilize the output signal. It is hoped that the reader will have gained some insight into the workings of some of these instruments and will feel more at ease using them. A suitable oscilloscope, such as the one shown in Figure 10-31, may, of course, be used to monitor a number of output signals when it is not convenient to see the recorder or, as in the case of the magnetic tape recorder, when there is no visible information. It would be fruitless to attempt to list all of the various oscilloscopes that may be used for this purpose.

Many university and hospital research laboratories have such a tremendous need for biomedical instrumentation that rooms such as the surgical suite shown in Figure 10-32 are not uncommon. In a few years such rooms will be a necessary part of almost every clinical and basic science laboratory. Many such operating rooms, however, are designed so that all equipment except the transducers and electrodes are located in a separate monitor room. The surgeon maintains constant audio and visual communication with the monitor room and thus has continuous information about the condition of his patient.

11

Some Examples of Biomedical Instrumentation

bioelectric potentials

blood pressure

blood flow

In this chapter three large areas of biomedical instrumentation will be discussed: measurement of blood pressure, blood flow, and various bioelectric potentials, such as ECG and intracellular potentials. It is obviously impossible to include all available techniques or instruments. The techniques and instruments described in this chapter should therefore be considered examples of currently used instrumentation in each area.

BIOELECTRIC POTENTIALS

A bioelectric potential may be defined as the difference of potential or voltage developed across a cell membrane as a result of difference in the concentration of ions on each side of the membrane. The potential difference is generally about 90 to 120 millivolts with the inside being negative to the outside of the cell in most cases. The potential difference is maintained in part by the metabolic activity of the cell and in part by a combination of the *semipermeable* properties of the cell membrane and differences in the rate of diffusion of various ions. The term "semipermeable" means that not all ion species are free to diffuse through the membrane. The accumulation of different ion species on each side of a membrane results in an asymmetrical charge distribution. The cell membrane is thus referred to as being *polarized*, and the potential is called the *resting potential.*

When nerve and muscle cells are sufficiently depolarized in a local membrane segment the muscle or nerve cell is said to be *stimulated.* Stimulation may occur from either chemical or physical phenomena. The minimum amount of depolarization necessary to stimulate the muscle or nerve is called the *threshold.* When the nerve or muscle is adequately stimulated the local depolarization resulting from the stimulus propogates in both directions from the point of stimulation at a velocity which depends on the properties of the nerve or muscle cell (not the stimulus). This "moving depolarization" is called an *impulse* or *action potential.* Following depolarization the cell immediately does metabolic work to re-establish the resting potential. Although the impulse travels in both directions from the point of stimulation, it is rectified (propagation in one direction is canceled) when the impulse travels from one nerve or muscle cell to another. The connection between two nerve cells is called a *synapse.* The properties of the bioelectric potentials generated by nerve and muscle cells have been studied extensively and may be reviewed by consulting any of the references on physiology at the end of this chapter, especially the textbook by Ruch and Fulton.

There are several bioelectric potentials which are of special interest to the physician and the basic medical scientist. Among these are: The direct measurements of the resting potential across almost any membrane in which the nerve impulse (action potential) is important to the basic scientist investigating the genesis of the impulse and to the physician who uses the same data as a diagnostic tool. The study of the electrical activity of the muscle fiber is important and is called *electromyography.* The nature and conduction velocity of the impulse generated by the pacemaker of the heart are extremely important to the cardiologist. The study of the cardiac membrane potential is called *electrocardiography.* The study of the electrical activity of the brain is called *electroencephalography*, and of the electrical activity of the retina, *electroretinography.*

A variety of special electrodes have evolved for study of bioelectric phenomena. Some of these electrodes are shown in Figures 11-1 and 11-2; others will be

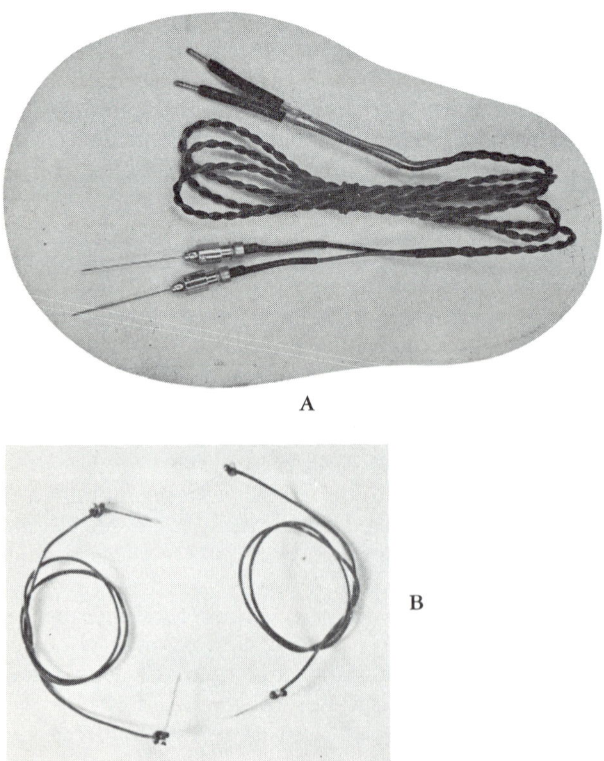

A

B

Figure 11-1. Small electrodes A, Needle electrode; B, pin electrode. (Courtesy of E & M Instrument Co., Inc., Houston, Tex.)

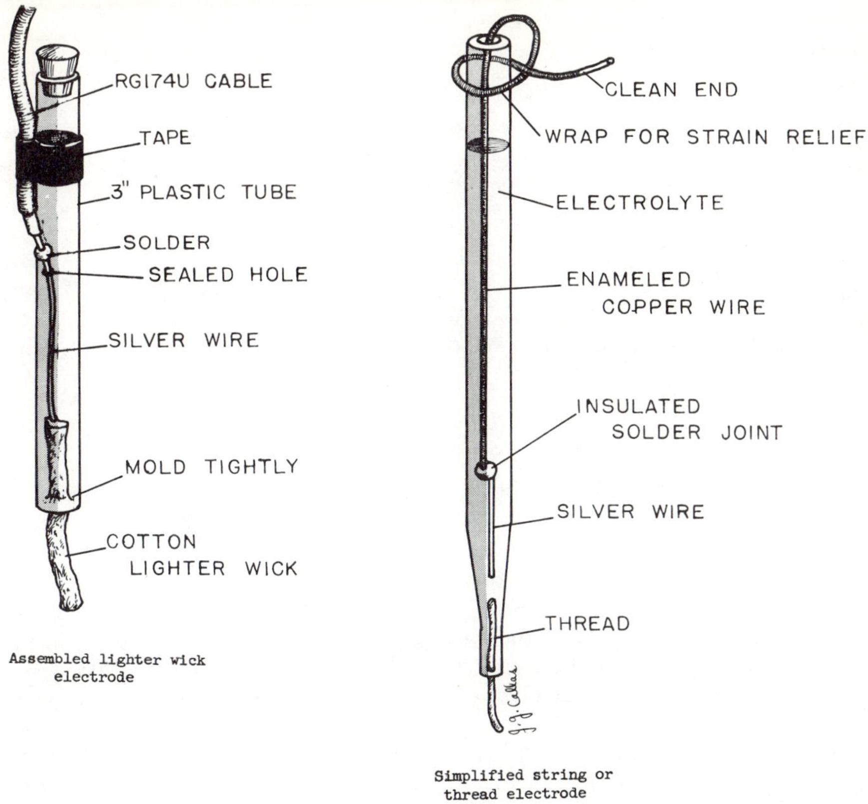

Figure 11-2. The wick electrode. (Courtesy of Project IMPS, University of California Extension, Berkeley, Calif.)

shown later in the chapter. In some cases the electrodes are applied to the surface of the skin; these may be metallic or wick-type, while others have very fine tips (about 0.5 microns in diameter) which can puncture the membrane without injuring it.

The amplifier used to make the measurement must have an extremely high input impedance (over 10^8 ohms, more often over 10^{10} ohms). The amplifier must draw a minimum of current from the source that is the site of measurement. The amplifier must be capable of rejecting a considerable amount of noise which is generated in the pickup leads because of the high-input impedance. It is often necessary to work in a shielded room in order to obtain usable signals. The "electrometer"-input stage has been used to insure high-input impedance, and, when it is, the inherent drift must be minimized. It is a tetrode vacuum-tube amplifier with positive feedback used to increase the input impedance of the tube. The electrometer is basically an unstable amplifier and tends to drift. Recently, completely transistorized amplifiers suitable for measuring biopotentials have been commercially introduced. Some of the best of these have a differential input and an input impedance of over 500 megohms (Zeltex, Inc., Concord, Calif., and Burr-Brown Research Corporation, Tucson, Ariz.).

It is often necessary to apply a controlled electrical stimulus to a muscle or nerve preparation in order to elicit an action potential. Figure 11-3A is a block

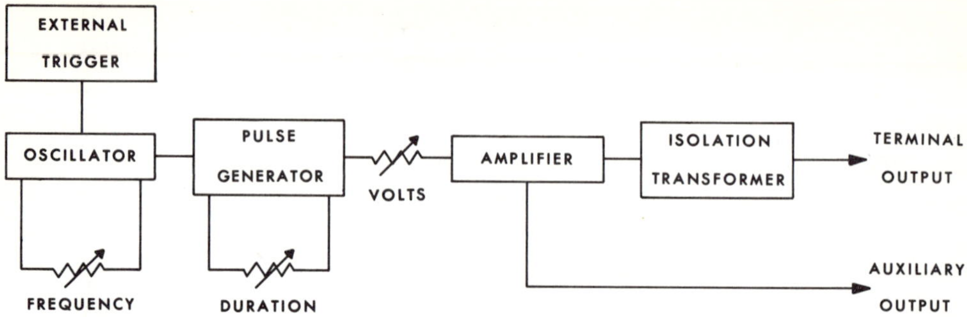

Figure 11-3A. Block diagram of a stimulator. (Courtesy of E & M Instrument Co., Inc., Houston, Tex.)

diagram and Figure 11-3(B) is a photograph of the *stimulator* sold by the E & M Instrument Company. Note that an oscillator is arranged so its frequency may be adjusted from 2 to 200 cps, or it may be triggered externally. The output of the oscillator is used to trigger a multivibrator, the duration of which may be controlled. This rectangular pulse is then amplified so that the output has the desired voltage and current. An isolation transformer is employed to avoid shocking the animal or test subject if a short circuit should occur and to help eliminate the noise resulting from the stimulator circuit which would affect the circuitry used to register the action potential. A more sophisticated stimulator manufactured by the Grass Instrument Company is shown in Figure 11-4.

An instrument called the IMPScope, which was developed under Project IMPS (Instrumentation Methods for Physiological Studies) at the University of California in Berkeley, combines a stimulator with an oscilloscope designed for the measurement of membrane potentials. The Heathkit IMPScope, as manufactured by the Heath Company, is shown in Figure 11-5. A block diagram of the instrument is shown in Figure 11-6, and Figure 11-7 illustrates the function of each part of the instrument.

The change of potential between the inside and the outside at a particular location on a nerve fiber with time results in the passage of the impulse along the nerve. This can be visualized by using the moist nerve chamber shown in Figure 11-8. A small amount of saline solution is placed on the bottom of the chamber and a nerve (*e.g.,* the sciatic nerve carefully removed from a pithed frog) is laid across the silver electrode wires. The first two at the left are connected to a stimulator such as the one shown in Figure 11-4. The next electrode to the right is usually reserved for grounding the nerve. Grounding the nerve may or may not be necessary; it may even increase the amount of noise picked up by completing a ground loop (see Chapter 12). Two of the other terminals are then connected to a suitable push-pull amplifier such as the IMPScope. The *action potential* or impulse may then be visualized following adequate stimulation. If two of the additional terminals are connected to another amplifier and displayed along with the first impulse, the *conduction velocity* of the impulse may be determined (see Figure 11-17).

One of the most common electronic measurements used by the modern physician is the electrocardiogram (ECG or EKG). The bioelectric potential developed by some specialized heart muscle fibers in the right atrium, called the *sinoatrial*

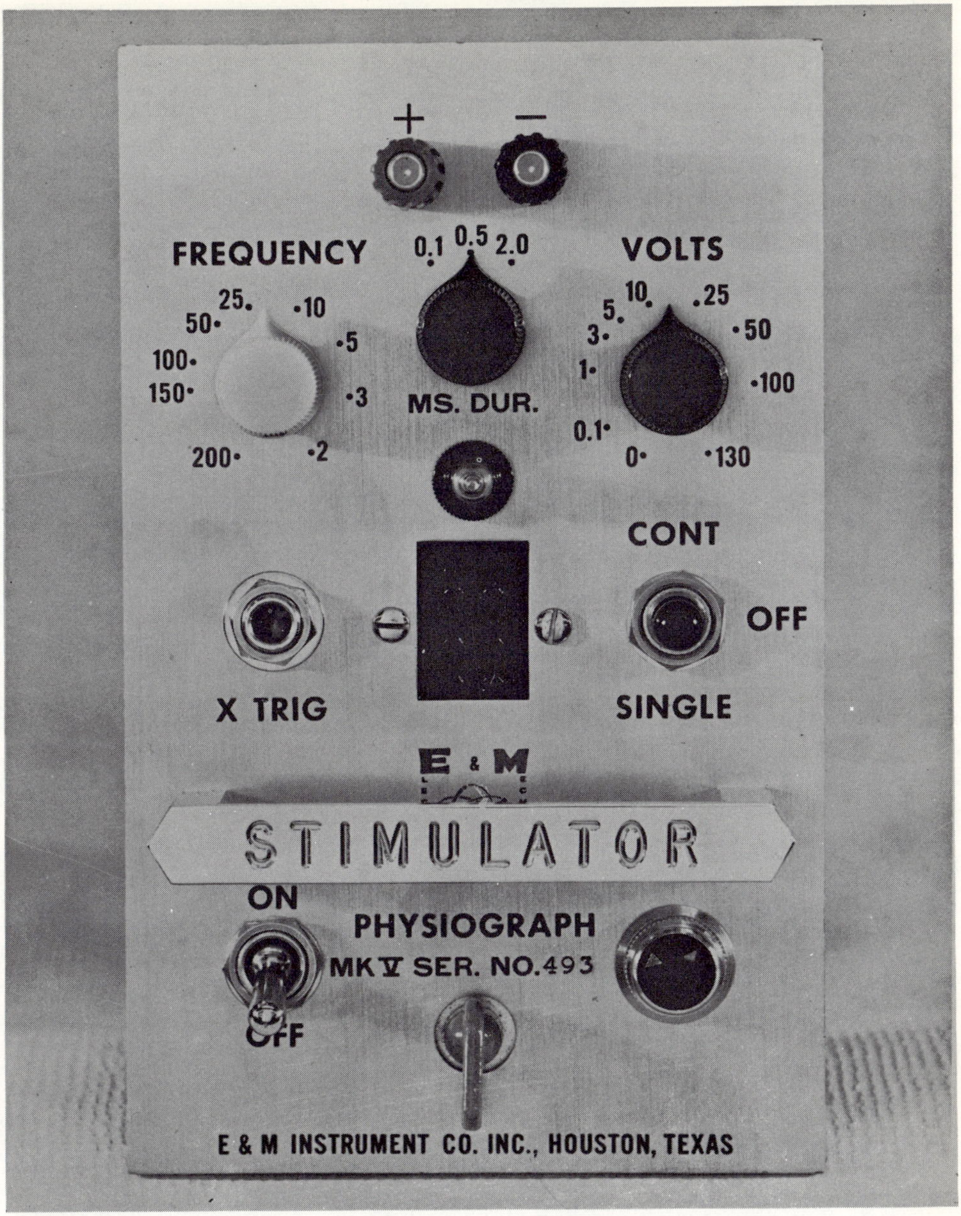

Figure 11-3B. Front panel of the Physiograph stimulator. (Courtesy of E & M Instrument Co., Inc., Houston, Tex.)

node (S-A node), spreads through the atria of the heart to the *atrioventricular node (A-V node)*, from whence it spreads through the ventricles of the heart. As this impulse travels through the heart it causes the heart muscle to contract. Therefore, the number of such impulses per second determines the rate of contraction of the

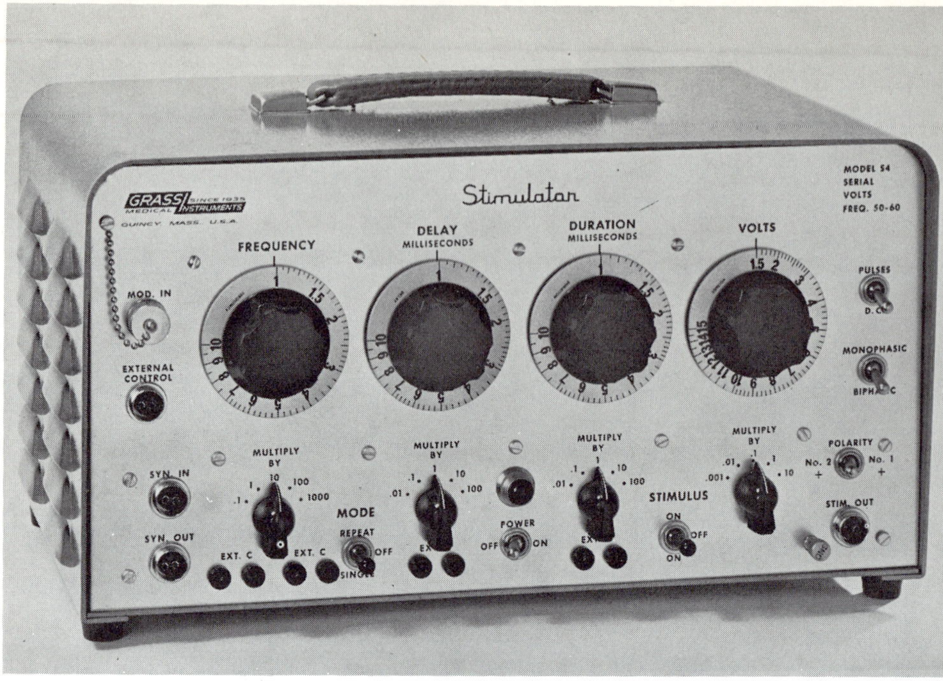

Figure 11-4. A physiological stimulator. (Courtesy of Grass Instrument Co., Quincy, Mass.)

Figure 11-5. The IMPScope (Courtesy of Project IMPS, University of California Extension, Berkeley, Calif.)

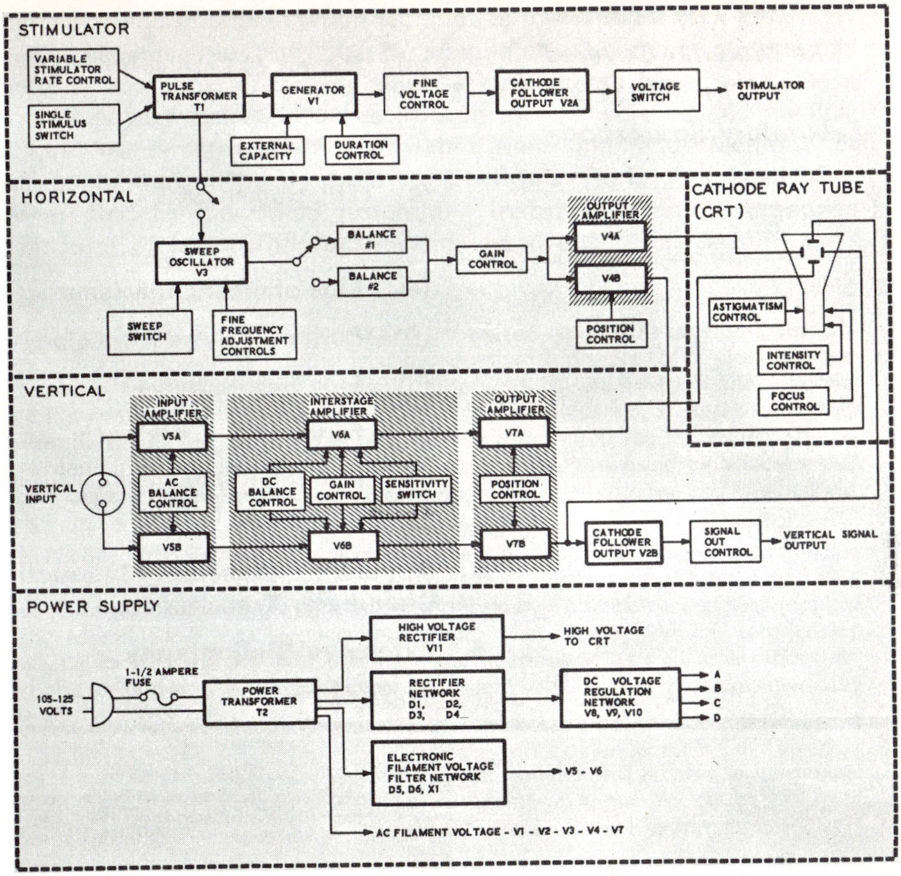

Figure 11-6. Block diagram of the Heathkit IMPScope. (Courtesy of Heath Co., Benton Harbor, Mich., and Norman N. Goldstein, University of California Extension, Berkeley, Calif.)

heart. For this reason, the S-A node (where the stimulus normally originates) is called the cardiac pacemaker.

The heart may thus be represented as a bar with definite polarity (called a dipole) located in a large bag of salt water (the animal's chest). The impulse generated by the pacemaker is relatively large (the order of magnitude of millivolts) and can therefore be detected at the surface of the skin. The position of the electrodes on the surface of the skin and the relative position of the dipole within the chest determine the shape of the potential measured on the surface of the skin. Several standard positions for the electrodes have evolved and are shown in Figure 11-9. The plate electrodes commonly used are shown in Figure 11-10.

A newer, more efficient electrode utilizes a small stainless steel wire cloth mounted in a semi-rigid cup. The wire cloth is held at a fixed distance from the skin. Standard electrode paste is used as a coupling agent. This electrode, shown in Figure 11-11, decreases artifacts due to movement.

Almost all electrodes used to measure bioelectric potentials tend to become polarized at the tissue-electrode interface. That is, as current flows through the

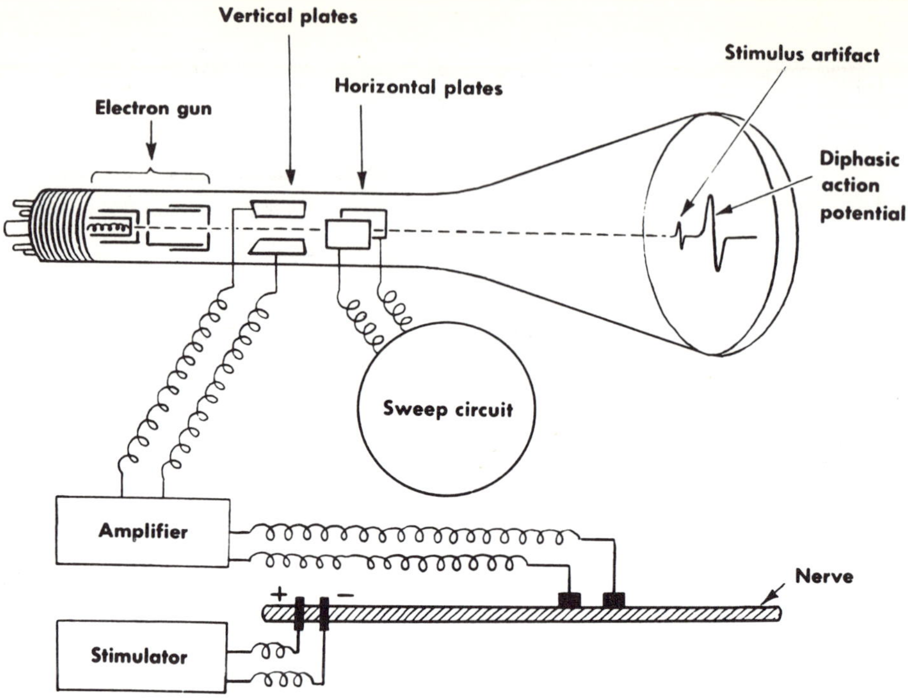

Figure 11-7. Functional diagram of an oscilloscope set-up to demonstrate nerve action potentials. (Courtesy of W. W. Tuttle and B. A. Schottelius, Textbook of Physiology, 14th ed., C. V. Mosby Co.)

electrode the number of ions flowing one way is greater than the number of ions flowing in the other direction. The result is that there is a DC potential or DC offset voltage which tends to mask the signal. Polarization may be thought of as an additional impedance which opposes the useful signal.

Polarization of the electrode can be avoided if a proper *electrode system* is used. When a metal is used as the electrode, some of the ions of the metal go into solution at the junction of the metal and the tissue. These ions tend to diffuse away and change the interface impedance, thus causing *polarization.* This phenomenon accounts for movement artifacts as well as providing an explanation for the DC offset voltage.

One electrode system which utilizes a silver-silver chloride pellet has recently been developed by the Beckman Instrument Company. The silver electrode is surrounded by a layer of silver chloride which acts as a stable source of silver ions. The silver chloride is especially suitable as it is relatively insoluble in aqueous solution, and in addition there is a high concentration of chloride ion in the tissue itself. The Beckman electrode is shown in Figure 11-12.

The electronic circuitry used to record the electrocardiogram is shown in block form in Figure 11-13. A differential or push-pull amplifier is generally used to minimize the noise pickup which is due to the high-input impedance. The gain is usually arranged so that the final read-out is 1 millivolt per cm of amplitude. A

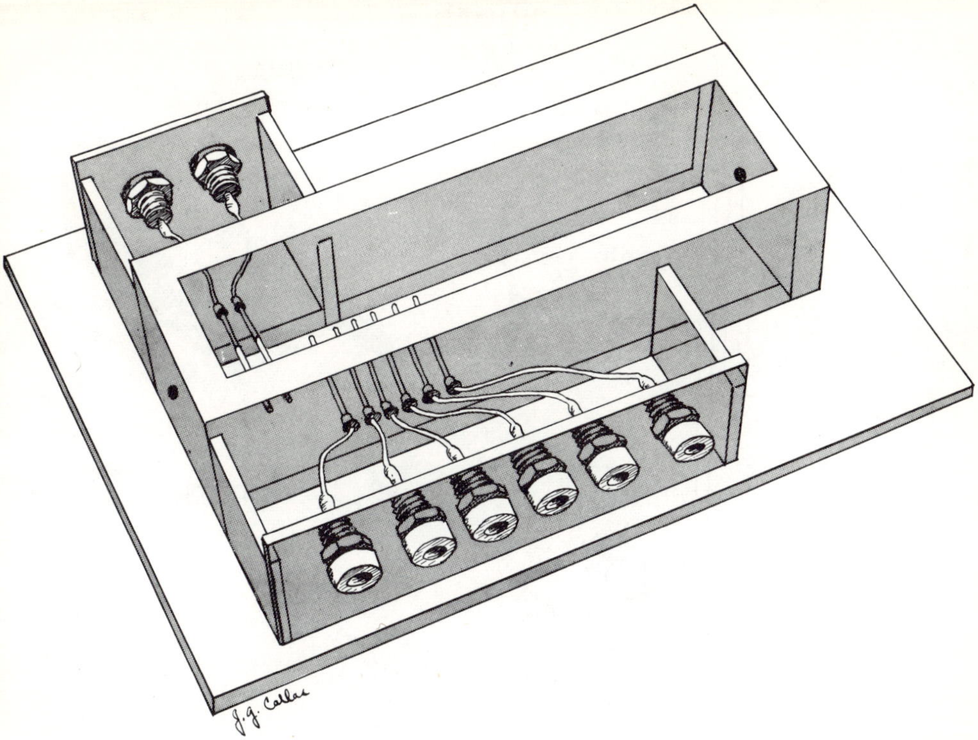

Figure 11-8. Perspective view of the moist nerve chamber (cover not shown). (Courtesy of Project IMPS, University of California Extension, Berkeley, Calif.)

rather large RC time constant is used (about 2 seconds). A great many electrocardiographs are commercially available.

The measurement of potentials on the surface of the scalp is called *electroencephalography.* These electroencephalogram or EEG potentials are generated in the brain and are related to the activity of the nerve cells which constitute the brain. An example of an EEG record is shown in Figure 11-14. These potentials are generally about 15 microvolts and from almost DC to 60 or 70 cps. Normal EEG signals are from 15 to 100 microvolts and from 0.5 to 30 cps.

Recording an EEG signal is an especially difficult task. The input impedance of the amplifier must be high. Therefore, small currents which may be induced in the "input circuit," including cables and electrodes, generate much larger voltages ($V = IR$) if the input resistance is very high than would be the case if the input resistance were lower. Here is an example where a high input impedance circuit must be used to measure microvolt signals. In order to minimize this effect it is necessary to carefully shield the electrode leads. The patient or subject tends to act like an antenna and must also be carefully grounded. It is often necessary to put the patient and the equipment in a specially constructed and properly grounded shielded room. One commercially available electroencephalograph is shown in Figure 11-15.

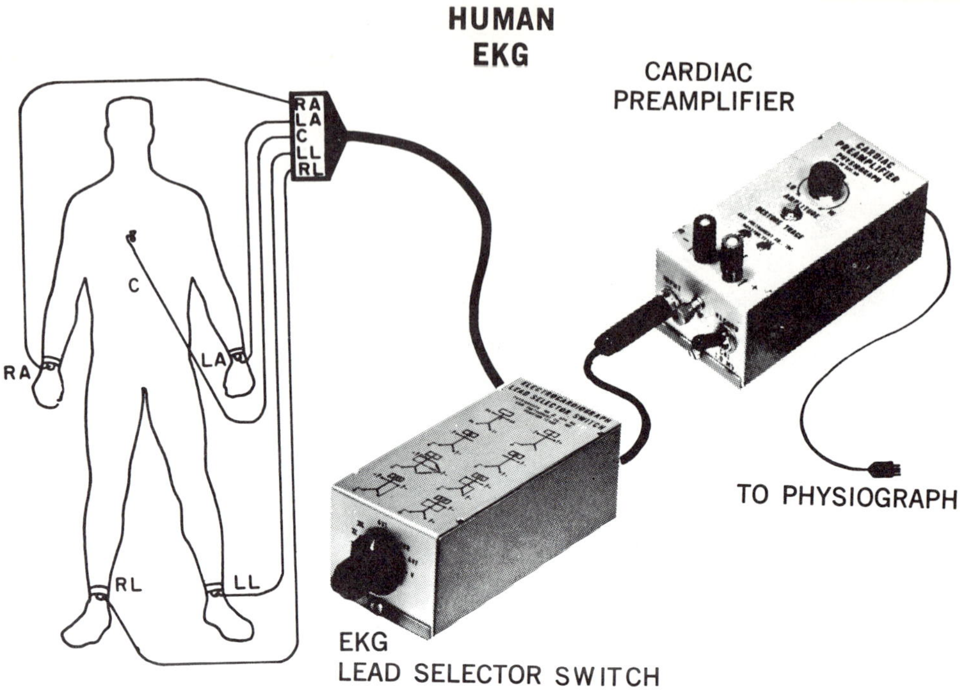

Figure 11-9. Measuring the ECG. (Courtesy of E & M Instrument Co., Inc., Houston, Tex.)

Figure 11-10. The ECG plate electrodes. (Courtesy of E & M Instrument Co., Inc., Houston, Tex.)

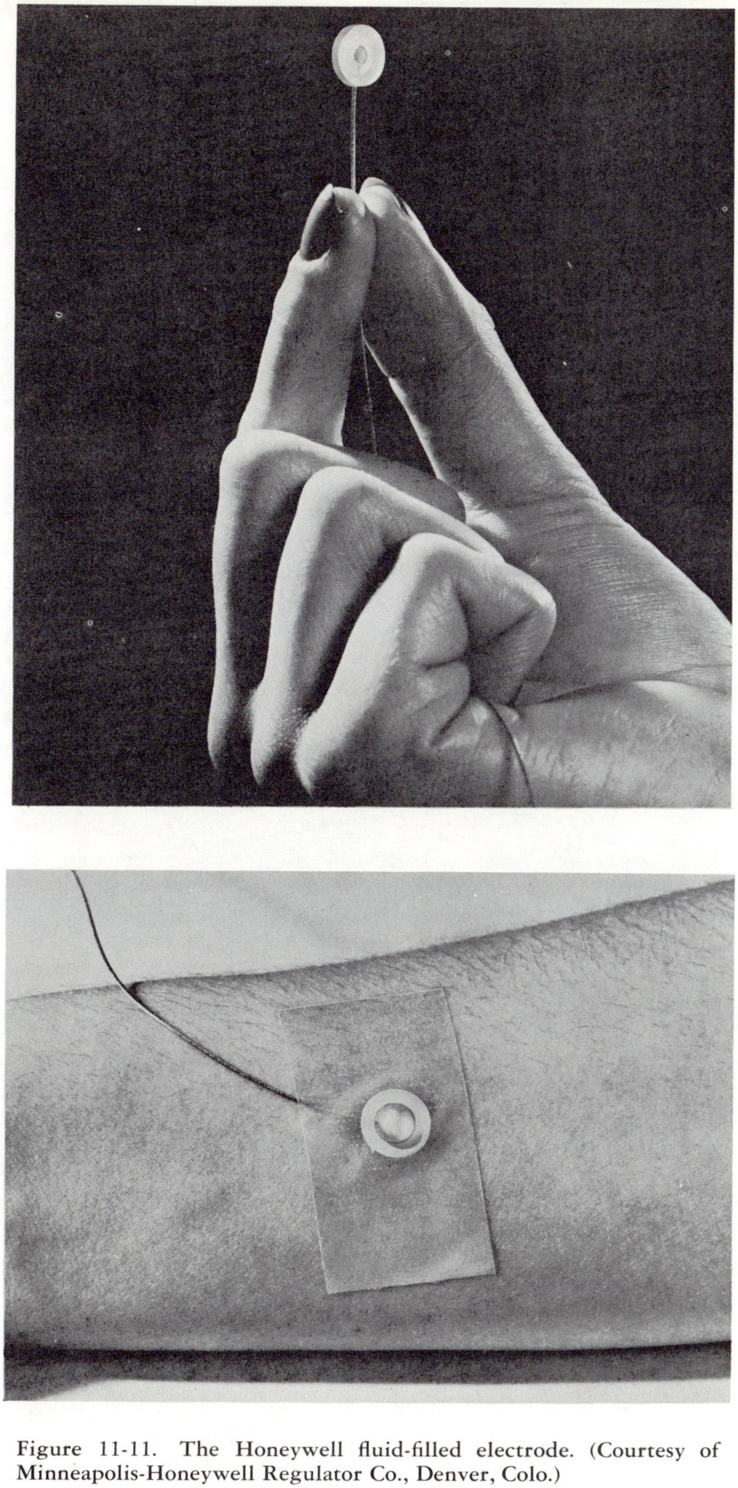

Figure 11-11. The Honeywell fluid-filled electrode. (Courtesy of Minneapolis-Honeywell Regulator Co., Denver, Colo.)

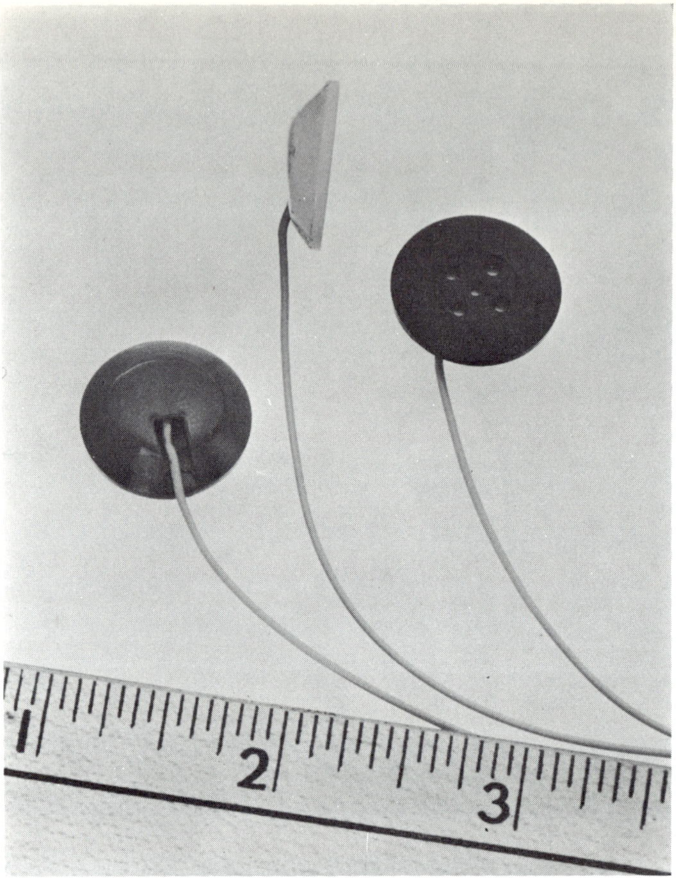

Figure 11-12. The Beckman silver-silver chloride surface electrode. (Courtesy of Beckman Instruments, Inc., Palo Alto, Calif.)

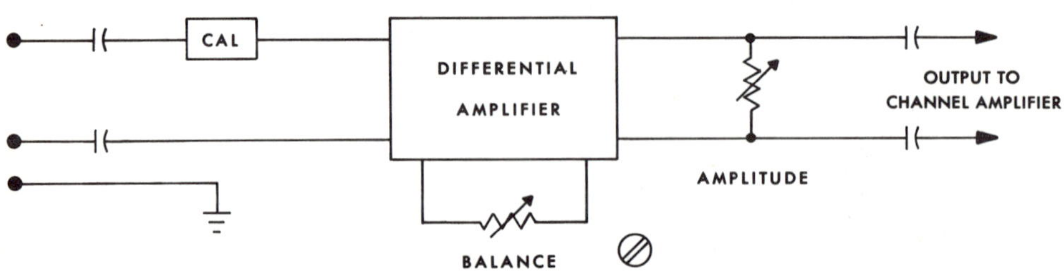

Figure 11-13. A block diagram of an electrocardiograph. (Courtesy of E & M Instrument Co., Inc., Houston, Tex.)

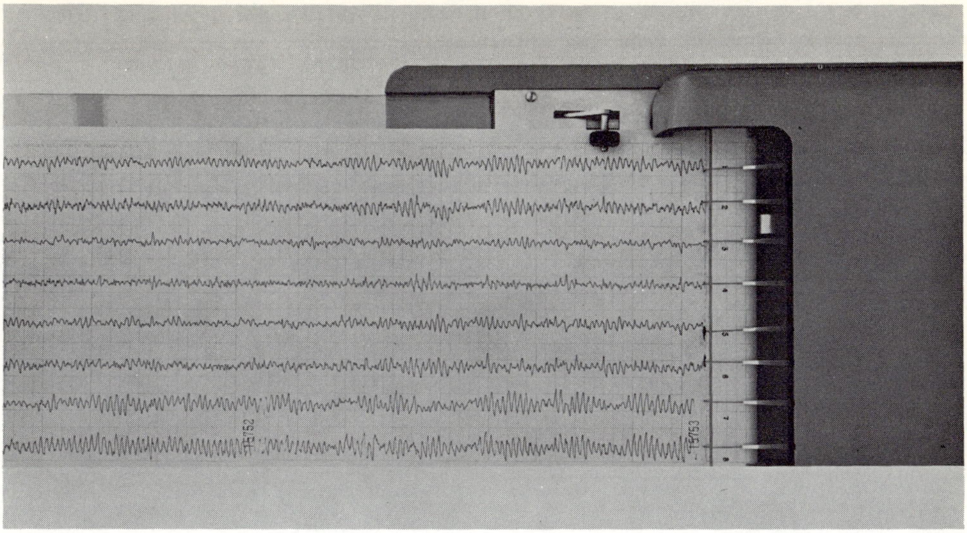

Figure 11-14. An example of an EEG. (Courtesy of Beckman Instruments, Inc., Palo Alto, Calif.)

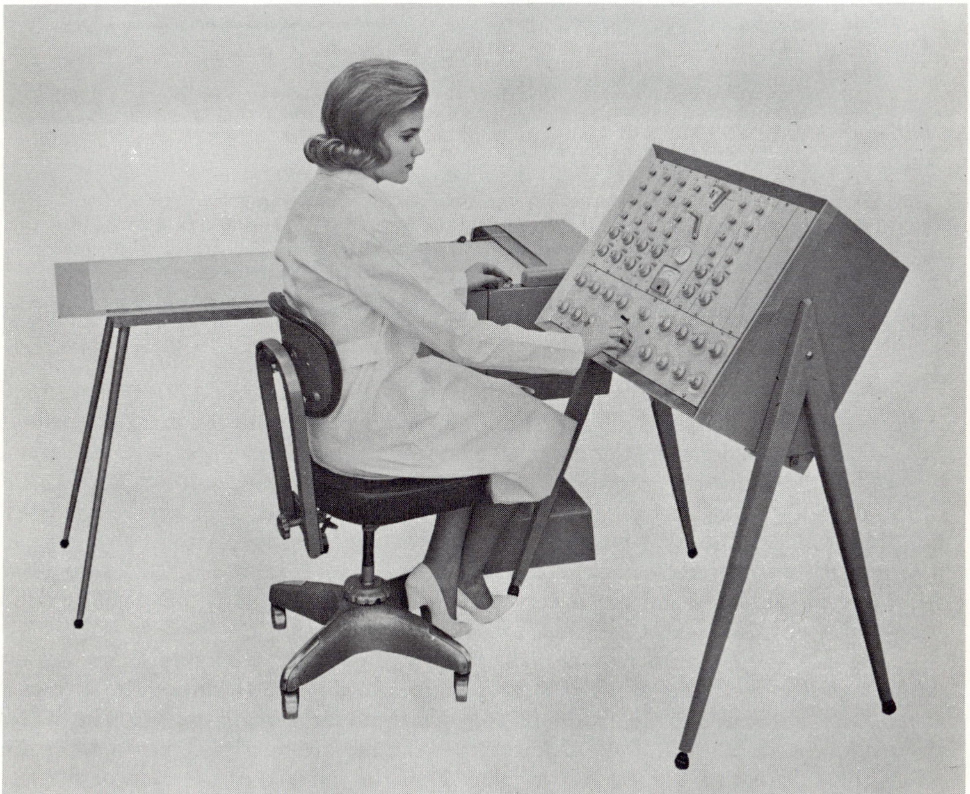

Figure 11-15. The Beckman/Offner type TC electroencephalograph (Courtesy of Beckman Instruments, Inc., Chicago.)

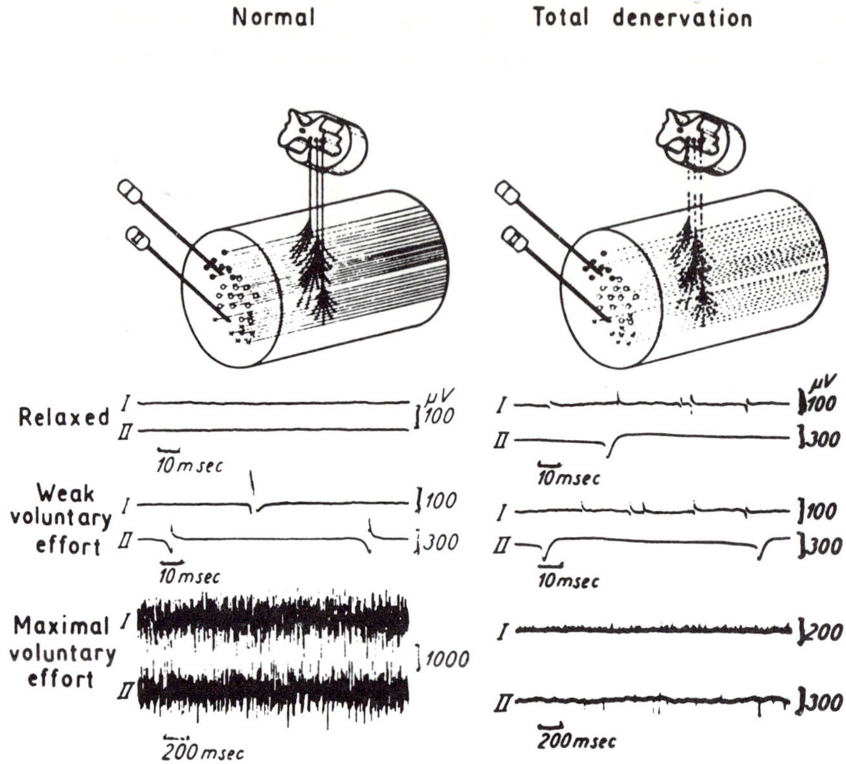

Figure 11-16. Electromyogram from normal and denervated muscle with a schematic of the muscle fibers from three motor units (marked •, 0, and X). Two recording electrodes I and II were used, and the records marked I and II correspond to these electrodes. Up is negative. (Courtesy of Fritz Buchtal, An Introduction to Electromyography, Gyldendalske Boghandel.)

The impulse or action potential which travels along a muscle fiber is essentially the same as the impulse which travels along the nerve fiber. In the muscle, however, the action potential causes the muscle to contract. The muscle action potential may be recorded by needle electrodes inserted directly into the muscle. A recording of the muscle potential is called an electromyogram or EMG. Figure 11-16 shows a bundle of muscle fibers which is divided into three groups or motor units. Each motor unit may be traced back to a single nerve fiber coming from the spinal cord shown at the top. The normal EMG is shown on the left, and the EMG which is recorded following total denervation is shown on the right.

The electromyogram may also be used to determine the conduction velocity of an impulse traveling along a nerve. Figure 11-17 shows three EMG tracings taken with the needle electrode in the right abductor digiti quinti. The needle electrode used as a ground was inserted into the muscle nearby. Tracing A resulted when the stimulating electrode was placed on the surface of the skin of the right arm near the ulnar nerve; B resulted when the stimulator was placed in the vicinity of the elbow, and C resulted when the electrodes were over the ulnar nerve in the wrist. The distance separating positions A and B is 0.15 meters and from B to C is

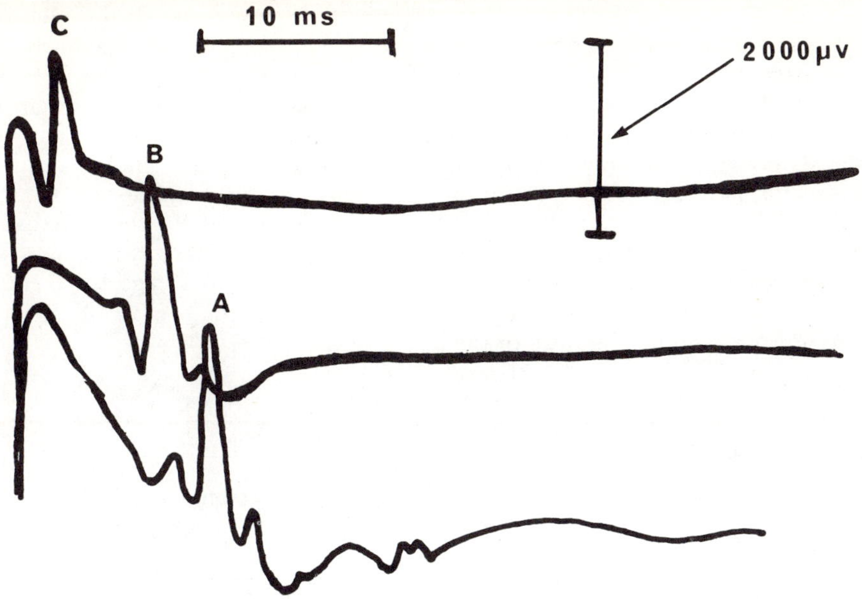

Figure 11-17. Measuring nerve conduction velocity in the right ulnar nerve. (Courtesy of Harry T. Zankel, Columbia, S.C.; modified for clarity by the author.)

0.25 meters. It may be seen from Figure 11-17 that the time difference for the peak of the tracings is 0.003 seconds from A to B and 0.0045 seconds from B to C. The impulse velocity or nerve-conduction velocity from A to B is therefore $0.15 \div 0.003$ or 50 meters/second and from B to C is 55.5 meters/second. The difference in impulse velocity may result from the fact that the nerve is somewhat smaller in diameter near the wrist than it is in the upper arm. The conduction velocity is related to the diameter of the nerve fiber. The difference in conduction velocity may also very well be an error in the reading of the record.

BLOOD PRESSURE

The circulatory system of an animal is a closed system of elastic tubes. The blood is forced to circulate by contractions of the heart muscle which develops a mean driving pressure of about 100 mm Hg above atmospheric pressure in man. The peak pressure is called the *systolic pressure*, while the lowest pressure measured at the same point is referred to as the *diastolic pressure.* The difference between the systolic and the diastolic pressures is called the *pulse* or *pulsatile pressure.* The systolic, the diastolic, and the pulse pressure will vary, depending on the location within the animal where the measurement is made and whether the animal is standing up or lying down, *etc.* In man, the heart beats approximately once every second, whereas in other animals the heart rate is faster or slower. The biological pressure pulse, which results from the heart beat, does not vary sinusoidally with time. In order to appreciate the true shape of the pressure wave, it is necessary to record faithfully not only the fundamental frequency, which is the heart rate, but also as many as five harmonics. The frequency of the second harmonic is twice that of the first; the third is three times that of the first harmonic, and the fifth harmonic is five times that of the fundamental frequency. If one measures two pressures along a particular vessel and plots the difference between the two pressures per unit length, this is called the *pressure gradient.* Figure 11-18 shows a harmonic analysis of a single pressure-gradient pulse. In this only four harmonics have been shown. The conclusion here is that if the heart rate is 180 beats/minute or 3 beats/second, it is necessary that the pressure manometer, amplifier circuitry, and read-out device have an over-all frequency response of at least 15 cps. The response should be unattenuated or *flat* to 15 cps. If the amplitude of the response has increased (or decreased) not more than 5 per cent from 0 to 15 cps, the frequency response is said to be *flat* to 15 cps. Pressure measured at the root of the aorta may require more harmonics because of the shape of the pulse.

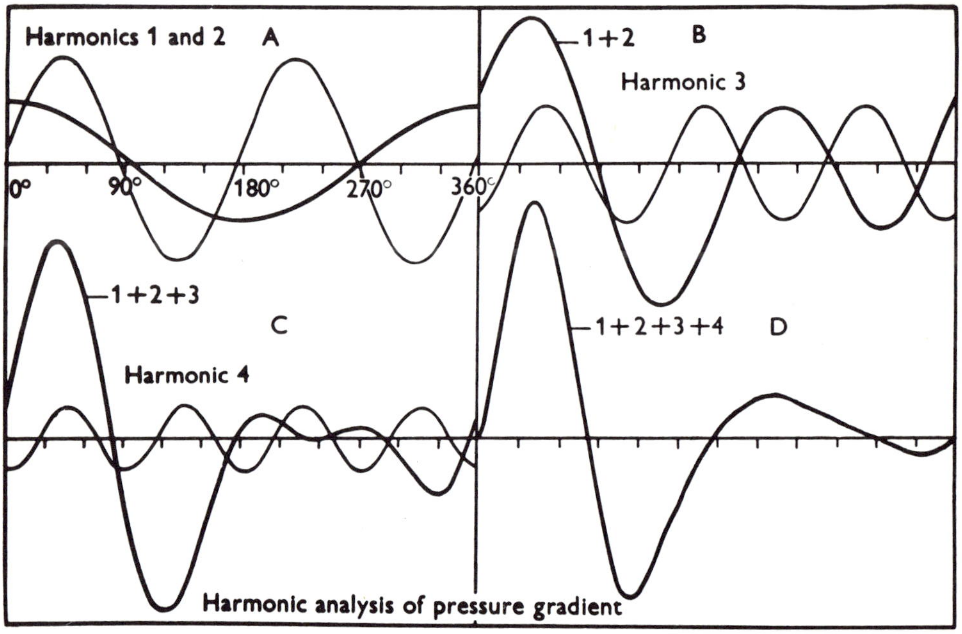

Figure 11-18. Harmonic analysis of a pressure gradient pulse. (Courtesy of D. A. McDonald, Blood Flow in Arteries, Edward Arnold Ltd.)

Pressure is defined physically as the force exerted per unit area. It is measured by the displacement of a constrained membrane, usually covering a fluid-filled chamber, to which is attached a device which can transduce the movement or bending of the membrane to a usable signal. Prior to the use of pressure gauges based on electrical or electronic principles, blood pressure was measured with a water- or mercury-filled U-tube manometer. A float was connected to a pen arrangement. In this way, the changes in pressure were represented directly as changes in the height of the pen. In a later development, shown in Figure 11-19(B), the fluid was confined to a chamber which was covered with a tightly stretched membrane (usually rubber). The pen was attached to the membrane so that an increase in the volume of the chamber caused the pen to move. It was also found that a mirror cemented to the membrane could be used to deflect a beam of light. The transducer illustrated in Figure 11-19(A) is constructed by blowing a large glass bubble onto the end of a glass tube. When the bubble is large enough, a slight negative pressure is applied to make a spoon-shaped transducer. The mirror is applied as shown. This transducer would be expected to have a higher frequency response than one in the rubber membrane. All pressure transducers have parts equivalent to those shown in Figure 11-19. In general, as indicated in Chapter 9, the measurement of pressure is merely a special case of the measurement of displacement. Even though present commercial pressure transducers may be far superior to the transducers described above, a great deal of useful physiological information can, nevertheless, be obtained by using these manometers.

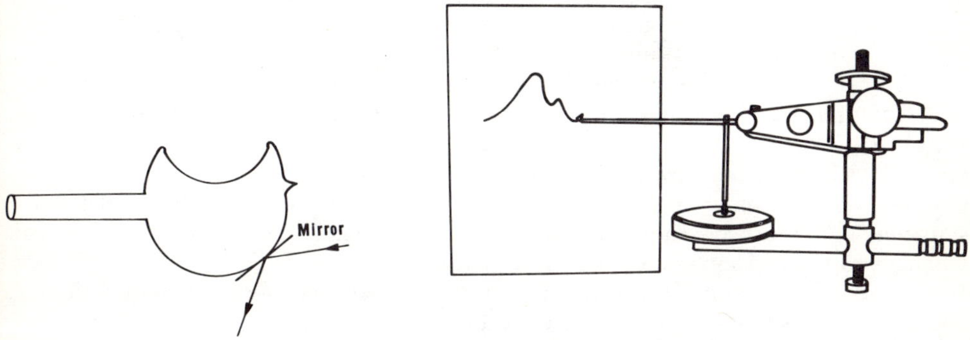

A B

Figure 11-19. Two varieties of mechanical manometers: A, glass bubble manometer; B, stretched rubber membrane.

Each manometer chamber contains a particular volume of fluid, and the movement of the membrane due to pressure fluctuations may be expressed in terms of a change in this volume. The sensitivity of the gauge is proportional to that change in volume per unit pressure change. This is called the *volume displacement* and is a characteristic of the gauge. The best of the Statham pressure transducers has a volume displacement of 0.01 mm^3 per 100 mm Hg pressure. The larger the amount of pressure necessary to distort or bend the membrane, the *stiffer* the membrane and the higher the frequency response.

Since the measurement of a pressure is in reality the measurement of a small displacement, it is not surprising that any principle that can be applied to the design of a displacement transducer can be applied to the design of a pressure transducer. A short list of physical principles that have been used includes changes in chamber volume, resistance, capacitance, mutual inductance, etc.

A transducer in which the pen is attached directly to a rubber membrane is shown in Figure 11-19(B). The volume displacement is rather large, and hence the sensitivity and the frequency response are low. The frequency response is, however, slightly better than the U-tube manometer filled with mercury where the pen floats on the surface of the mercury. The advantage of this type of manometer is that it is inexpensive to build and does not need electronic equipment.

The frequency response of the system described above could be substantially increased if the weight of the pen were to be decreased to almost nothing. This can be done if a tiny mirror is attached to the membrane. A small movement of the mirror deflects a light beam, and the deflections are recorded photographically. Since a very minute movement of the mirror can produce a readable deflection, it is possible to reduce the volume of the chamber and to use a much stiffer membrane. An example of this *optical manometer* has already been shown in Figure 11-19(A). The Bourdon tube manometer illustrated in Figure 9-12 is another example of this type of manometer. Recently manometers have been suggested in which a very tiny manometer is built onto the end of a glass fiber bundle. In this way, fiber optics can be used to transmit the light beam to and from the mirror and the entire transducer can be implanted in the animal. This type of pressure transducer is capable of extremely high sensitivity and high frequency response if it is properly designed. The use of extremely long light beams (2 to 6 meters long) and photographic registration is being replaced with photocells that detect changes in the amount of light flux.

A very small change in the separation of two plates with a dielectric material between them, as diagramed in Figure 9-20(top), forming a capacitor will change the capacitance of the unit. This change in capacitance can be detected with electronic circuitry. *Capacitance manometers* are extremely sensitive, and their frequency response is extremely high. This type of manometer is, however, temperature sensitive and must be thermally stabilized immediately before and during use. Changes in the capacitance of the cables with movement also add to the instability. It is more difficult to use this type of manometer than the strain-gauge manometer, and for that reason it is not being used as often as it was in the past.

There are several types of manometers in current use that may be classified as *inductance manometers.* In all of these, the distortion of the membrane causes a change in the inductance of a coil which results in a change in the voltage across the coil. Review of equation 1.19 reveals that a change in inductance may result from an increase or decrease in the number of turns (inductance varies with the square of the number of turns), the magnetic permeability of the core, and the dimensions of the coil. In some cases, one primary and one secondary coil are used, and when the spring-loaded core which couples the primary and the secondary coils is moved back and forth, the voltage induced across the secondary changes in proportion to the pressure applied. An example of this scheme is the Dallon-Telco catheter-tip manometer shown in Figure 11-20. It will be seen later that it is a distinct advantage to have the manometer at the source of the pressure measurement.

The differential transformer, as shown in Figure 9-15, has also been used as a physiological pressure transducer. In this case the two secondary coils are oppositely wound, so that if the spring-loaded core is symmetrically positioned, the voltage across one secondary coil will oppose the voltage across the other. Movement of the core can thus change this symmetry, and the result is a signal developed across the combined secondary coils. The core may be spring loaded to accept pressures from one side ("single-ended") or both sides at the same time, thus measuring the *differential pressure* between two different sites.

The Sanborn differential pressure transducer, designated 267B, is shown in Figure 11-21(A). The Sanborn 267A is the single-ended equivalent transducer. The Sanborn 267 series has a range of -100 to $+400$ mm Hg pressure and volume displacement, reported by the Sanborn Company, of 0.03 mm^3 per 100 mm Hg

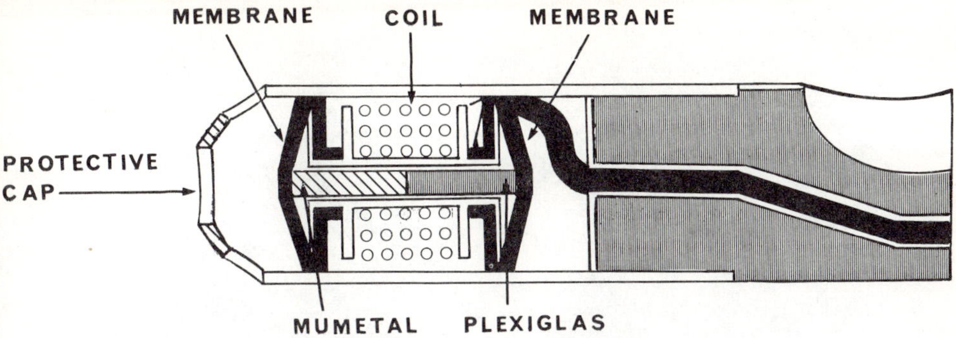

Figure 11-20. The Dallon-Telco catheter-tip manometer.

pressure. The Sanborn 268 series has a smaller range, −40 to +40 mm Hg, and its volume displacement is 0.3 mm^3 per 100 mm Hg pressure. Figure 11-21(B) is the frequency response curve reported by the Sanborn Company. The company does not indicate how the measurements were made except that distilled water was used. How many experimental points were obtained before these curves were drawn? Note that some form of catheter, cannula, or needle must be used. It is of paramount importance to report the length, diameter, and type of material used in the fabrication of the catheter or tubing between pressure source and the manometer membrane. The wall thickness should also be stated. The frequency response, as reported by the Sanborn Company, is flat (which means that the relative amplitude has increased not more than 5 per cent above the mean pressure registration) to no more than 8 or 9 cps.

McDonald reported that five harmonics are needed to accurately record most physiological pressures, but in some cases when the pressure changes abruptly, as in vessels near the heart, ten harmonics would be better. This means that the Sanborn transducer is not accurate enough to register fine detail if the heart rate of the animal is more than 96 beats per minute. When the frequency response is lower than it should be for a particular measurement, some *harmonic information* is lost. The moving parts of the differential transformer type of manometer, in general, have more mass than the moving parts of other varieties of pressure transducer. This contributes to their lower frequency response.

As has been previously stated, any transducer that can be used to measure displacement can be used to monitor blood pressure. Materials which demonstrate piezoelectric properties have been used to make sensitive transducers with an extremely high frequency response. The frequency response at the low end, from 0 to 20 cps, is very poor. Although the piezoelectric crystal pressure transducer has industrial application, its use as a physiological pressure transducer is not as yet practical because of the nonlinearity of the response to pressure.

The resistance strain gauge has attained wide use as the transducer of choice for the measurement of biological pressures. It is simple to use and requires a minimum of associated electronic equipment. Two strain-gauge manometers are shown in Figure 11-22. The manometer shown in Figure 11-22(A) is the Statham P23Dd transducer which has a range from −40 mm Hg to +750 mm Hg. The manometer in Figure 11-22(B) is a Statham P23H; it consists of two matched

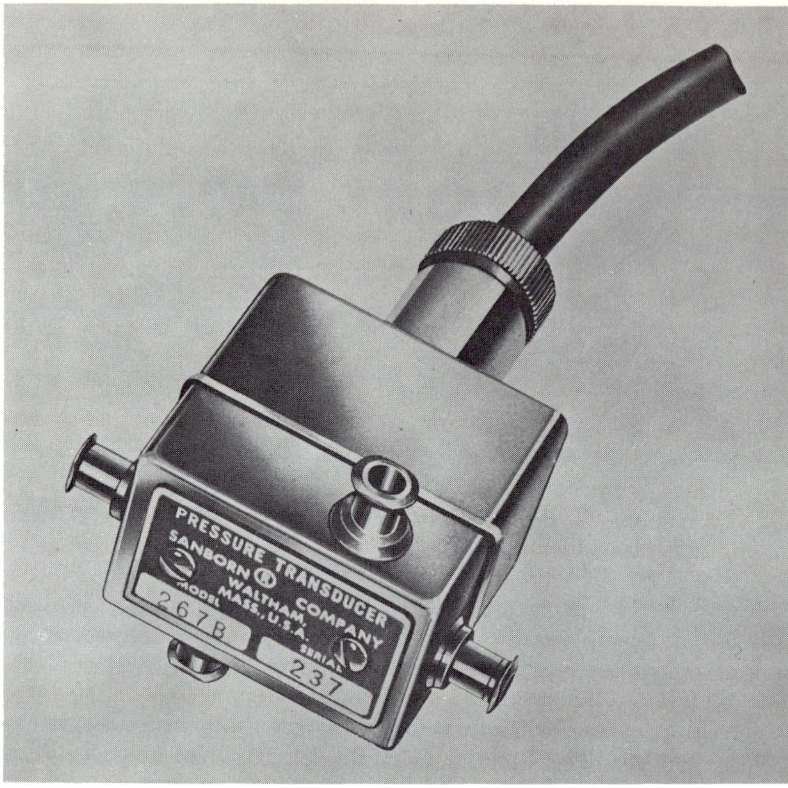

Figure 11-21A. **The Sanborn differential pressure transducer. (Courtesy of Sanborn Co., Waltham, Mass.)**

P23Dd transducers arranged so that the pressure applied to the unit on the left is subtracted from the unit on the right. The P23H also provides an output for each separate pressure as well as the differential pressure.

The P23Dd transducer has a volume displacement of 0.02 mm^3 (1×10^{-5} ml) for an applied pressure of 100 mm Hg. Statham Transducers, Inc., also manufactures a type P23Da and a type P23Db, both of which have a volume displacement of 0.04 mm^3, and type P23Gb, which has a volume displacement of 0.01 mm^3 for an applied pressure of 100 mm Hg. The volume displacement of the Statham P23Gb transducer leads to the conclusion, knowing the dimensions of the active area of the diaphragm, that the maximum excursion of the strain wire is only 2.6×10^{-7} cm. From purely electrical considerations, knowing the characteristics of the strain wire, the excursion estimated by J. Burns (Statham Physiological Transducers) is 3.0×10^{-7} cm. The corresponding membrane excursion for the P23Dd transducer would be about 3.4 times greater for the same applied pressure. This information leads to the conclusion that the sensitive membrane in the P23Gb transducer has an elastance which approaches the bulk modulus of water. In other words, the transducer membrane is almost as incompressible as water.

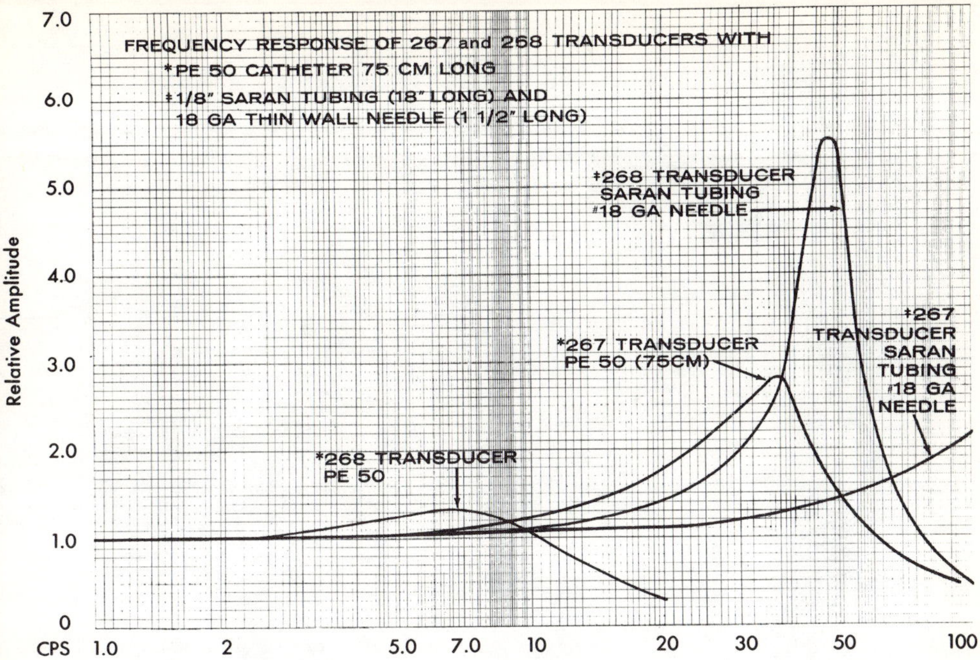

Figure 11-21B. The frequency response of the Sanborn transducer. (Courtesy of Sanborn Co., Waltham, Mass.)

It is important to remember that a pressure manometer, like those shown in Figure 11-22, is always used with a catheter, a cannula, a piece of tubing, or a needle between the sensitive membrane and the source of the pressure (for example, an animal's artery). The ultimate frequency response of the manometer is a combination of the frequency response of not only the transducer but also the catheter. Factors affecting the frequency response of the manometer-catheter system will be discussed later in this chapter.

If the pressure measurement could be made at the pressure source, that is, within the artery or heart, much of the difficulty associated with the use of a catheter or cannula, such as removal of air bubbles, would be eliminated. Figure 11-23(A) is a photograph of the Statham SF-1 transducer. The details of construction are shown in Figure 11-23(B). Notice that the strain wire is wound around a fluted cylinder. This fluted cylinder holds a captured pressure which acts as a spring against the strain wire. The SF-1 transducer element is extremely difficult to manufacture as it is extremely fragile, but I have found it to have a flat response from 0 to 90 cps. Air bubbles in the catheter are not important because the pressure measurement is made at the source.

An electrical network which can be used with the strain-gauge manometer is shown in Figure 11-24. The power supply may be either AC or DC although DC is preferred. The resistor R_3 is used to adjust the excitation voltage of the bridge. The resistor R_c in the calibrating circuit is arranged so that when the switch is

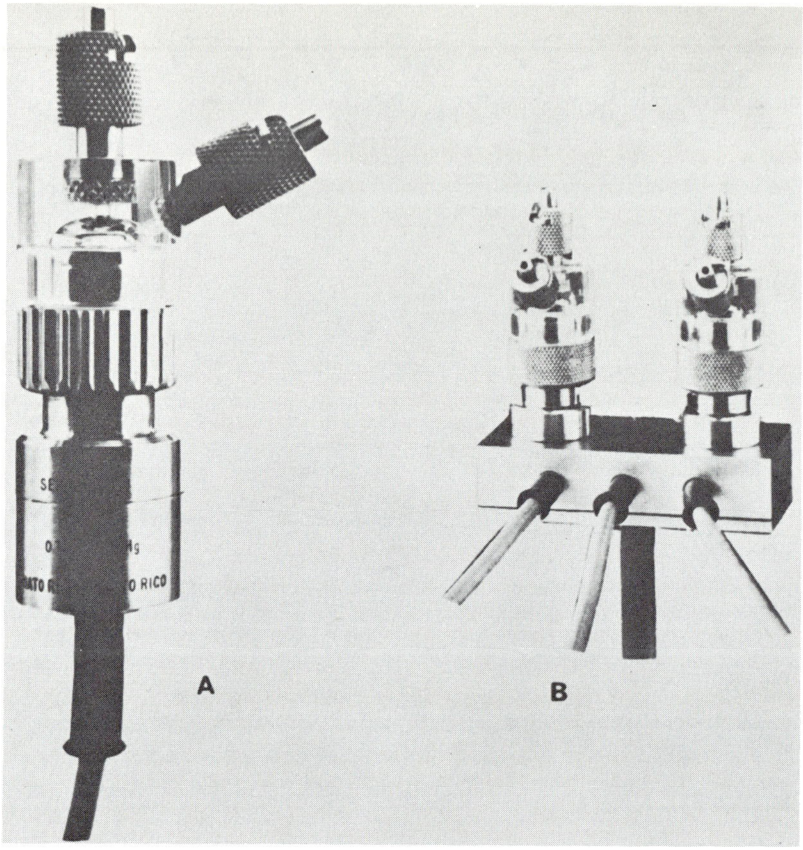

Figure 11-22. Statham physiological transducer: A, type P23Dd; B, type P23H. (Courtesy of Statham Transducers, Inc., Puerto Rico.)

closed it is in parallel with one of the active elements of the gauge. The value of R_c is chosen so that it represents an imbalance of the bridge equivalent to a particular pressure. The zero adjusting network consists of R_1, which is a wire-wound potentiometer with a value between 20K and 30K Ω, and R_2 which should be about 100 times the transducer bridge resistance measured between terminals 2 and 3.

The voltage V_g presented to the receiving instrument is equal to:

$$V_g = \frac{FNE}{\frac{R}{R_g} + 1} \qquad 11.1$$

where: F = calibration factor given by the manufacturer in microvolts per volt excitation per cm Hg pressure.
N = pressure in cm Hg
R_g = input resistance of the recording instrument
E = excitation voltage
R = output resistance (between terminals 2 and 3 in Figure 11-24)

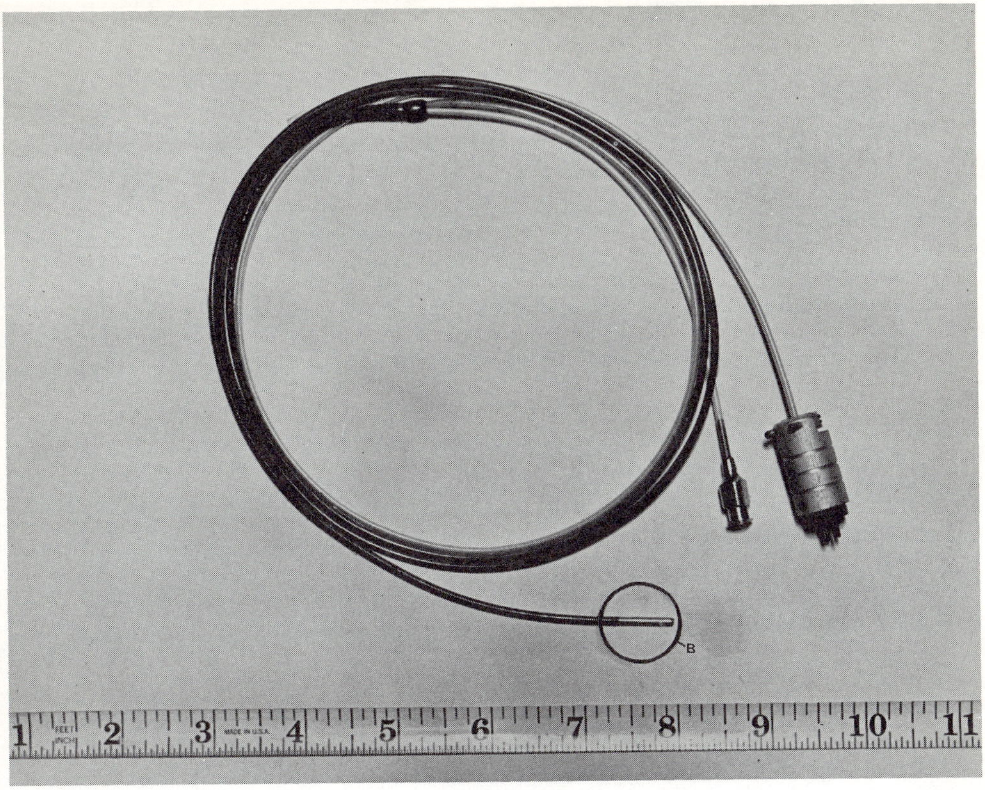

Fig. 11-23A

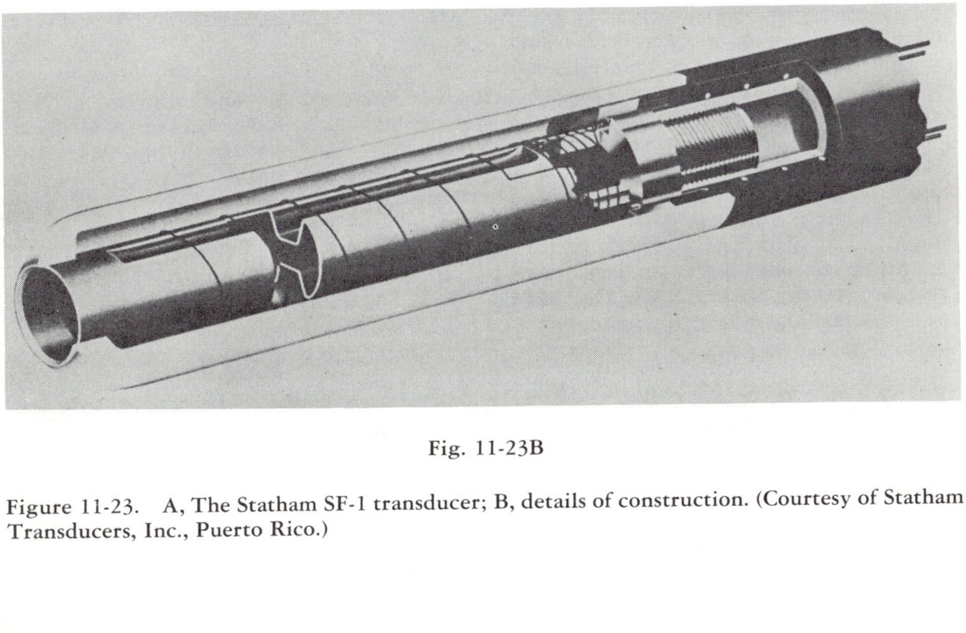

Fig. 11-23B

Figure 11-23. A, The Statham SF-1 transducer; B, details of construction. (Courtesy of Statham Transducers, Inc., Puerto Rico.)

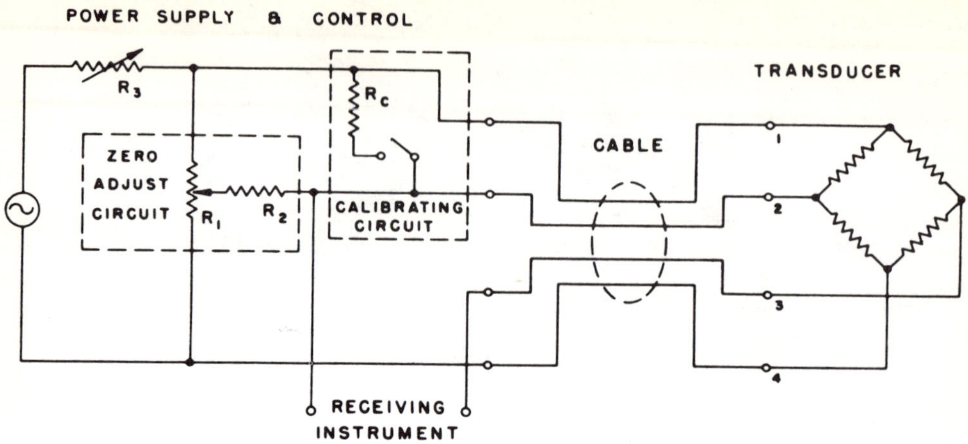

Transducer terminals 1 and 4 = power input

Transducer terminals 2 and 3 = output to receiving instrument

Figure 11-24. The electrical network used to operate the strain-gauge manometer. (Courtesy of Statham Transducers, Inc., Puerto Rico.)

If the input resistance of the receiving instrument is very high compared to the gauge resistance ($R_g \gg R$) then equation 11.1 reduces to:

$$V_g = FNE \qquad 11.2$$

The current (I_g) in microamperes through the input circuit of the recording instrument is given by:

$$I_g = \frac{FNE}{R + R_g} \qquad 11.3$$

Equation 11.3 is merely an expression of Ohm's law applied to equation 11.2.

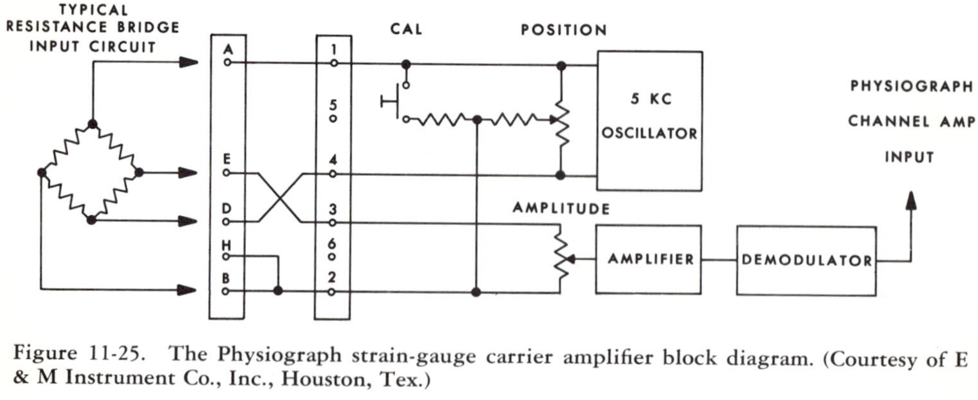

Figure 11-25. The Physiograph strain-gauge carrier amplifier block diagram. (Courtesy of E & M Instrument Co., Inc., Houston, Tex.)

Figure 11-25 is the block diagram for a complete carrier amplifier. This particular instrument itself is shown in Figure 11-26. It uses a 5 kilocycle/second sinusoidal voltage to excite the bridge of the transducer. The position control is equivalent to the zero adjusting network in Figure 11-24. The signal amplitude is controlled by dividing the output signal, using a potentiometer, but it could have been adjusted by altering the excitation voltage. Following further amplification, the signal is then demodulated to remove the carrier, and the signal representing changes in pressure is fed to the driver amplifier of the recording system.

No matter what instrumentation is used to measure pressure, the user should be able to evaluate parameters such as the frequency response necessary for his application. To do this, he must calibrate not only the transducer, but the entire manometer-catheter system and the electronic circuitry as well. This particular discussion does not include evaluating the frequency response of the electronic circuitry (the amplifiers and the recorder), but is limited to the hydraulics system. A number of systems have been developed to calibrate the manometer-catheter system, and more complete reviews have been written by Hansen (1949), Fry (1957), McDonald (1960), and Yanof, Rosen, McDonald, and McDonald (1963).

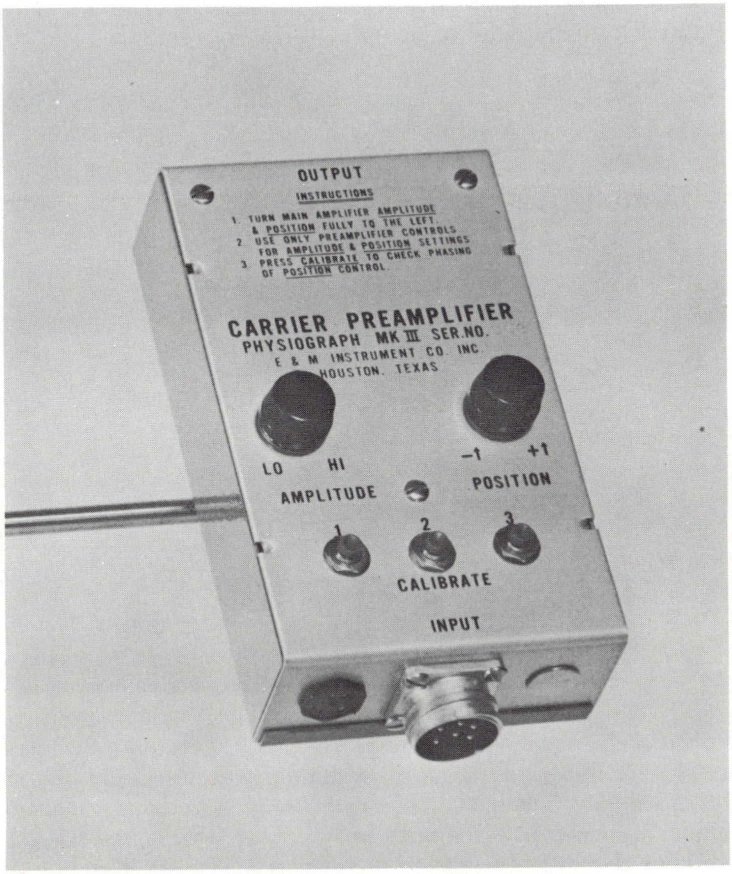

Figure 11-26. A photograph of the Physiograph carrier amplifier. (Courtesy of E & M Instrument Co., Inc., Houston, Tex.)

It is generally agreed that the best method to evaluate a manometer-catheter system is to attach the entire device to a fluid-filled pressure generator, being careful to calibrate the system exactly as it is to be used. There are two general classifications of pressure calibration which seem to fit this discussion. The first of these is called a *transient analysis* because, as the name suggests, some rapid change is made. There are certain rapid changes which are routinely used, such as the analysis of the response resulting from the application of a square-wave voltage, current, pressure, etc. The mathematical details necessary to completely understand the transient analysis are well beyond the scope of this book.

It is possible, however, to apply a transient analysis to a pressure system without understanding all of the details. The following method was first described by Hansen in 1949. It is based on his own mathematical description of the behavior of a catheter-manometer system based on Frank's theoretical analysis published in 1903. Hansen's technique uses a syringe with a small hole bored through the plunger. This syringe is attached to the manometer-catheter system, and the plunger is withdrawn, a finger closing the hole on the plunger. This develops negative pressure within the system. When the finger is removed from the hole, air rapidly rushes in and the entire system vibrates. Analysis of these free vibrations, using the basic mathematical theory, enables the user to calculate not only the natural frequency but also the amount of damping of the system.

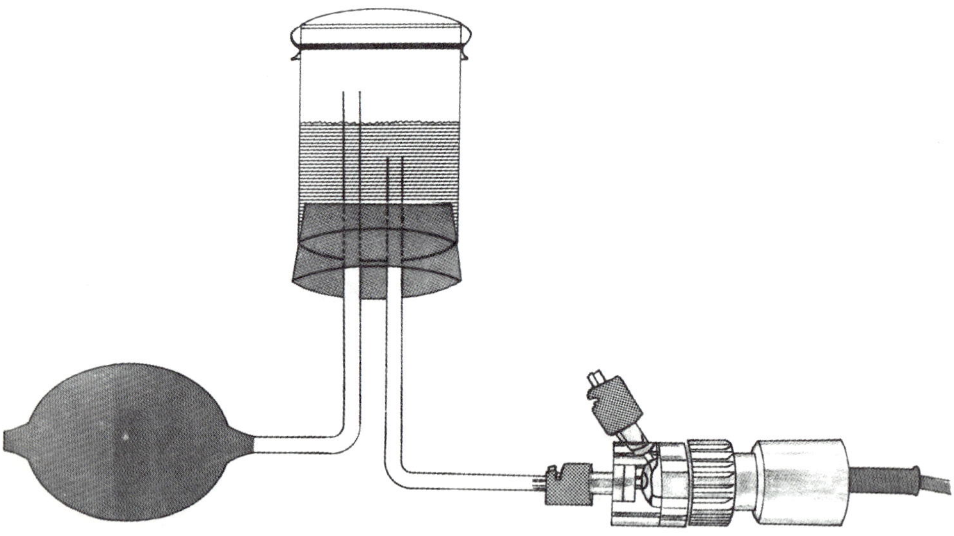

Figure 11-27. The apparatus used for transient analysis of a catheter-manometer system.

Figure 11-27 illustrates a more commonly used system for the transient analysis. It consists of a 1-inch o.d. lucite cylinder. A very tight-fitting rubber stopper with two holes is sealed into the bottom side. One hole contains a glass or plastic tube which goes almost to the top of the cylinder. The other end of this tube goes to a rubber bulb and a one-way valve. The catheter part of the catheter-manometer system is sealed into the other hole, but the catheter extends just beyond the rubber

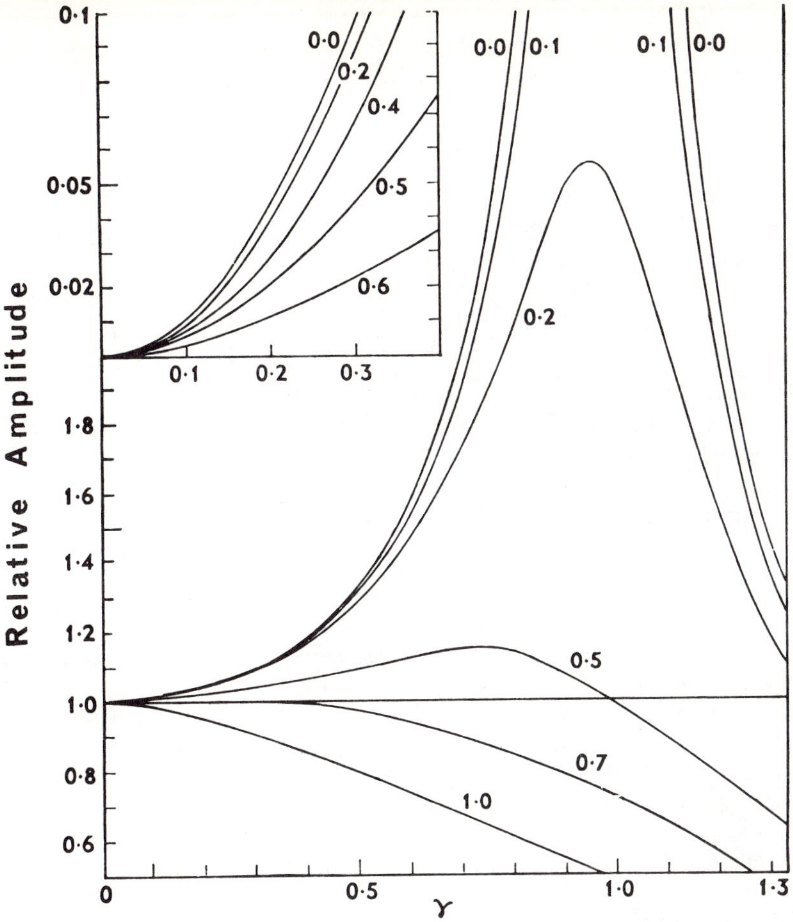

Figure 11-28. The amplitude response of a manometer under different conditions of damping. Inset is a more detailed display of the behavior at relatively low driving frequencies. (Courtesy of D. A. McDonald, Blood Flow in Arteries, Edward Arnold Ltd.)

stopper. The catheter-manometer system is thoroughly flushed with previously boiled distilled water (boiled to remove the air in solution) and the lucite cylinder is also filled to a level shown in Figure 11-27. A thin rubber membrane, *e.g.,* tip of the finger of a surgeon's glove, is attached to the top of the cylinder. The pressure within the cylinder is increased, using the rubber bulb, until the membrane is stretched almost to its limit. The valve is then shut to maintain the elevated pressure. Next, the recording system and associated circuitry are turned on and are allowed to record the pressure when the rubber membrane is burst with a hot soldering iron or cigarette. The pressure suddenly decreases, and the system vibrates for a time and then reaches an equilibrium position.*

*The reader should review the discussion of damping found in Chapter 2; see particularly Figure 2-11 and 2-12 at this time (pages 37 and 38).

The amplitude of the first cycle (measured from the final equilibrium position to the peak) is called ϕ_1. The amplitude of the second cycle measured in the same way is called ϕ_2. The number of seconds between the first peak and the second peak is the period of one complete damped oscillation, T. The *logarithmic decrement* (Λ) is defined as:

$$\Lambda = \log_e \phi_1 - \log_e \phi_2 \qquad 11.4$$

The hydraulic damping β (assuming that the frequency of the electronic circuitry exceeds the hydraulic frequency response) is:

$$\beta = \frac{\Lambda}{\sqrt{4\pi^2 + \Lambda^2}} = \frac{2\pi\beta}{\sqrt{1 - \beta^2}} \qquad 11.5$$

The frequency at which the system would oscillate in *simple harmonic motion* if it were disturbed is called the *natural frequency* of the system and, in this analysis, is given by:

$$\omega_o = 2\pi f = \frac{\sqrt{4\pi^2 + \Lambda^2}}{T} \qquad 11.6$$

(where f is the frequency in cps)

In order to compare a variety of catheter-manometer systems, the frequency of the pressure oscillation ω may be compared to the natural frequency ω_o of the system. The amplitude of the pressure response may be compared to the response of the instrument to a steady (DC) pressure. This is called *relative amplitude,* and is shown in Figure 11-28, for systems with damping from 0 to 1. The phase-lag relationship as a function of relative frequency when damping varies from 0 to 1 is shown in Figure 11-29. The symbol γ denotes ω/ω_o.

Let us now consider the factors which affect the dynamic response of the catheter-manometer system used in physiological studies. Among these factors are the radius (r) and the length (λ) of the catheter; the viscosity (η) and the density (ρ) of the fluid medium ($\rho = 1$ in this analysis); and the volume elasticity (E) of the system.

The resonant frequency ($\omega_o = 2\pi f_o$) may be expressed in terms of the volume elasticity of the system,* with the density $\rho = 1$, as:

$$\omega_o = r\sqrt{\frac{\pi E}{\lambda}} \qquad 11.7$$

Where E is the volume elasticity which is defined as a change in pressure per unit change in volume and is expressed in units such as dyne/cm^5. If the damping factor (β) is written as the ratio of ω_o to the critical damping, which is equal to ω_o, then equation 11.7 may be rewritten:

$$\beta = \frac{\beta_o}{\omega_o} = \frac{4\eta}{r^3}\sqrt{\frac{\lambda}{\pi E}} \qquad 11.8$$

When the Hansen damping is expressed in this way, the relative contribution of the various parameters can be evaluated. One can see from equation 11.8 that the

*See McDonald, p. 303.

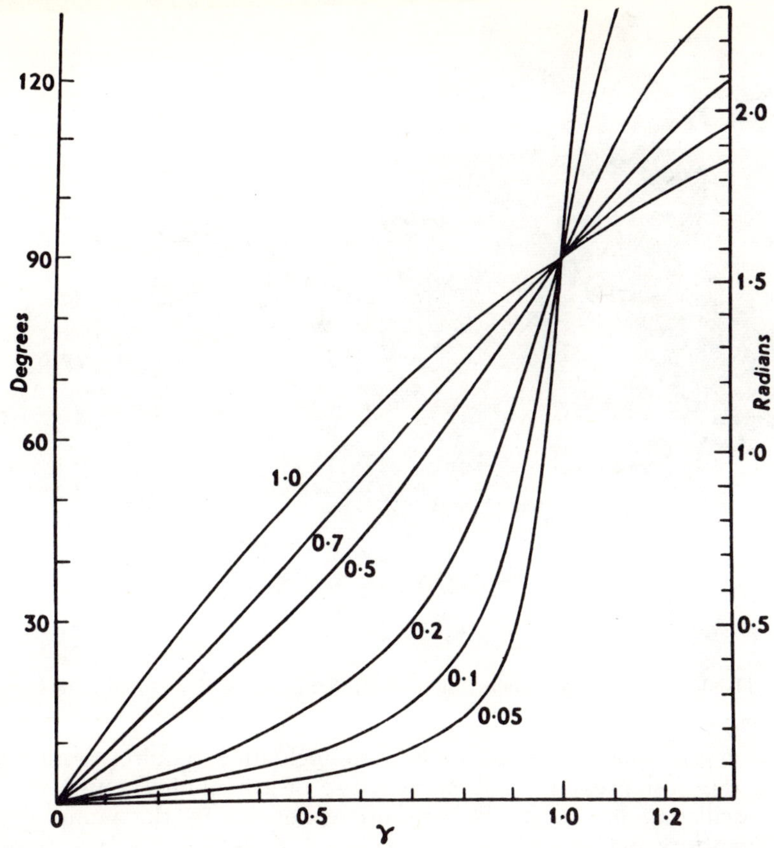

Figure 11-29. The phase-log in a manometer driven by an oscillatory force of constant amplitude. (Courtesy of D. A. McDonald, Blood Flow in Arteries, Edward Arnold Ltd.)

diameter of the catheter is much more important than its length in determining the frequency response of a catheter-manometer system, and:

$$\beta_o = \frac{4\eta}{r^2} \qquad 11.9$$

Since the forcing function is oscillatory in nature, the fluid resistance derived by Womersley* should more appropriately be used:

$$\beta_o = \frac{\eta\, \alpha^2 \sin \epsilon}{2r^2 M'} \qquad 11.10$$

where $\alpha = r\sqrt{2\pi n f/\eta}$; M' and ϵ are solutions of the equation of motion and are a function tabulated by Womersley.† The effect of equation 11.10 is to increase β_o with frequency, especially when α is large.

*See McDonald, p. 303.

†See McDonald, Table VI.

Figure 11-30. A pump for generating a sinusoidal pressure in a closed-fluid system.

Although the Hansen "pop" calibration technique requires a minimum of equipment, the best pressure calibration is use of a fluid-filled pump which generates an oscillating pressure. The pressure generated by the pump should be sinusoidal in form and should be able to generate sufficiently high frequencies (to about 90 cps). Figure 11-30 is a photograph of such a pump. Basically, the pump consists of a syringe mounted in a plastic block (Fig. 11-31), the plunger of which is driven by a modified Scotch-yoke assembly shown in Figure 11-32. It is, of course, important that the stroke be of the order of 0.001 inch or less so that excessive pressures are not generated. The pump itself is described in more detail in an article by Yanof and associates (1963).

The pump is used by filling the clean pump block (Fig. 11-31) with previously boiled water and a very small amount of detergent or ethyl alcohol. The catheter is fitted onto the Linden fitting on the top right side. The front orifice (used in the flow studies) is blocked with a brass plug. The valve on the upper left is used to control the mean pressure within the block. The system is completely filled with fluid and flushed long enough to remove *all bubbles* and then flushed some more just to be certain. This flushing must be done with freshly boiled distilled water containing a small amount of detergent. After a few days, the water must be reboiled. When all flushing is complete the system is closed and the pump is slowly brought to full speed. Then the pump is turned off and the pressure recording system is simultaneously turned on. Figure 11-33 illustrates records made in this fashion when the Statham P23Dd transducer is used with a variety of intracardiac cannulas. If a bubble were present, the resonant frequency (peak) would occur at the low-frequency end (on the left). This "quick-check" system may be employed when the catheter-manometer system is in place in the animal.

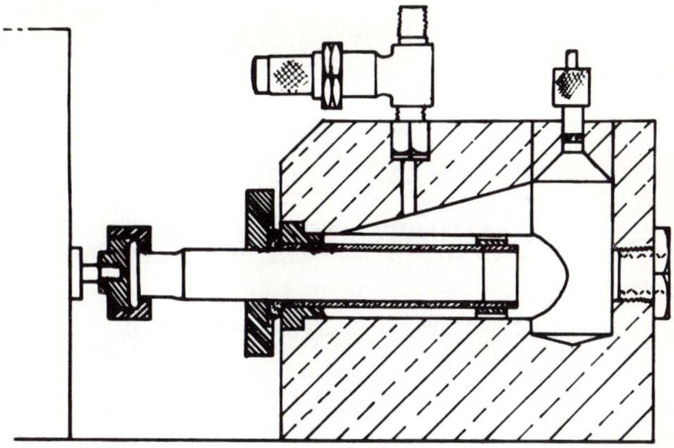

Figure 11-31. The plastic pump block containing the syringe piston). (Yanof, H. M., *et al.,* Phys. Med. Biol. 8:407-422, 1963. Courtesy of Taylor & Francis, Ltd., publisher.)

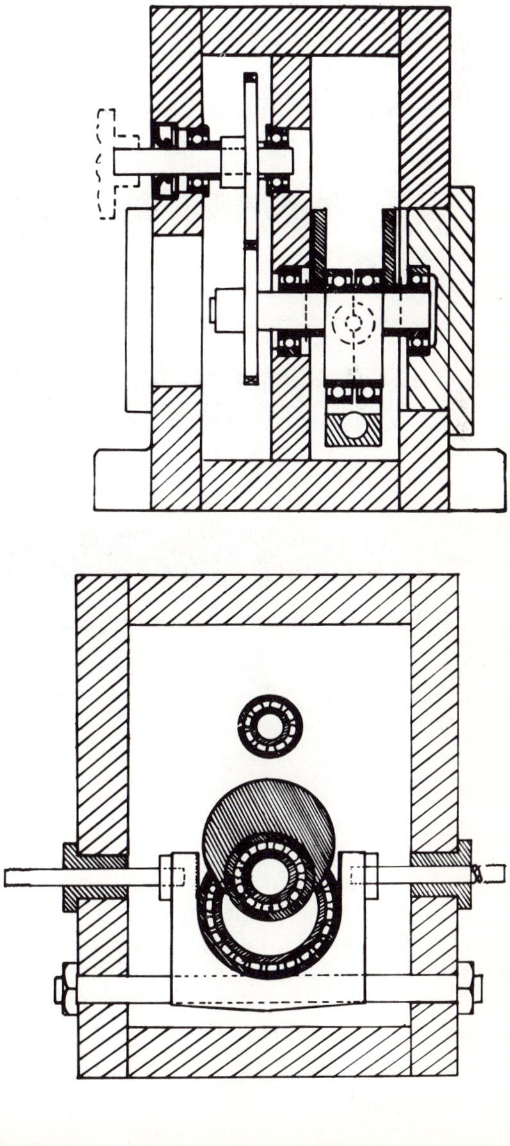

Figure 11-32. The drive mechanism of the pump shown in Figure 11-31. (Yanof, H. M., *et al.,* Phys. Med. Biol. 8:407-422, 1963. Courtesy of Taylor & Francis, Ltd., publisher.)

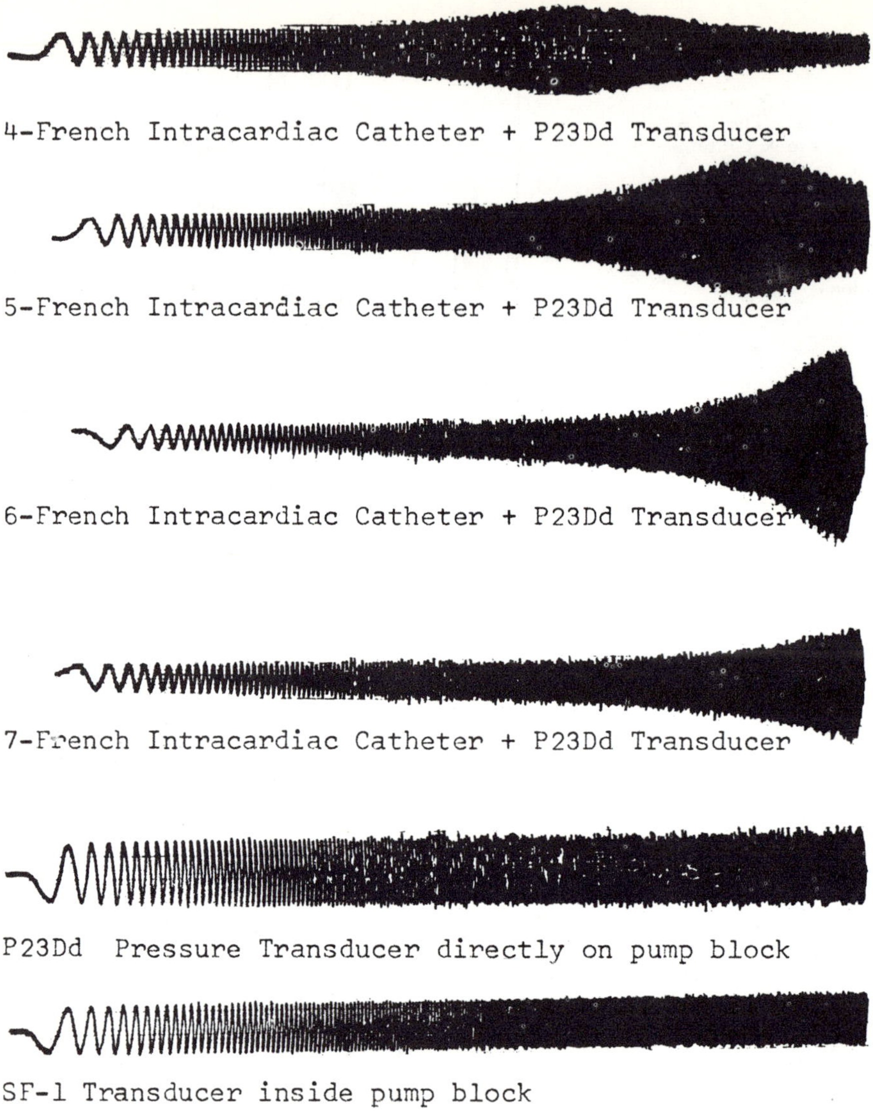

Figure 11-33. A "quick check" of the frequency response of a catheter-manometer system.

The actual frequency response of the catheter-manometer system must, however, be measured in the *steady state.* This is done by allowing the servo-motor to reach a certain speed and recording the pressure, then increasing the servo-motor speed slightly and again recording the pressure. The results, with a variety of intracardiac catheters and two sizes of polyethylene tubing, are shown when the Statham P23Dd transducer is used (Fig. 11-34), and when the Statham P23Gb transducer is used (Fig. 11-35). The user must be careful to measure the frequency

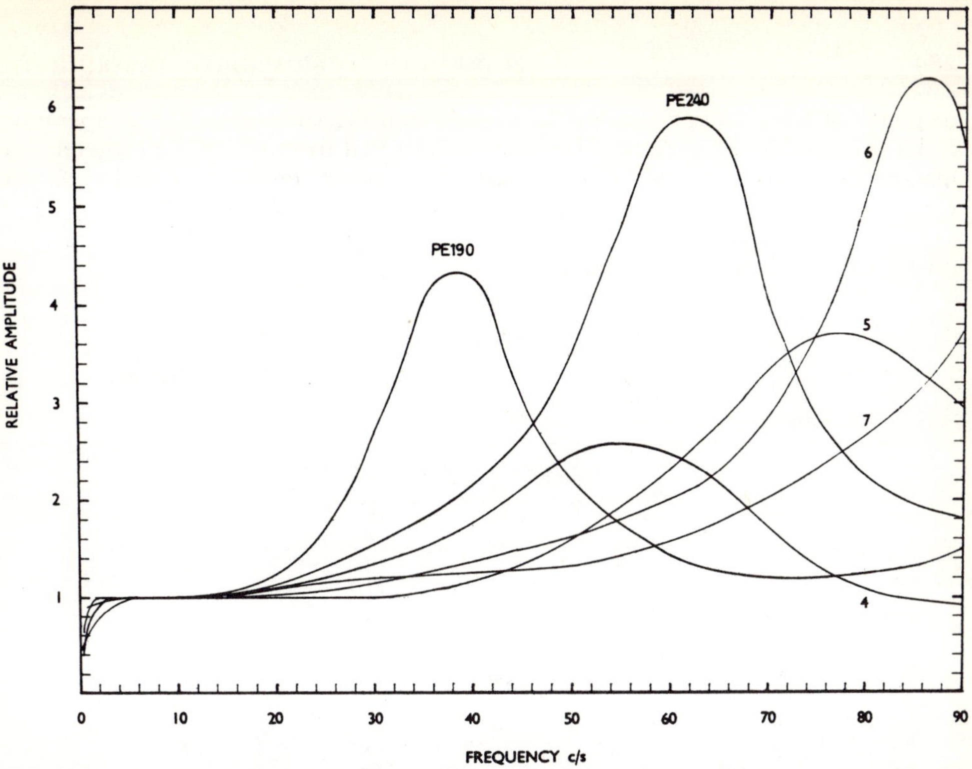

Figure 11-34. The frequency response, using the Statham P23Dd transducer and a variety of catheters (e.g., 4=4-French, P.E.=polyethylene; sizes are Clay-Adams Co. sizes), manufactured by the United States Catheter Instruments in Glens Falls, N.Y. (Yanof, H. M., *et al.,* Phys. Biol. Med. 8:407-422, 1963. Courtesy of Taylor & Francis, Ltd., publisher.)

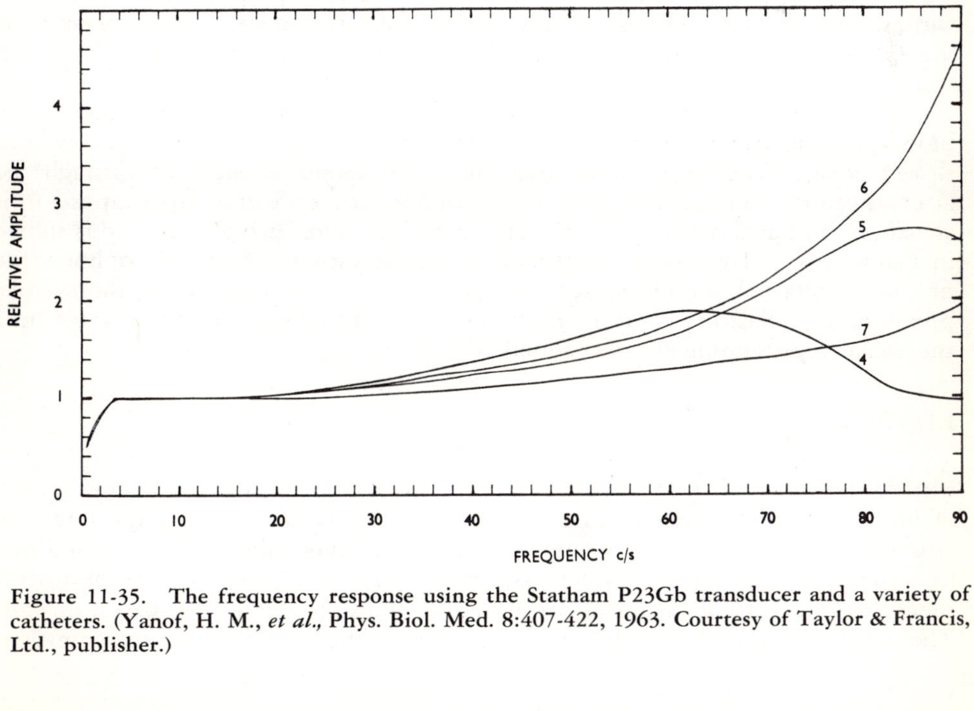

Figure 11-35. The frequency response using the Statham P23Gb transducer and a variety of catheters. (Yanof, H. M., *et al.,* Phys. Biol. Med. 8:407-422, 1963. Courtesy of Taylor & Francis, Ltd., publisher.)

response of the catheter-manometer system with the catheter at the temperature to be used (body temperature) because the material from which the catheters are manufactured may have a different elastance at room temperature and at animal temperatures. Table 11-1 gives some working values. In Figures 11-34 and 11-35 and Table 11-1 the numbers given for intracardiac catheters are French sizes, and polyethylene tubing is indicated by P.E.

TABLE 11-1*

Test Catheter	*Internal radius cm*	f_o *c/s*	+5% *c/s*	*Volume Elasticity of system* (E) 10^{10} *dyne/cm*^{5}
	P23Dd Transducer + Test Catheter			
4F	0.023	56	15	0.83
5F	0.033	78	28	0.78
6F	0.046	86	24	0.49
7F	0.058	—	28	—
PE-190	0.060	39	11	0.06
PE-240	0.084	62	14	0.08
	P23Gb Transducer + Test Catheter			
4F	0.023	65	23	1.12
5F	0.033	85	21	0.93
6F	0.046	—	24	—
7F	0.058	—	34	—

*From Yanof, H. M., *et al.,* Phys. Med. Biol. 8:407-422, 1963.

Thus far, this chapter has dealt with the *direct* measurement of blood pressure. This requires the vessel to be punctured and a catheter or needle inserted. Instruments exist, however, to measure the systolic and the diastolic pressures in man without opening a blood vessel. This means of measurement of blood pressure is the *indirect* method. Instruments of this type are manufactured by the Beckman Instrument Company and the E & M Instrument Company.

Figure 11-36 illustrates the basic principles used in this measurement. The blood pressure cuff is automatically inflated to about 220 mm Hg and allowed to slowly deflate. The microphone transduces the sound of the flow through the artery near the surface just below the compression cuff and superimposes this signal on the signal representing the falling cuff pressure. A typical record is shown in Figure 11-37. By proper adjustment of the sensitivity of the microphone, the first superimposed sound crosses the cuff pressure curve, indicating the systolic pressure. The diastolic pressure is the point on the falling pressure curve where the signal representing the last sound is seen.

BLOOD FLOW

The instantaneous measurement of blood flow has interested biological workers for many years. Ideally the physiologist would like to determine the amount of blood flowing through a vessel at any instant without disturbing the flow, the vessel, or the physiology of the animal. It has not been possible, to the present time, to accomplish this to the complete satisfaction of research workers. Some instruments require cutting the vessel to insert the transducer into the stream of flow. This procedure grossly changes certain aspects of the physiology of the animal as,

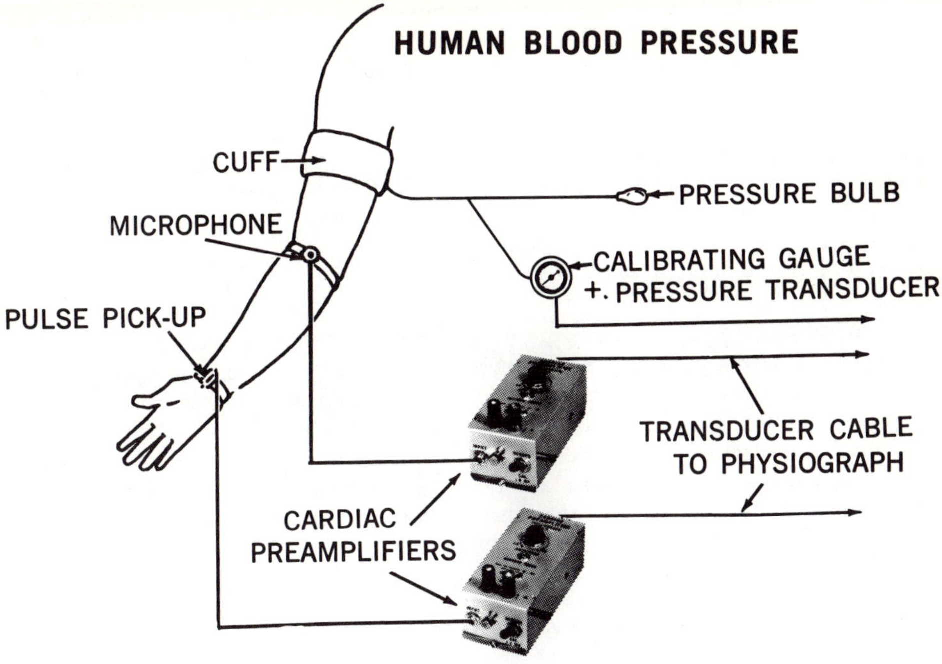

Figure 11-36. The instruments used to measure blood pressure indirectly. (Courtesy of E & M Instrument Co., Inc., Houston, Tex.)

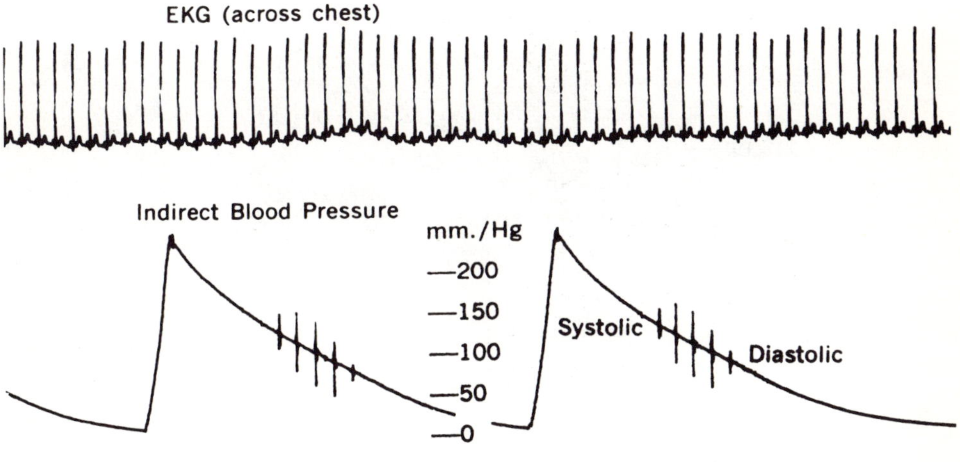

Figure 11-37. A sample record of an automatic indirect blood pressure registration. (Courtesy of E & M Instrument Co., Inc., Houston, Tex.)

for example, cutting a vessel destroys the fibers of the sympathetic system which are in the wall of the vessel. Inserting a transducer disturbs the laminar character of the flow. Adaptations of purely mechanical devices, such as the rotameter, are useful in some experimental arrangements in which mean flow is to be measured, but in the past decade the ultrasonic and electromagnetic flowmeters have become widely used in the research laboratory. It would appear, however, that an electromagnetic flowmeter is the instrument of choice for the measurement of instantaneous volume flow through a particular vessel. Most blood flow instruments measure only velocity and do not measure flow unless the diameter of the vessel is held fixed (measuring velocity through a fixed area is the same as measuring volume flow.)

None of the instrument types mentioned thus far can be used to determine the blood flow through a particular region, such as the flow of blood through the skin or through an entire limb, unless the flow is supplied by a single vessel. This is especially true for clinical studies in which human subjects are used or when surgery is not indicated. Another class of instruments may be used for these clinical measurements, but most of them cannot as yet be quantitatively calibrated. This section will survey some examples of flowmeters in current use, but it cannot do justice to every instrument that is available.

The ideal instrument for measuring the flow of blood and other fluids in animals must have the following characteristics:

1. A range between 0.1 ml/minute and 8 to 10 liters/minute
2. Linearity of response
3. A signal-to-noise ratio of at least 50
4. Equal sensitivity to both positive and negative flows
5. Temperature stability
6. A frequency response flat (± 5 per cent of response at zero frequency) to 50 cps (for some applications).

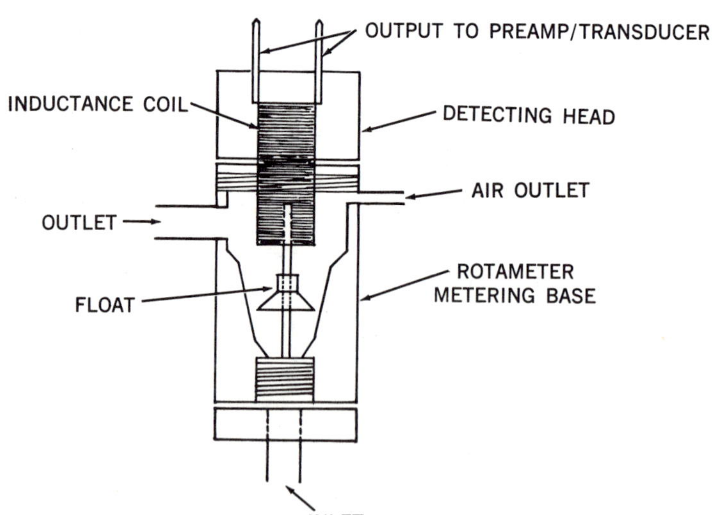

Figure 11-38A. The blood flow rotameter transducer. (Courtesy of E & M Instrument Co., Inc., Houston, Tex.)

In addition to the above electronic criteria, the flowmeter should not interfere with the flow being measured. Cutting into an animal, and especially into the vessel carrying the flow, obviously should be avoided. It is not possible, or at least it has not been possible thus far, to design the ideal flowmeter described above. Whatever instrument is used, the investigator must be aware of its limitations. He must have calibrated the instrument not only statically, the voltage output equivalent to the number of milliliters per minute of flow, but also dynamically, that is, response to a sine wave flow of constant amplitude as a function of frequency. These calibration techniques should be carried out *in vivo.* Many investigators have been fooled when using an external device because a transducer which may work very well in a beaker of saline or a plastic bath holding the artery often may be expected to work differently *in vivo* as a result of the corrosive properties of blood, noise problems, etc.

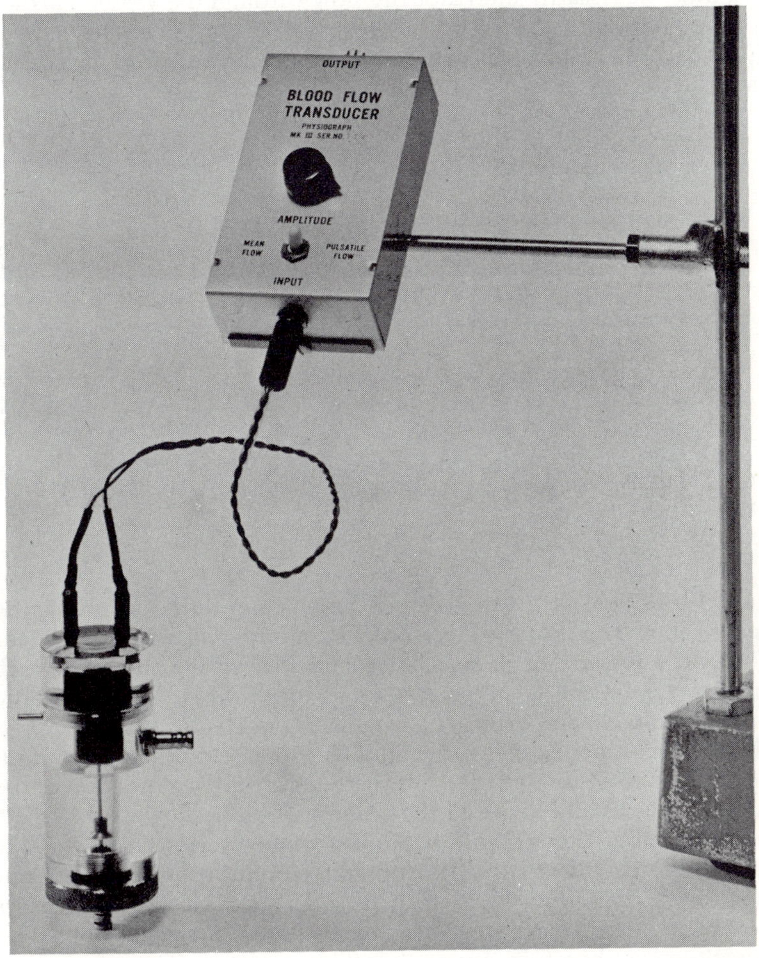

Figure 11-38B. Photograph of the instrument in Fig. 11-38A. (Courtesy of E & M Instrument Co., Inc., Houston, Tex.)

The simplest method of determining the flow of blood through a vessel is to expose the vessel, cut it, and allow the blood to flow into a graduated cylinder for a specific time. This is a relatively unquestionable method for measuring blood flow and is quite commonly used to calibrate any flowmeter device. A variety of mechanical units, called *stromuhrs*, have been designed to fit in series with a vessel. In all of these units, one chamber is allowed to fill, and the time necessary to fill it is noted. The flow to that chamber is then stopped and diverted to another chamber while the first chamber is allowed to empty into the blood stream. The physiological application of such a mechanical device is, however, severely limited.

Another type of mechanical device is shown in Figure 11-38. In this case, the transducer consists of a chamber with a specially shaped compartment containing a float. The greater the flow, the higher the float rises in the chamber. The amount of flow can thus be determined by the height of the float. This type of flowmeter is called a *rotameter*. In this particular rotameter a variable inductance displacement transducer is used to sense the height of the float. This rotameter, like the mechanical stromuhr, must necessarily interrupt the flow through the vessel. Furthermore, the inertia of the float causes the frequency response of the rotameter to be low. Figure 11-39 is a block diagram of the instrument described above as it is used with the Physiograph recorder.

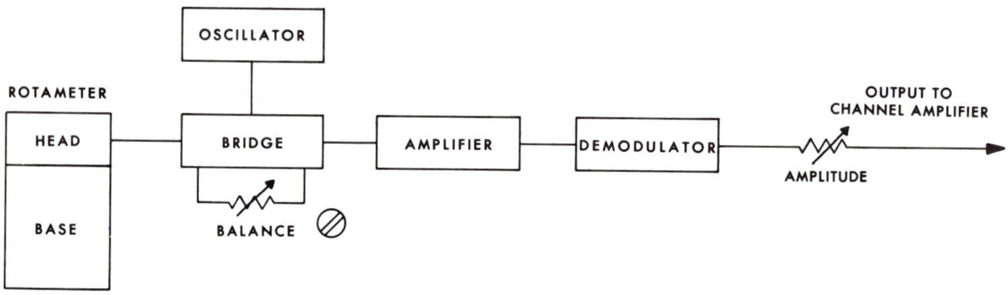

Figure 11-39. The block diagram for the rotameter shown in 11-38. (Courtesy of E & M Instrument Co., Inc., Houston, Tex.)

Figure 11-40 is a mechanical device based on a different principle. This flow transducer, called the Potter electroturbinometer, consists of a stainless steel turbine which is suspended in the flow by circular spring clips. The turbine blade is designed to stay in the lucite chamber and rotate when the blood flows past the blades. A two-pole alnico permanent magnet is encased in the body of the turbine. As the blades turn, the magnet turns, inducing an alternating current in the pickup coil. The frequency of the induced alternating current is a function of the speed of rotation of the blades and hence of the volume of blood flowing past the blades and through the chamber. The output of the instrument varies linearly as a function of flow when the rate of flow is high (over 1 liter/minute) but becomes non-linear at lower flow rates. This transducer with its associated circuitry has been used for estimation of the cardiac output, but it necessitates cutting an opening in a major blood vessel and is therefore not an "ideal" instrument. The electroturbinometer, like other basically mechanical flowmeters, does not indicate the instantaneous flow, but rather provides an indication of the mean flow.

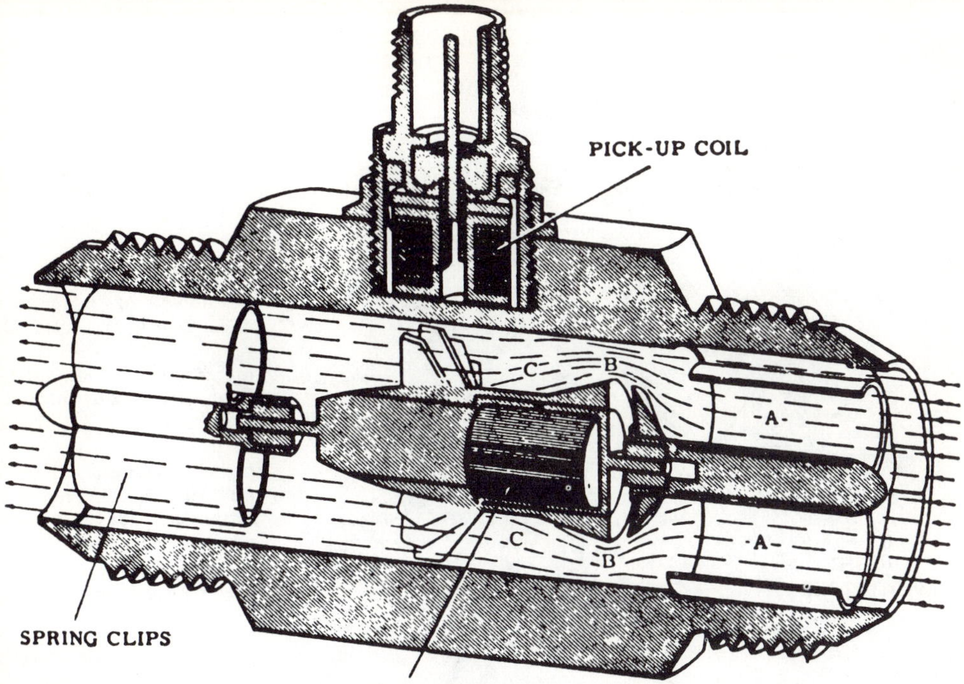

Figure 11-40. The Potter electroturbinometer. (Courtesy of S. J. Sarnoff and E. Berglund, The Potter electroturbinometer: an instrument for recording total system blood flow in the dog. IRE Transactions on Medical Electronics, vol. ME-6, no. 4, p. 270; published by the Professional Group on Medical Electronics of the IRE, 1959. Used by permission of The Institute of Electrical and Electronics Engineers.)

A more sensitive flowmeter based on the mechanical displacement of a bristle projecting into the stream of flow is shown in Figure 11-41. The basic component of this transducer is the RCA 5734 vacuum tube shown in Figure 9-22. The output of the transducer is basically non-linear, but electronic circuitry has been designed to make the output a linear function of flow. The bristle flowmeter, unlike the rotameter and the electroturbinometer, does respond to rapid changes in flow as well as both positive and negative flows. This type of instrument is, however, sensitive to temperature variation and in some instances appears to respond to changes in pressure as well as flow. Transistor or semiconductor photosensitive components may be of value in the improvement of this device in the future.

It is possible to use a principle based on the heat loss to sense the flow of a liquid through a tube. A small heating coil may be placed in the blood stream to cause local elevation of the blood temperature. Then, if a heat-sensing element, for example, a thermistor, is placed downstream to the heating coil, the thermistor will change temperature as a function of the velocity of the blood moving past the heating coil and the thermistor. Descriptions of devices of this sort appeared in the literature as early as 1928;* they are generally referred to as *thermostromuhrs.*

---

*Rein, H., Die Ausnützung der Wärmekonvektion zur Messung der Durchstromung uneroffneter Bludgefasse. Aberhalden Handbuch der Biologeschen Arbeitsmethoden, Volume 8, Chapter 5, pp. 693-716 (1928-35).

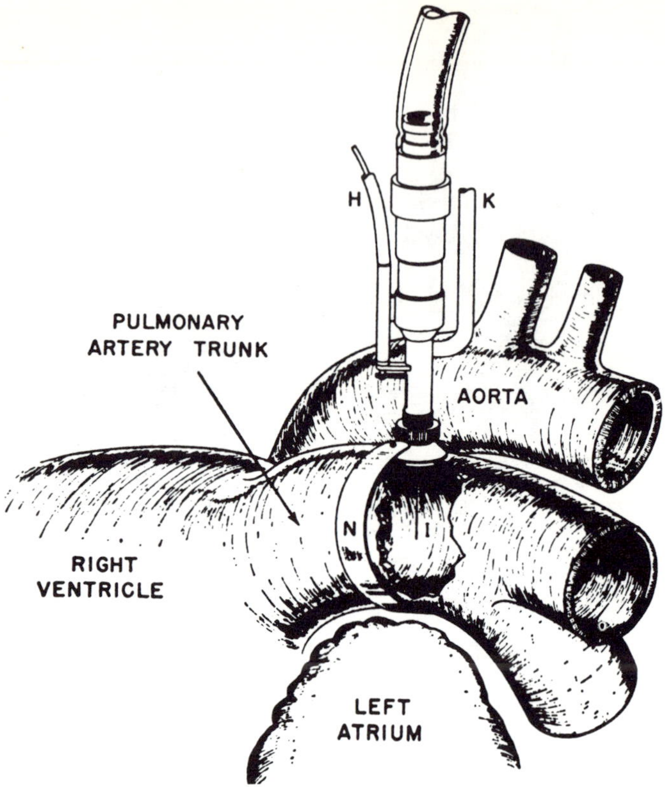

Figure 11-41. The bristle flowmeter. (Courtesy of G. Brecher, A critical review of bristle flowmeter techniques. IRE Transactions on Medical Electronics, vol. ME-6 no. 4, p. 270; published by the Professional Group on Medical Electronics of the IRE, 1959. Used by permission of The Institute of Electrical and Electronics Engineers.)

Another thermal device, similar in principle, is called an *isothermal* velocity probe. This transducer consists of a small thermistor which is inserted into the flow and maintained above the blood temperature. As blood flows past the heated thermistor, it is cooled. The heat-sensitive element, in this case, changes resistance as a function of temperature. The thermistor is used as one leg of a Wheatstone bridge, and the changes in temperature then appear as changes of potential across the bridge. A block diagram and circuit schematic for the isothermal flowmeter are shown in Figure 11-42.

The thermostromuhr and the isothermal flowmeters which have been described have two major defects: One, this flowmeter design will not measure reverse flow and, two, the output is not a linear function of flow. It is possible that by using two thermistors and a multiple bridge arrangement, both forward and reverse flow could be measured. The non-linearity of the instrument could be corrected by building electronic compensation into it. Thermistors can be obtained which are very small and these can be built into the tip of a needle or onto the end of a long intracardiac catheter. This type of flowmeter would appear to be quite feasible for clinical use in human subjects, although a high-frequency response should not be expected.

The use of ultrasound to measure the velocity of blood flowing within a vessel is a method which approaches the ideal flowmeter previously described. There are two major types of ultrasonic flowmeters in use, the transit-time and the Doppler back-scatter instruments. The principle on which the transit-time instrument is

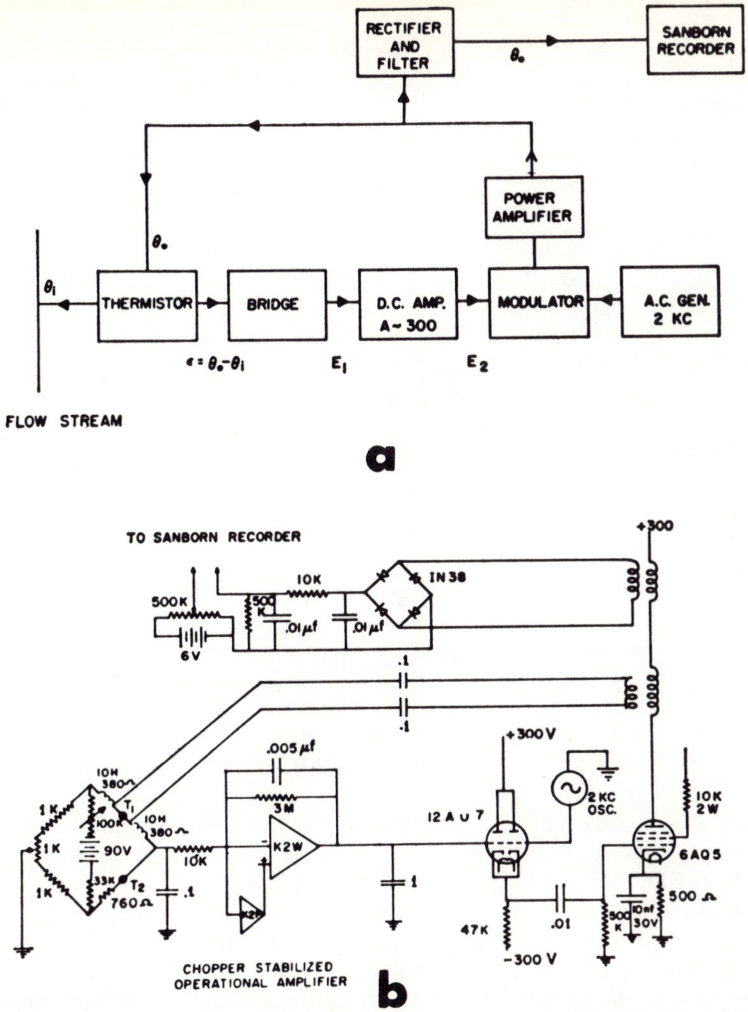

Figure 11-42. a, Block diagram of the isothermal flowmeter; b, circuit schematic. (Courtesy of S. Katsuro, R. Weiss, D. Baker, and R. F. Rushmer, Isothermal blood flow velocity probe. IRE Transactions on Medical Electronics, vol. ME-6 no. 4, p. 270; published by the Professional Group on Medical Electronics of the IRE, 1959. Used by permission of The Institute of Electrical and Electronics Engineers.)

based depends on the fact that the velocity of an acoustic wave (in the ultrasonic frequency range in the order of 380 kc) is increased as it moves in the direction of the flow and decreased as it travels against the flow. A cuff or cannula, on which two ultrasonic transducers are mounted 1 cm apart, is placed around a blood vessel (or in the case of a cannula, in series with the vessel). Barium titanate crystals when driven to resonance emit a burst of sound. These crystals also develop a voltage across them when a sound wave causes them to resonate. The ultrasonic flowmeter causes one transducer to transmit a burst of ultrasonic waves which is received by the other transducer. The other transducer then becomes the transmitter and emits

its own burst of ultrasound, which is detected by the first transducer which is now acting as a receiver. The frequency of this alternation is about two times per second. The difference in transit time (Δt) between the upstream and the downstream signals is then electronically subtracted, and the result after considerable electronic circuitry is a voltage signal proportional to the flow. Mathematically this is expressed as:

$$\Delta t = \left(\frac{\ell}{c - v \cos \theta}\right) \text{upstream} - \left(\frac{\ell}{c + v \cos \theta}\right) \text{downstream} \qquad 11.11$$

where: c is the velocity of the sound and
v is the blood velocity
θ is the angle between the axis of sound propagation and the axis of flow
ℓ is the distance between crystals

which can be expressed:

$$\Delta t = \frac{2\ell v}{c^2 - v^2 \cos^2 \theta} \qquad 11.12$$

but c^2 is very much larger than v^2 so equation 11.12 can be reduced to

$$\Delta t = \frac{2\ell}{c^2} v \cos \theta \qquad 11.13$$

A block diagram which illustrates the principles of operation of the ultrasonic flowmeter as well as the arrangement of barium titanate crystals in the flow cuff is shown in Figure 11-43.

The ultrasonic flowmeter is capable of meeting almost all of the necessary conditions for an ideal flowmeter. The transducers are lightweight and can be implanted in the animal, causing a minimum of physiological change. The electronic circuitry employed in this type of flowmeter is relatively complex. The circuitry must be ultrastable so as to avoid changes in transit time which would appear as erroneous flow signals. This type of flowmeter is difficult to calibrate because there is some uncertainty as to the diameter of the fluid column, the velocity of which is being measured by the ultrasonic flowmeter. An instrument that measures velocity cannot be relied upon to measure volume flow unless the diameter of the vessel is constant or at least known throughout the measurement. The average user could not be expected to "trouble shoot" if anything should go wrong with the complex circuitry.

Figure 11-44 illustrates the placement of the emitter and sensor crystals of the Doppler flowmeter. The piezoelectric emitter crystal is excited with a 5-megacycle oscillator. The 5-megacycle ultrasonic beam is aimed at the blood (or other liquid) flowing within the sleeve. Some of the ultrasonic energy is reflected toward the sensing piezoelectric crystal—a signal, the frequency of which is dependent on the flow of the moving reflector. A portion of the 5-megacycle signal from the emitter crystal also strikes the sensor crystal. The two signals are mixed and the algebraic combination of the frequencies (beats) is converted to a signal which modulates the frequency of a 100-megacycle FM telemeter transmitter. The resulting signal may be received some distance from the animal. The total power necessary to run this equipment is less than 100 milliwatts. The Doppler flowmeter

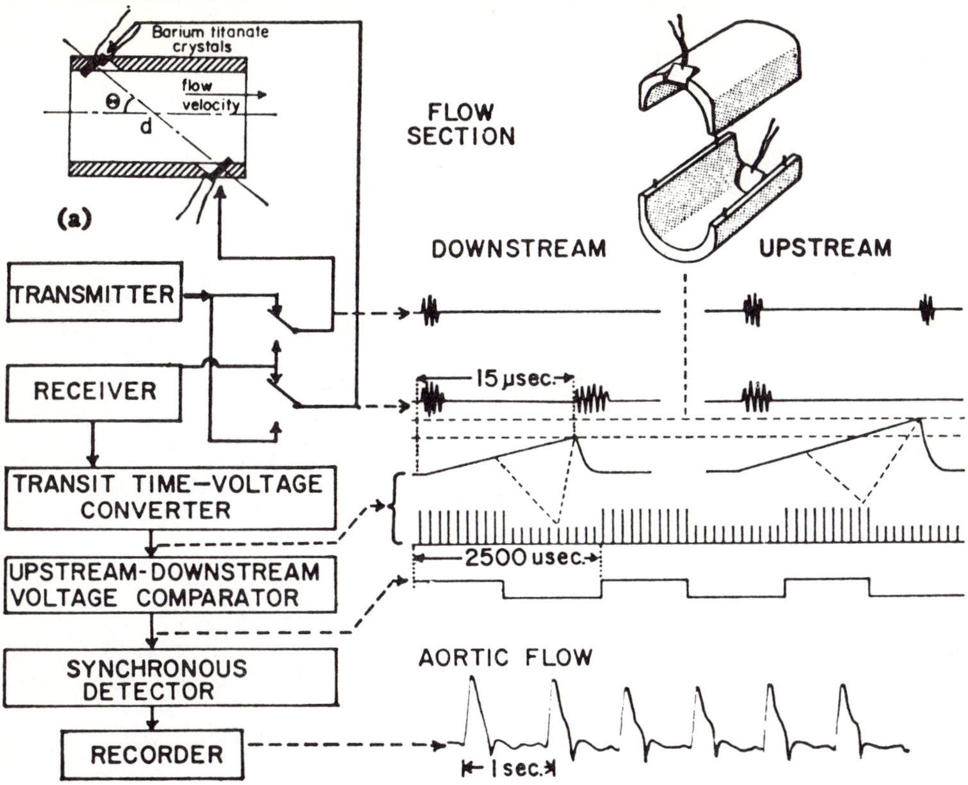

Figure 11-43. Block diagram of the ultrasonic flowmeter. (Courtesy of William R. Farrall, Design Considerations for Ultrasonic Flowmeters. IRE Transactions on Medical Electronics, vol. ME-6, no. 4, p. 201; published by the Professional Group on Medical Electronics of the IRE, 1959. Used by permission of the Institute of Electrical and Electronics Engineers.)

or radar flowmeter, which may be a better name, measures velocity of fluid in only one direction. A bidirectional radar flowmeter is under development at the present time and will soon be available.

The transit-time and the radar or Doppler ultrasonic flowmeter are best suited to measuring relative changes in the velocity of a moving fluid. The conversion of velocity of flow to volume flow requires a constant diameter, and there is some evidence that the chronically implanted transducer tends to cause scar tissue formation which decreases the diameter of the vessel, thus making the calibration with respect to volume flow less accurate. Nevertheless, the ultrasonic flowmeter is an instrument which should be given careful consideration by anyone wishing to measure blood flow.

The most commonly used flowmeter in the United States, as of the writing of this book, is the electromagnetic flowmeter. The principle on which it operates was initially described by Faraday in 1832, and the first actual measurements were made by Young and his associates in 1920. A successful attempt to measure velocity of flow in tubes was made by Williams in 1930. Use of the instrument for blood-flow measurement was then suggested by Fabre in 1932. First applications of the electromagnetic flowmeter to blood flow were reported independently by Kolin and Wetterer in 1936 and 1937 respectively.

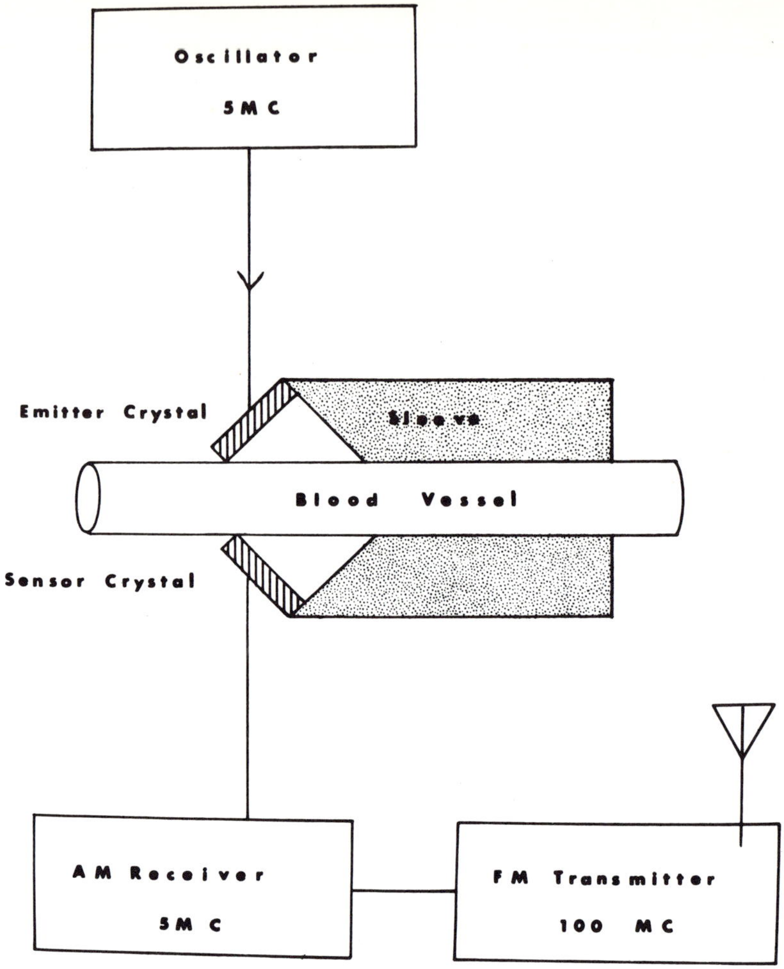

Figure 11-44. The Doppler or radar flowmeter.

The ideal blood flowmeter can be placed around an intact vessel in the animal and the cables brought through the skin. The blood flow can then be measured with the animal unanesthetized and free to move about without restriction. A small, lightweight transducer has been described by Kolin. The amplifier must be capable of amplifying signals of 1 to 50 μv with a noise level of less than 0.5 μv referred to the input. (See Chapter 12 for a discussion of noise.) The frequency response of the instrument should be flat to 50 cps. The instrument must also faithfully record flow signals in spite of annoying potentials present in the surrounding tissues, such as those arising from the heart (ECG) (see Fig. 12-3).

There are three basic types of electromagnetic flowmeters which use alternating current to energize the magnet coil. These flowmeters are classified according to the shape of the magnet-current pulse, namely, sine wave, square wave, or trapezoidal wave. A fourth type uses a sine-wave magnet current, but the demodu-

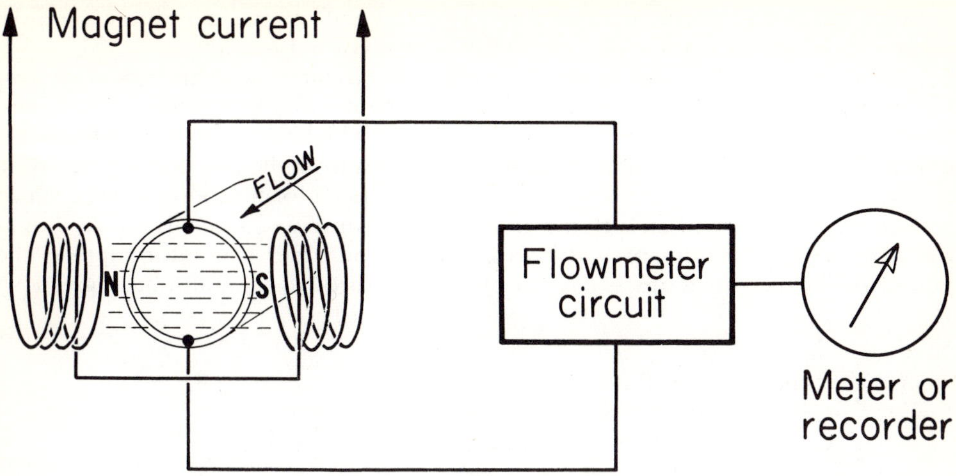

Figure 11-45. Principle of electromagnetic induction. (Courtesy of Lawrence Radiation Laboratory, Berkeley, Calif.)

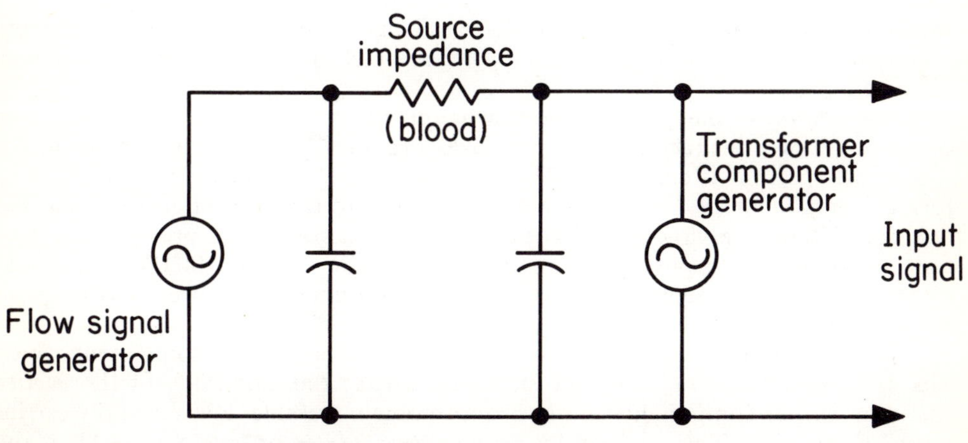

Figure 11-46. Analog of potentials forming the input signal of an electromagnetic flowmeter. (Courtesy of Lawrence Radiation Laboratory, Berkeley, Calif.)

lating system is sufficiently different to merit a separate classification. This last type, which uses phase-sensitive demodulation, was designed by Olmsted and will be discussed later in this chapter.

In the operation of the electromagnetic flowmeter, an alternating current energizes the magnet coils, as shown in Figure 11-45, resulting in an alternating magnetic field perpendicular to the field.

The energizing current may be a sine wave, square wave, trapezoidal wave, or some other time-variable shape. The flow of fluid through the vessel induces a voltage across the electrodes, which are perpendicular to both the field and to the vessel. This voltage is a measure of the velocity of the fluid in the vessel. Since the vessel is held to a specific diameter by the artery cuff of the transducer, the flow measured in units of volume per unit time can be recorded directly.

The voltage (V) induced in the electrode circuit is a combination of two separate voltage signals generated in the electrode circuit of the transducer. Figure 11-46 is a schematic diagram to explain the signals which are generated in the transducer. The transformer component generator results from the fact that the electrode circuit of the transducer element (including the blood) forms a loop of wire which acts as the secondary of an *air** transformer. This unwanted signal is referred to as the *transformer component* or *quadrature voltage.* The other generator supplies the potential V_f which develops as a result of the flow through the transducer, given by:

$$V_f = B\ell\, v \sin\theta \times 10^{-8} \qquad 11.14$$

where: V_f is the flow signal (volts)
ℓ = diameter of the blood vessel (cm)
$v \sin\theta$ = the time varying velocity of fluid in the flow probe
B is the magnetic flux density

Figure 11-46 shows how the flow signal and the transformer component combine to form the actual input signal. The *source impedance* is the impedance between the pickup electrodes in the transducer. Notice that the *source impedance* is capacitively coupled to ground by *stray capacitance.* The stray capacitance has a tendency to shift the flow signal with respect to the transformer component. This will become important when the differences among types of electromagnetic flowmeters are discussed.

The sine-wave electromagnetic flowmeter was first described by Kolin. Two of Kolin's flow transducers are shown in Figure 11-47. In this type of flowmeter the magnet current is sinusoidal. The electrode signal contains a sinusoidal transformer component (90° out of phase with respect to the magnet current) and the flow signal. Since the flow signal is of the order of microvolts the present sine-wave flowmeters employ a push-pull amplifier with a gain of over 100,000. Band-pass filters and negative feedback are used to limit frequency response so as to avoid noise signals.

The flow signal must be separated from the transformer component which acts as a carrier system. The flow signal modulates the amplitude of the carrier (the transformer component) and in order to obtain a usable flow signal the carrier must be removed. This is done by using a *demodulator* or *gate* circuit. Thus, the signal is allowed to pass through the network only when the gate is open. There are many circuits and schemes that can be used as a gate.† Commercial versions of the sine-wave flowmeter have used either the vacuum-tube *gated amplifier* or a simple switching transistor. The *gated amplifier* is a dual triode in which a signal (called a gating signal), usually derived from the magnet supply, drives the amplifier from cut-off to some usable portion of its characteristic curve, thus allowing the flow signal to appear at its output. The main problem with this type of circuit is that no matter how accurate the gating signal is with respect to the flow and transformer component signals, it cannot take into account the change in phase which results from inevitable changes in source impedance. These small changes in source impedance then appear as drift in sine-wave instruments.

---

*Blood and tissue have essentially the same magnetic permeability as air.

†A complete discussion of these gate circuits may be found in *Pulse and Digital Circuits* by J. Millman and H. Taub (McGraw-Hill Book Co., 1956). It would be worth-while for the reader to investigate gating circuits thoroughly as they have many uses in biomedical instrumentation.

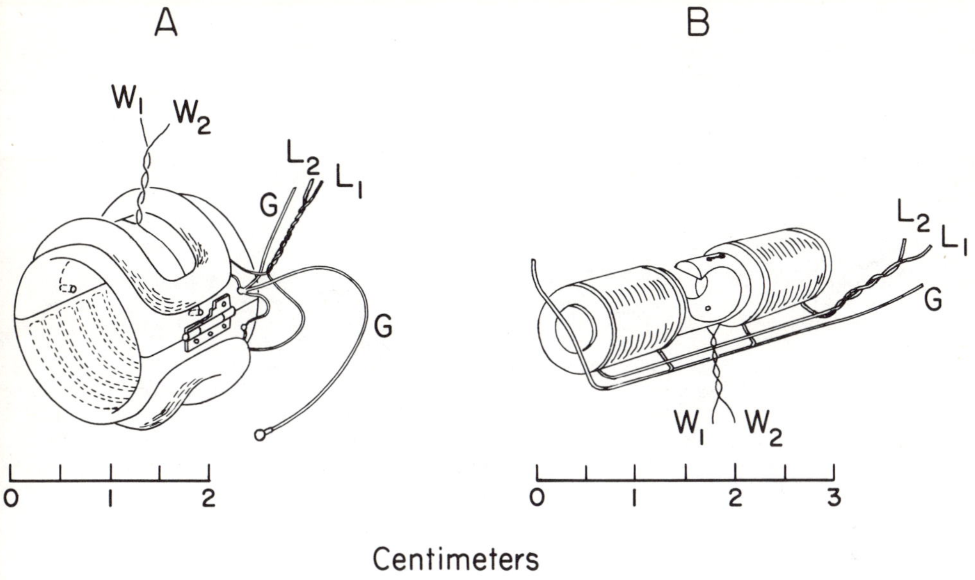

Figure 11-47. Two examples of miniature flow transducers designed by Alexander Kolin. (Courtesy of Lawrence Radiation Laboratory, Berkeley, Calif.)

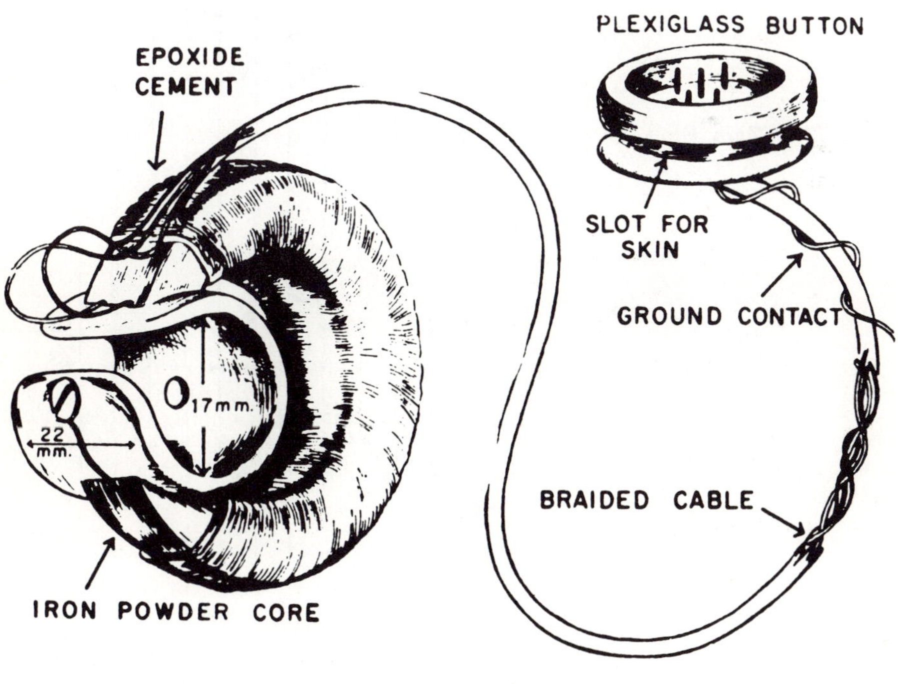

Figure 11-48. An Olmsted flow transducer. (Courtesy of Frederick Olmsted, Measurement of cardiac output in unrestrained dogs by an implanted electromagnet meter. IRE Transactions on Medical Electronics, vol. ME-6, no. 4, p. 211; published by the Professional Group on Medical Electronics of the IRE, 1959. Used by permission of The Institute of Electrical and Electronics Engineers.)

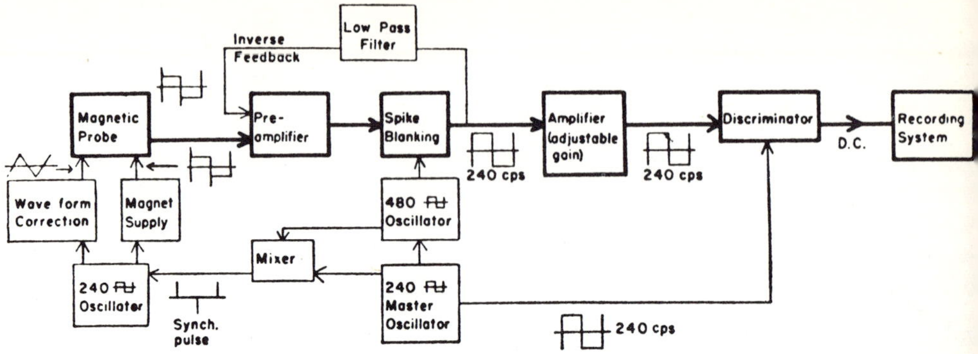

Figure 11-49. Block diagram of the square-wave flowmeter circuit. Courtesy of M. P. Spencer, and A. B. Denison, Jr., The square-wave electromagnetic flowmeter: theory of operation and design of magnetic probes for clinical and experimental applications. IRE Transactions on Medical Electronics, vol. ME-6, no. 4, p. 221; published by the Professional Group on Medical Electronics of the IRE, 1959. Used by permission of The Institute of Electrical and Electronics Engineers.)

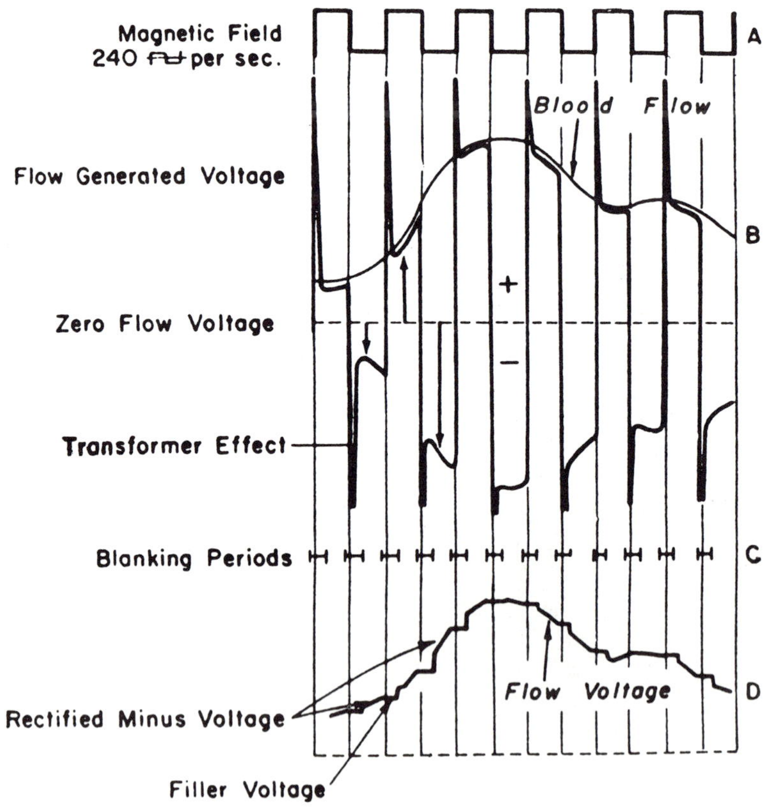

Figure 11-50. Waveforms associated with the square-wave flowmeter. (Courtesy of M. P. Spencer and A. B. Denison, Jr., The square-wave electromagnetic flowmeter: theory of operation and design of magnetic probes for clinical and experimental applications. IRE Transactions on Medical Electronics, vol. ME-6, no. 4, p. 221; published by the Professional Group on Medical Electronics of the IRE, 1959. Used by permission of The Institute of Electrical and Electronics Engineers.)

To avoid this problem a flowmeter using sine-wave magnet excitation and a phase-detection system was designed by Olmsted. In this circuit a portion of the magnet signal is squared by means of a bistable multivibrator *squaring circuit.* The transformer and flow signals from the transducer are also squared and mixed with the squared magnet signal. The end result is a square wave in which one vertical edge moves right and left as a function of flow. The movement of the edge is detected with a circuit called a *phase detector.* This circuit works very well except that since the magnet must be on in order to use the circuitry, it is a problem to design a method for obtaining a *zero flow* signal. An example of a flow transducer designed by Olmsted is shown in Figure 11-48. Notice the wire loops on the top of the artery cuff. These are adjusted to *buck out* or cancel some of the transformer component so as to bring it into the same amplitude range as the flow signal.

The square-wave flowmeter uses circuitry similar to the sine-wave flowmeter except that the magnet *current* is a square wave. In order to provide a square magnet current it is necessary to employ current feedback in the magnet amplifier rather than voltage feedback. Figure 11-49 is a block diagram of a square-wave flowmeter, and Figure 11-50 shows the waveforms associated with each block.

The use of a square-wave current to energize the magnet would seem to be a good idea. The transformer component should be instantaneous, and any shifts in the phase between the transformer component and the flow signal would not affect the circuit. The only problem is that in square-wave flowmeters designed to date, the instantaneous transformer component has an amplitude of 50 to 80 volts. This is excessive when one considers that the flow signal is only in the microvolt range. Also, as shown in Figure 11-51, the transformer component does not decay instantaneously, as it should, but instead decays exponentially with a finite time constant.

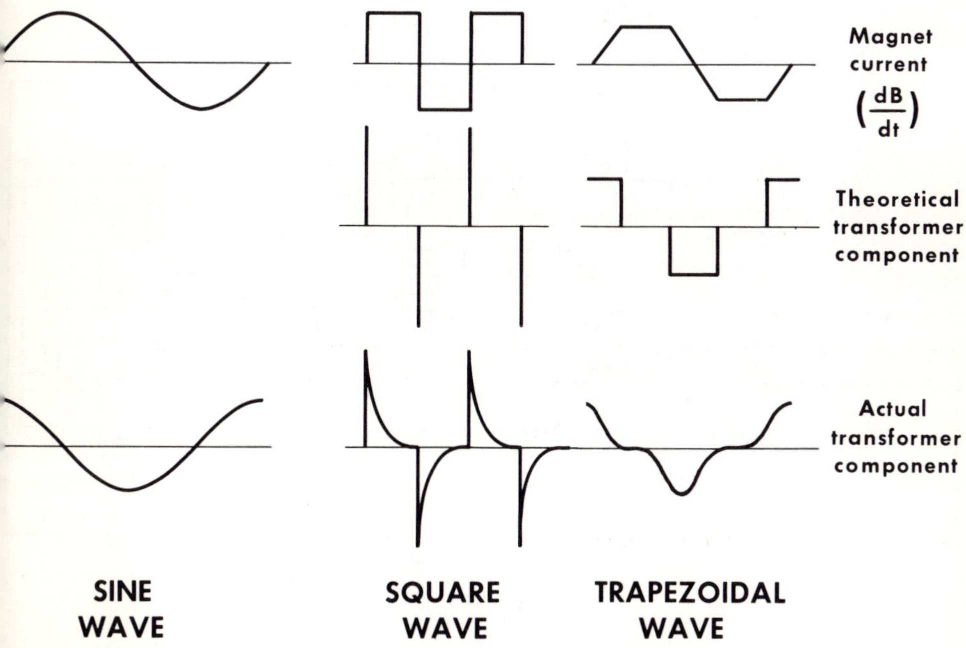

Figure 11-51. Transformer component in the sine-wave, square-wave, and trapezoidal-wave flowmeters. (Courtesy of Lawrence Radiation Laboratory, Berkeley, Calif.)

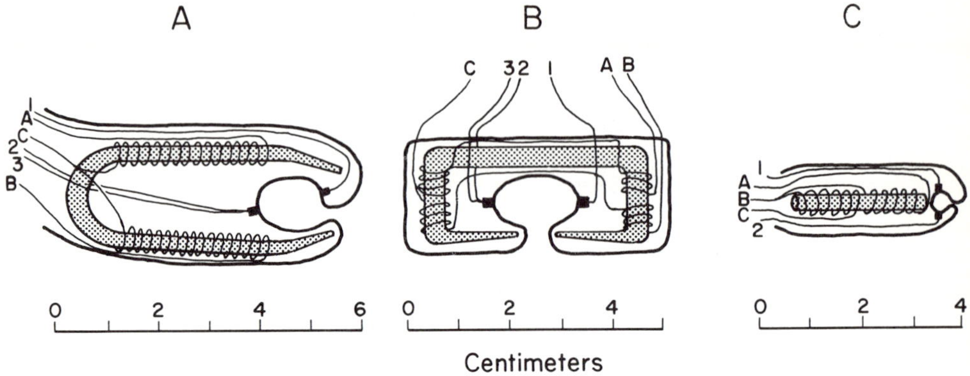

Figure 11-52. Transducers designed for the square-wave flowmeter. (Courtesy of Lawrence Radiation Laboratory, Berkeley, Calif.)

This is due to the use of capacitors, resistors, vacuum tubes, and other components in the circuitry. Therefore, for both of these reasons, an attempt is made to turn the entire amplifier off during the time that the transformer component is extremely large. This technique has met with some success as the square-wave instrument has been used in many laboratories for some time. Its greatest problem appears to be a need for large transducers, some of which are shown in Figure 11-52. It should be possible, however, to design a small transducer that would have a transformer component of 50 to 100 microvolts.

The *trapezoidal-wave flowmeter* was designed to avoid the problems which seemed to exist in the flowmeters previously described. Figure 11-53 is a block diagram showing the various components of the instrument. The commercial

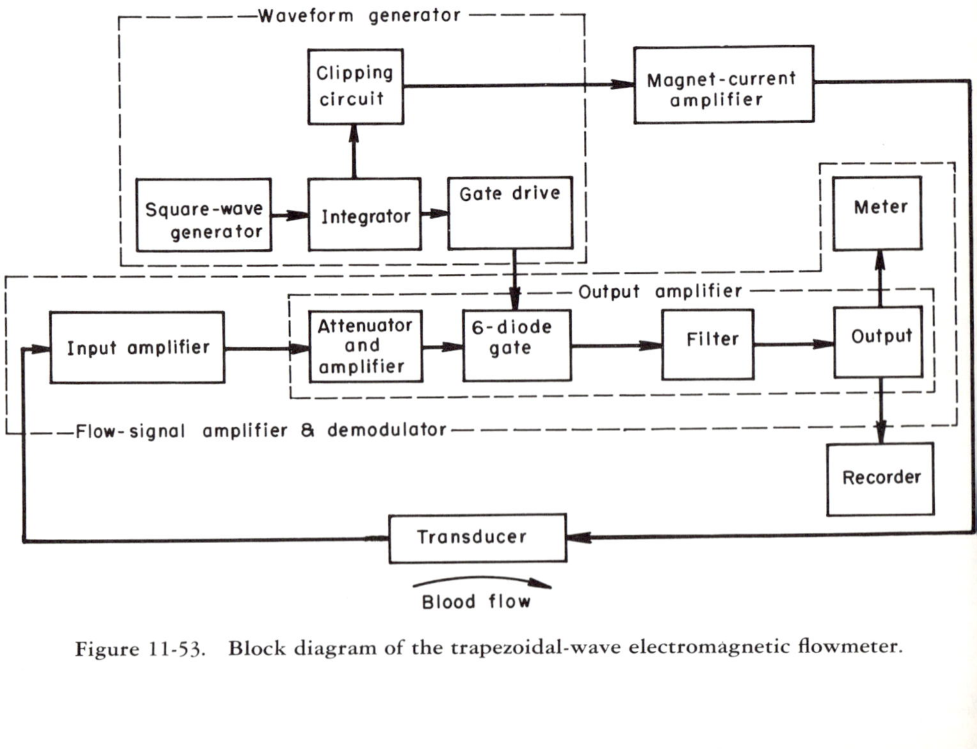

Figure 11-53. Block diagram of the trapezoidal-wave electromagnetic flowmeter.

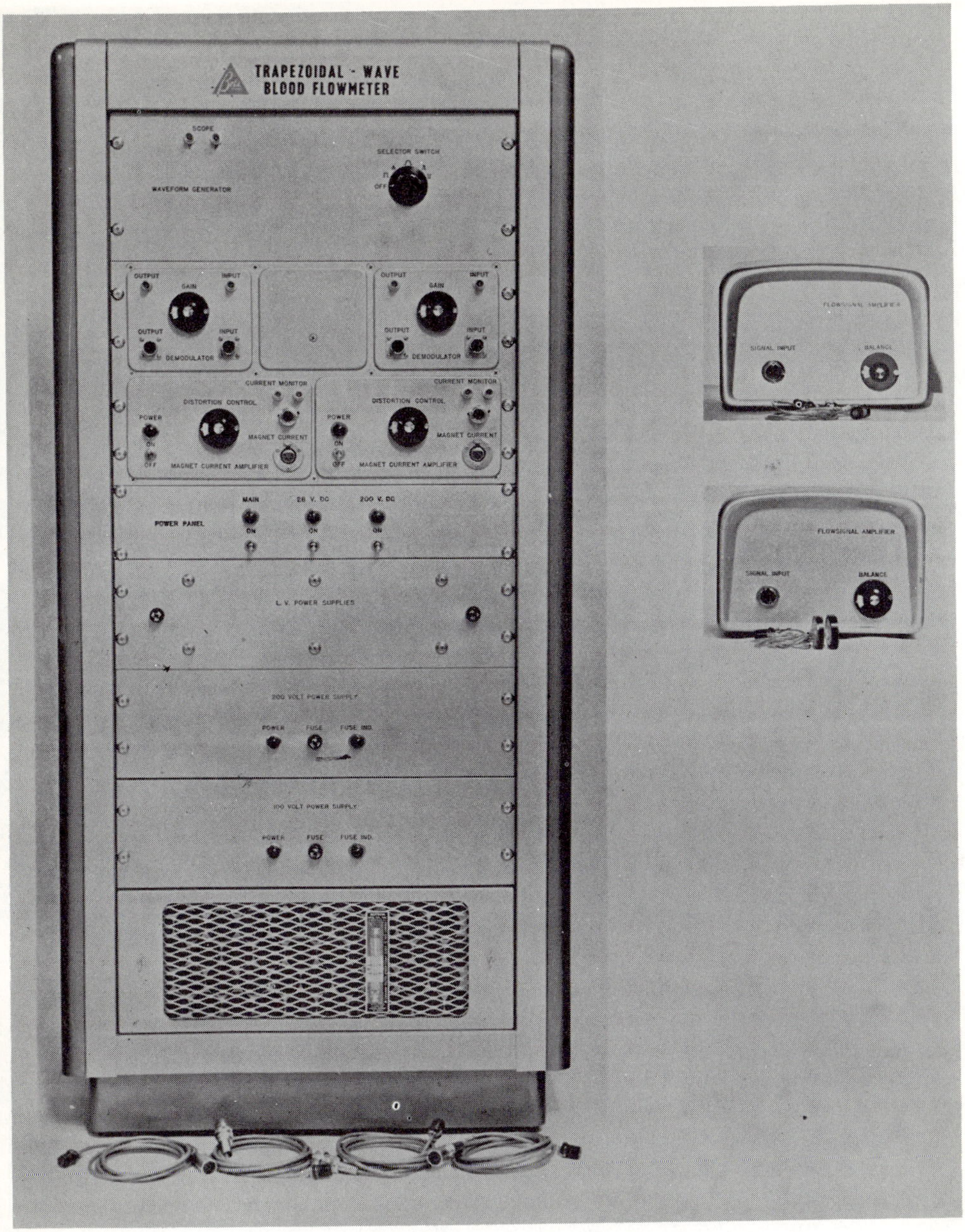

Figure 11-54. The two-channel trapezoidal-wave electromagnetic flowmeter.

version of the instrument is shown in Figure 11-54, and some transducers designed for use with this commercial flowmeter are shown in Figure 11-55.

The trapezoidal wave is generated by integrating a square wave to form a triangular wave which is then clipped by a pair of zener diodes to form the trapezoid-wave signal for the magnet current amplifier. The transistorized magnet current

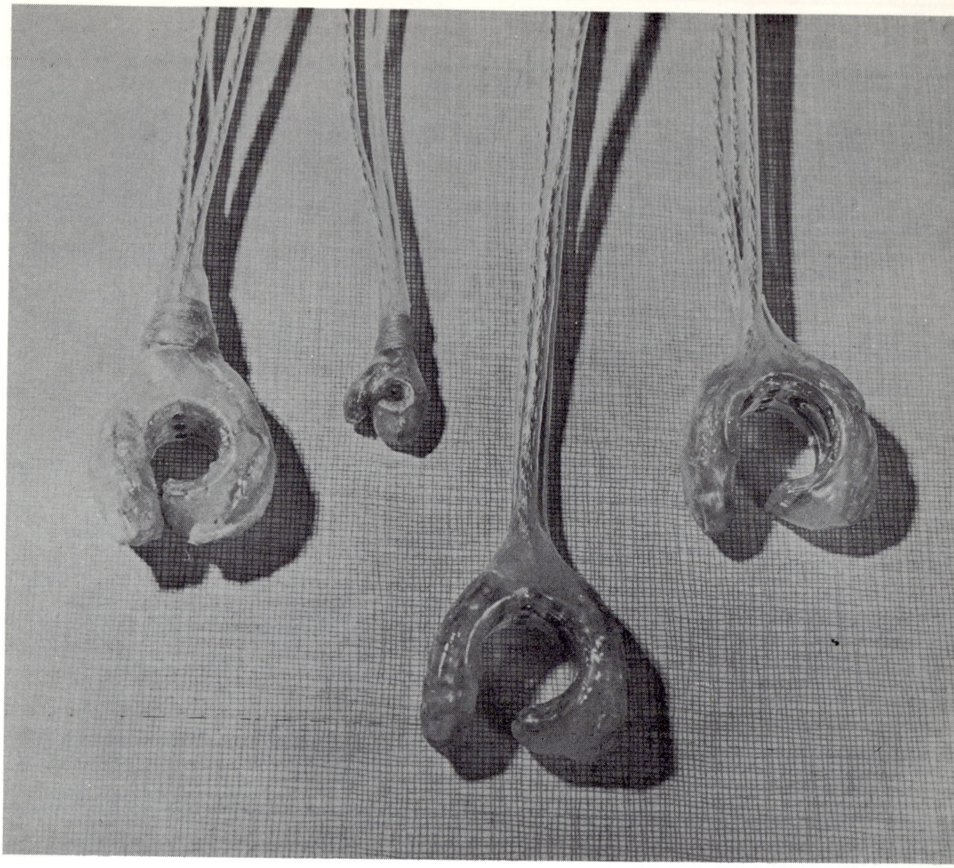

Figure 11-55. Transducers designed for the trapezoidal-wave flowmeter.

amplifier uses an adjustable current feedback (*distortion* and *shape* controls shown in Figure 11-56) to insure that the magnet current supplied to the magnet coils is, indeed, trapezoidal. In order to achieve stability, any circuit requires the use of a highly regulated power supply with as low a ripple as possible. The use of good power supplies will improve the stability of any instrument. All electromagnetic flowmeters need this stability for use in chronic experimentation. The reader should pay particular attention to this point when examining commercial instruments.

The input amplifier is a low-noise push-pull amplifier with a high-input impedance. The high-input impedance is required to minimize the effect of slight changes in source impedance. An input transformer with a very high reflected impedance and excellent magnetic shielding as well as electrostatic shielding has been used, but this has been replaced by eliminating the input transformer, resulting in an even higher input impedance. The transformer component generated with this instrument is shown in Figure 11-53. A feedback network has been designed to alter the amplitude and direction of the transformer component by adjusting a ten-turn potentiometer on the front panel. Each transducer requires its own initial

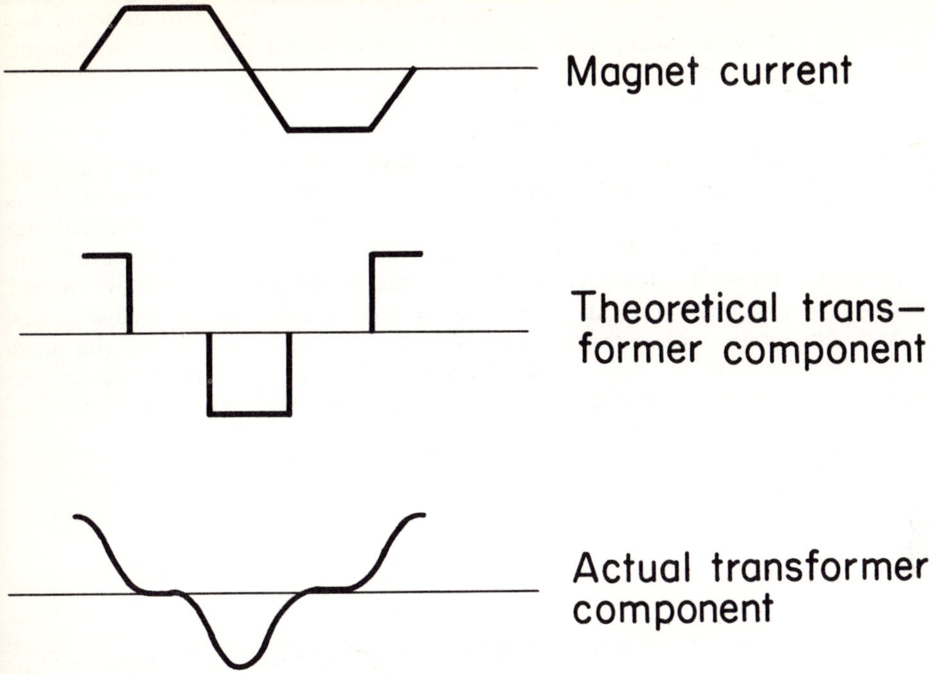

Figure 11-56. The transformer component generated by the trapezoidal-wave flowmeter. (Courtesy of Lawrence Radiation Laboratory, Berkeley, Calif.)

setting of this potentiometer. By adjustment of the transformer component in this way, the signal when the magnet is switched off can be adjusted to be equivalent to *zero flow*. This is done by occluding the vessel *once* and making the necessary adjustments in the input circuit before releasing the vessel. This procedure can be carried out *in vivo*.

Calibration of any electromagnetic flowmeter is difficult. It is especially important that the system be calibrated in the same way as when used for the flow study itself. The transducer of the electromagnetic flowmeter is particularly sensitive to noise, and the entire system must be properly grounded. In some flowmeters, however, this is not obvious because the noise signals have been filtered out. These noise signals do appear to affect the circuit enough to alter the calibration even though they do not appear in the output. Thus, while the frequency response of a flowmeter may be tested by using an electronic oscillator and other instruments which will supply a modulated signal to the input of the flowmeter signal amplifier, it is still important to make these measurements *in vivo*. This is especially true for the biological investigator who is less experienced in the "fine art of proper grounding." (See Chapter 12.)

The electromagnetic flowmeter transducer must be chosen so that it is the correct size for the vessel in which it is to be used. If the transducer is to be used for an acute measurement it should be tight fitting, but if, on the other hand, the transducer is to be implanted more or less permanently it should be a loose fit. In this case the animal will fill the space between transducer cuff and the vessel with fibrous tissue which serves to hold the transducer in place as well as acting as a saline bridge. The transducer will tend to be unstable for a few days following "permanent" implantation.

Static calibration may be accomplished by putting the transducer around the femoral artery. The animal must be chosen so that its femoral artery is the correct size for the transducer. The artery is cannulated with polyethylene tubing, and the transducer is placed around the vessel proximal to the cannula. A screw clamp is then put onto the cannula. Blood is collected for 15 seconds, and the volume per unit of time is then determined. The amount of blood flowing through the system may be adjusted somewhat by adjusting the screw clamp. The above animal preparation is not "physiological," but it does suffice to calibrate the transducer *in vivo.* Static calibration of a transducer implanted around the root of the aorta may also be effected by using a dye dilution technique. Usually, however, the aortic flow pulse during diastole (heart relaxation) is taken as zero flow.

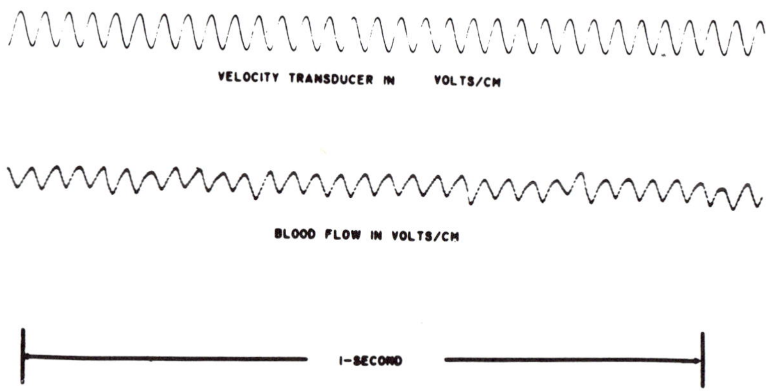

Figure 11-57. Dynamic calibration of the trapezoidal-wave flowmeter at 27 cps.

Frequency calibration is done by cannulating the common iliac artery with the nozzle from the pump shown in Figures 11-30 and 11-32. The transducer is placed around the femoral artery as before. A pediatric blood pressure cuff is inflated distal to the transducer to keep the pump from collapsing the vessel on its backstroke. Figure 11-57 is a record of the flow at 27 cps. The output of the velocity transducer which follows the movement of the piston is shown below the flow signal.

In summary, the electromagnetic flowmeter is a useful instrument for measuring blood flow in animals. There are as yet many problems to be worked out. An example of a problem which is universal to all flowmeter transducers, as well as a good many other instruments, is the encapsulation material. A variety of epoxy resins, silicone rubbers, and other plastics such as denture material have been used. In some cases the salty tissue fluid seeps into the transducer and causes an increase in the stray capacitance between the electrode circuit and ground or between the magnet supply and the electrode circuit. Salt leakage can also cause gradual shorting of the magnet coils, thus decreasing the magnet current and hence the magnetic field. *Tests in salt water do not begin to approach actual animal conditions because blood and tissue fluid are much more corrosive than salt water.* Compatibility of encapsulation materials with cables and wires is also a problem. New materials are being developed continually, and the search for a suitable encapsulation material will undoubtedly be carried on until the right combination is found.

REFERENCES:

A Critical Study of the Response of Manometers to Forced Oscillations, H. M. Yanof, A. L. Rosen, N. M. McDonald, and D. A. McDonald (Phys. Med. Biol. 8:407-422, 1963).

A Trapezoidal-Wave Electromagnetic Blood Flowmeter, H. M. Yanof (J. Appl. Physiol. 16:566-570 1961)

An Evaluation of Modern Pressure Recording Systems, D. L. Fry, F. W. Noble, and A. J. Mallos (Circ. Res. 5:40-46, 1957).

Blood Flow in Arteries, D. A. McDonald (Arnold Publishers. London, 1960).

Design of an Implantable Flowmeter Transducer Based on the Helmholtz Coil, H. M. Yanof, A. L. Rosen, and W. C. Shoemaker (J. Appl. Physiol. 18:227-230 1963).

III. *Flowmeters: Their Theory, Construction, and Operation,* E. Wetterer, Handbook of Physiology II (American Physiological Society, Washington, D. C., 1963).

Improvements in the Trapezoidal-Wave Electromagnetic Flowmeter, H. M. Yanof, P. Salz, and A. L. Rosen (J. Appl. Physiol. 18:230-232, 1963).

Instrumentation Methods for Physiological Studies, Vol. 1, Norman N. Goldstein (Operations Services Department, University of California Extension, Berkeley, Calif. 1964).

Medical Electronics in Cardiovascular Disease, Charles K. Friedberg, ed. (Grune and Stratton, New York, 1963).

Medical Physics, Vol. 3, Otto Glasser, ed. (Yearbook Publishers, Chicago, 1960).

Medical Physiology and Biophysics, ed. 18, T. C. Ruch and J. F. Fulton (W. B. Saunders Co., Philadelphia, 1960).

Pressure Measurement in Human Organism, A. T. Hansen (Acta Physiol. Scand. [supp. 68] 19:1-230, 1949).

12

Noise and Its Elimination

general remarks
source noise
noise inherent in the vacuum tube
noise inherent in other components
noise inherent in semiconductors
electrostatic interference
electromagnetic interference
noise generated in the animal
establishing a ground in the laboratory

GENERAL REMARKS

When one thinks of noise one usually thinks of some disturbing or irritating sound. Noise means different things to different people. To some it is a group of children playing when they are trying to concentrate; to others it is a loud radio or television, or a student practicing his first violin lesson. Similarly, *electronic noise* is an unwanted signal. It is a part of the output of an instrument not related to the measurement itself. Sometimes it is generated within the instrument, while at other times it is caused by external interference. This chapter will serve only as an introduction to the kinds of noise that are troublesome to the biomedical instrumentalist and will contain some ideas about how to avoid or eliminate them. There is no substitute for good solid experience. It is safe to add at this point that the major problem which confronts those making biological measurements with electron instruments is noise and its elimination. The instrument user is therefore encouraged to learn as much about noise elimination as possible.

In order to determine how much noise is present in the output of the instrument, the following procedure is used. The input terminals of the initial stage of amplification are shorted to ground, and either the peak-to-peak or the RMS output signal is noted. A known signal is then applied to the same input terminals, and either the peak or the RMS signal is noted. The signal-to-noise ratio is calculated by using both RMS values or both peak-to-peak values. The signal-to-noise ratio that is acceptable depends on the instrument being tested. An ultrasensitive low-level amplifier may do well with a signal-to-noise ratio of only 50, whereas other instruments should have a signal-to-noise ratio of 500 to 1000 or more. It is not uncommon for an estimation of the noise to be stated as 1 *microvolt referred to the shorted input.* In this case the gain of the amplifier or instrument is determined by dividing the output voltage by a given input voltage. The voltage output of the instrument when the input terminals are shorted is then divided by the gain. The resulting answer is called the *shorted-input noise referred to the input.* There are several other ways to describe the noise in a given instrument, but signal-to-noise ratio will suffice for the beginning instrumentalist.

SOURCE NOISE

This type of noise is introduced at the *source* of the measurement. It is associated with the electrical resistance of the material being measured and the temperature of the material at the source. In many animal measurements, especially those in which electrodes are in contact with tissue, the animal itself is the *source.* The amount of noise generated at the source is a function of the total resistance across the input terminals of the amplifier. (This includes the input resistance of the amplifier.) Small currents are generated from the random motion of electrons. When these small currents flow through a large resistance, the voltage developed is noise, and it may well be in the signal range. The noise signals which are generated may have any frequency. It should also be pointed out that the animal tends to act as an antenna. Therefore, a spectral analysis of the noise (the amount of noise signal at each frequency) would show a rather large 60-cps component. This is due to the fact that all electrical appliances use house current which in the United States is 60 cps (50 cps in most other countries), and this 60 cps is radiated from a variety of wires, etc.

Most biomedical electronic instruments are not sensitive to high-frequency signals as they are almost completely attenuated by the instrument circuitry. Other instruments which are designed to handle signals from DC to 100,000 cps or more do not attenuate high-frequency signals. The amount of source noise which appears in the output is related to frequency band-width of the instrument. Whitfield expresses the source noise as:

$$v_t = 126\sqrt{RB} \times 10^{-12} \text{ volts}$$

where R is the source resistance in ohms, B is the band-width in cps, and t is the temperature of the source. Higher temperatures cause more electrons to move, thus causing the source noise to increase. This type of noise can never be completely eliminated but it can be decreased by limiting the band-width of the electronic circuitry.

Measurements necessitating the use of metallic electrodes, such as needle electrodes, which come in contact with tissues, tend to generate a large amount of source noise. This is especially so when the electrode tends to move around. The use of dissimilar metals also generates large noise potentials.

NOISE INHERENT IN THE VACUUM TUBE

There are several types of noise generated within the vacuum tube which contribute to the over-all noise in the vacuum-tube amplifier. These may be divided into several categories:

1. shot noise
2. induced grid noise
3. partition noise
4. secondary emission noise
5. flicker noise
6. anomalous sources of noise.

The thermionic emission of electrons was described in Chapter 4. It will be recalled that the electrons are emitted from the cathode of a rectifier (or an amplifier tube) and form a cloud in the vicinity of the cathode. This density of this cloud is increased when the temperature of the cathode is elevated. The electron cloud, however, is not uniform, as some electrons are emitted which have considerably more energy than others. When a voltage is applied to the plate some electrons are accelerated toward the plate, but they do not arrive there at precisely the same instant. The arrival of the electrons at the plate may be likened to the firing of a shotgun at a barn door. Not all the shot arrives at the same instant nor is it uniformly distributed. The failure of all the electrons to arrive at the plate simultaneously gives rise to *shot noise.* This type of noise has a fairly broad frequency spectrum. It, like source noise, may be reduced by limiting the frequency response of the amplifier, or by heating the filaments of the vacuum tube with a low ripple-highly regulated DC power supply or with a storage battery. Shot noise can never be completely eliminated, especially in the low-frequency range.

The *induced grid noise* is caused by the motion of the electron past the grid. The electron induces a charge on the grid which results in a slight current in the grid-cathode circuit. The result is a small amount of noise which is not significant under 15 megacycles and is consequently not of considerable importance in biomedical instruments. Likewise, *partition noise* is caused by slight fluctuations in the potential of grids other than the control grid, such as in a tetrode or a pentode. As electrons pass these other grids, the partition noise is generated. When the moving electron strikes the plate it may cause emission of secondary electrons from the plate itself, thus giving rise to *secondary emission* noise. Shot noise, induced grid noise, partition noise, and secondary emission noise are very similar. The end result is the sum of the RMS of each individual type of noise.

Flicker noise is the result of an inhomogeneous potential distribution over the cathode surface so that there are spontaneous bursts of electrons boiled off which arrive at the plate and appear as signal. The term *flicker noise* arises from analogy to an unevenly burning material that causes the flame to waver unsteadily. Flicker noise or cathode noise decreases in amplitude with increasing frequency. It is an important source of noise in biomedical instruments since it occurs at low frequencies. This type of noise can be minimized by operating at a lower filament voltage than recommended. Well-designed low-noise amplifiers often have a separate high-quality power supply (minimum ripple and good regulation) which operates at about one half to three fourths of the rated filament voltage. It is especially important that the first one or two stages of amplification generate a minimum of noise because the noise in each stage is amplified as if it were signal.

There are a number of other causes of noise in vacuum tubes, such as the presence of a minute amount of gas in the tube. These gas molecules are likely to be present in a new tube. After a time they will be ionized and will travel out of the tube in the external circuit. Proper aging of the tube will help to eliminate these gas molecules. Another source of noise is a loose mechanical structure within the tube which increases when the tube is moved or jarred. This is called microphonic noise as the tube acts like a microphone. Some tube types are especially sensitive to movement, shaking, or even acoustic pressure. Such a tube is called a microphonic tube and should be replaced, especially in the first few stages of amplification. A good low-level amplifier will *shock mount* the input tube to avoid this problem.

NOISE INHERENT IN OTHER COMPONENTS

Almost every component used in electronic circuitry generates some noise. In some cases, changes in temperature alter the properties of a component. This is especially true of resistors. In other cases small undetectable shorts develop which change the value of the component and generate noise. The transformer is especially sensitive to the electric and magnetic fields which surround it.

Each component has its own problems, but modern technological advances have improved almost every variety of electronic element. A good example is the development of the metal film resistor. These high-quality resistors can be obtained which change less than one part per million per degree centigrade of temperature change. This means a 1-megohm resistor changes its resistance less than 1 ohm per degree centigrade of temperature change. It is important to use these high-quality (expensive) resistors in at least the input stage of any sensitive amplifier. The manufacturer should be asked if he has used metal film resistors and high-quality capacitors in his instrument. Considering the cost of instruments, these components do not increase the cost enough to be left out.

NOISE INHERENT IN SEMICONDUCTORS

The genesis of noise in the semiconductor is similiar to that in vacuum tubes. There are basically three sources of noise in semiconductors:

1. thermal noise
2. shot noise
3. flicker noise ($1/f$ noise).

Thermal noise is due to the random motion of electrons or holes and is present whether or not an electric field is present. The frequency spectrum of thermal noise is very broad. When viewed on an oscilloscope screen, thermal noise looks like grass, and it is often called just that. It may also be referred to as Johnson's noise after its discoverer F.B. Johnson (1925). Thermal noise is not confined to semiconductors. It is, as might be expected, found to be a contributor to source noise and to noise in vacuum tubes.

Shot noise in semiconductors is very similar to that in vacuum tubes except that the emission of electrons is not confined to the cathode but may occur anywhere in the crystal structure, but it cannot occur if no voltage is applied. Shot noise in transistors and diodes is caused by the carrier's drifting or diffusing through the semiconductor.

Flicker noise in semiconductors requires the presence of an electric field. The amplitude of the flicker noise is inversely related to frequency and is, on the whole, that noise which occurs over and above shot and thermal noise.

In general, transistor amplifiers are noisier than their vacuum-tube equivalents. The development of the transistor and multifunction semiconductor device is occurring so rapidly that such a comment may be out-dated by the time this book is printed.

ELECTROSTATIC INTERFERENCE

The effect of bringing a charged body close to a conductor was discussed in Chapter 1. If a negatively charged body is brought close to a conductor it will induce a charge in the conductor as shown in Figure 12-1(A). If the other end of the charged conductor is attached to ground through a resistor, the excess electrons

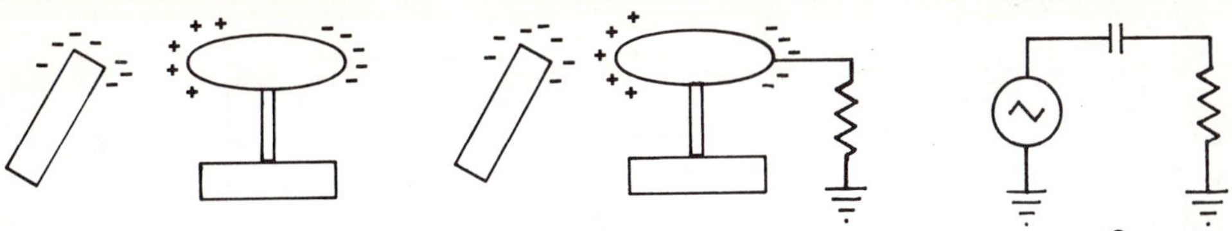

Figure 12-1. A, An induced charge on a conductor; B, leakage of the charge to ground through a resistance; C, the equivalent of B.

will flow to ground, and voltage will be developed across the resistor. This voltage is, in effect, a noise signal. Figure 12-1(C) is the electronic circuit which represents the physical situation described in Figure 12-1(B). The charged rod is the generator and the left plate of the capacitor, and the charged conductor is the right plate of the capacitor. Thus the signal from the generator passes through the capacitor and the resistor to ground.

In actual electronic circuitry the charged rod generator may be compared to any source of voltage such as a wire plugged into the wall (even if the switch is off and no current is flowing). The dielectric medium between the plates of the capacitor in this case is air and the other plate is any wire, chassis, or component in the circuit of the instrument. Wires and transducers in the input circuit are the most effective parts of the circuit for generating noise because they are usually unshielded, and in addition the noise is amplified by all stages. Careful placement of wires used to fabricate an instrument is necessary to avoid noise.

Electrostatic interference can be eliminated or substantially reduced by surrounding the input wires, transducers, etc., with a metal screen, mesh, or foil which is then connected to ground with a very low resistance wire. This prevents the electrostatic field from affecting those parts within the shield. Figure 12-2 illustrates the principle of electrostatic shielding. Notice that when the input wire is not shielded, the electrostatic signal is coupled directly to the wire. When, however, as shown in Figure 12-2(B), the wire is surrounded by a shield and the shield is grounded, the excess electrons travel directly to and from ground. In Figure 12-2(C) a plug was used to connect one part of the input lead, a transducer, for example, to the wire leading into the amplifier itself. Even though both sections of wire are shielded, if the plug is non-metallic or ungrounded and even though it is only 1/2 inch long, noise will be capacitively coupled to the input circuit, thus ruining the effectiveness of the shielded wires. A small amount of tinfoil wrapped around the plug and *properly* grounded will usually be sufficient to eliminate this source of noise.

It is also helpful to unplug all appliances and circuits not in use in the vicinity of the instrument being used to avoid sources of potential that cause electrostatic interference. Rooms used for sensitive measurements should not be located near diathermy or x-ray equipment. It is sometimes necessary to completely enclose the entire room in which certain measurements, such as the electroencephalogram, for example, are made. This may be done by surrounding the room with copper screening, soldering all edges together, and properly grounding the cage at one point. Fluorescent lights, telephones, and extraneous line cords must, however, be kept out. Doors and windows, like the plug in Figure 12-2(C), must also be shielded.

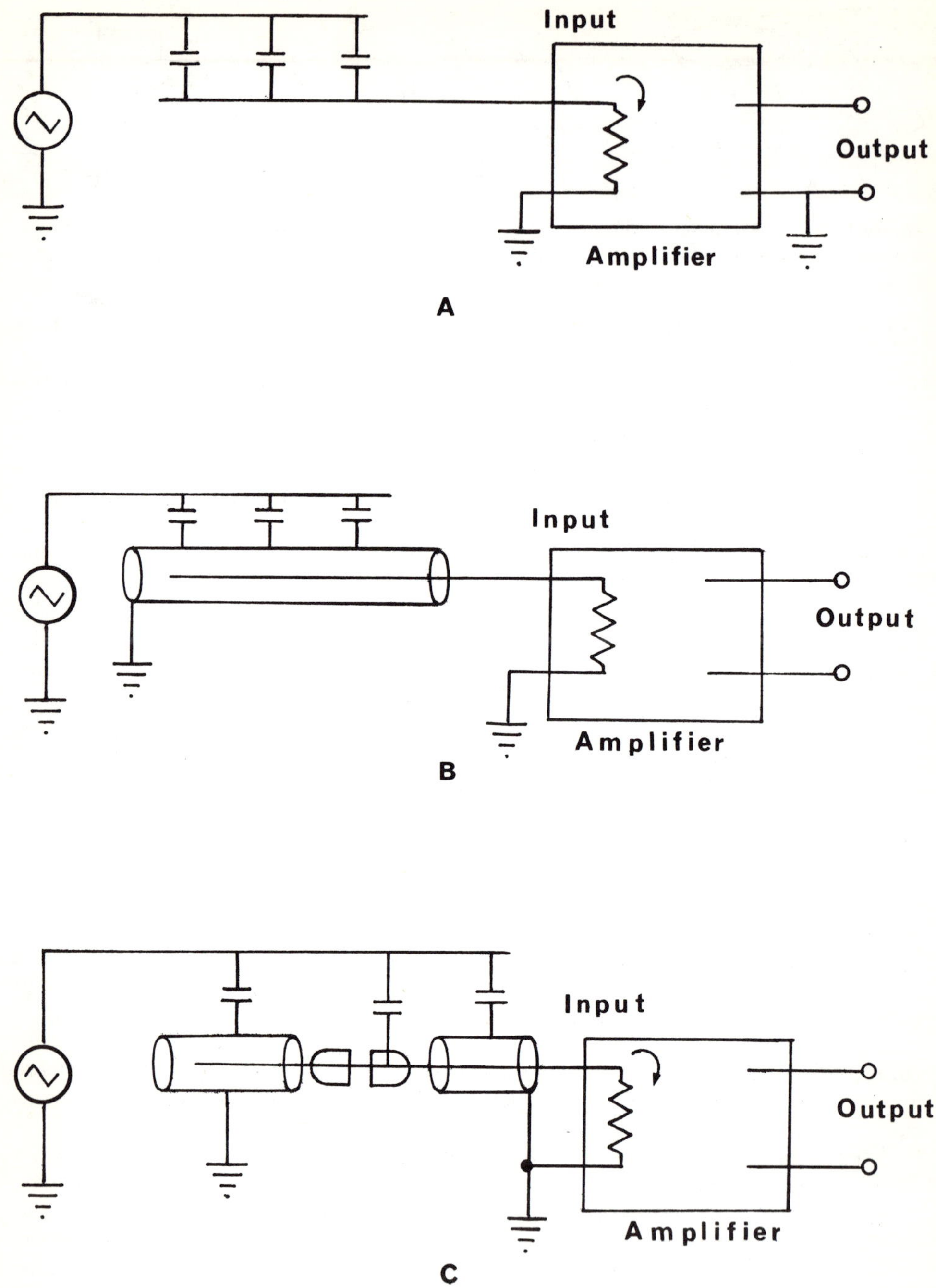

Figure 12-2. Principle of electrostatic shielding.

ELECTROMAGNETIC INTERFERENCE

A wire which carries a current generates a magnetic field as shown in Figure 1-13. Similarly, when a time-varying magnetic flux crosses a wire which is part of a circuit, current flows in the wire. This current is not due to any useful signal and is therefore noise. Electromagnetic interference may be compared to a transformer in which the current-carrying noise generator is the primary coil, and the conductors making up the instrument may be compared to the secondary coil. In this analogy the core material of the transformer is air.

Circuits and parts of circuits which are sensitive to electromagnetic interference are shielded in much the same way as for electrostatic interference except that the shield must be a material which has a high magnetic permeability like iron or a special alloy called mumetal. The magnetic shielding should then be grounded at a common point so that it serves as electrostatic shielding as well.

It is sometimes necessary to use equipment, such as a motor, in the vicinity of the instrument making a sensitive measurement. It is helpful to surround the motor with a high magnetic permeability shield to attenuate the magnetic field which it generates and thus keep the electromagnetic signal away from the instruments being used. If there is a source of magnetic flux in or near the laboratory that cannot be located, a search coil may be used to find it: Ten or twenty turns of fine insulated magnet wire are carefully wrapped around a cotton applicator stick. The leads from the search coil are attached to a shielded cable and connected to a sensitive oscilloscope. The search coil may then be faced in various directions. The signal viewed on the oscilloscope will be greatest when the open side of the coil is facing in the direction of the source of the magnetic field (90° to the magnetic flux).

NOISE GENERATED WITHIN THE ANIMAL

When a measurement is being made in an animal, the instrumentalist may be faced with a variety of noise generators which he did not expect. The biggest offender in this category is the ECG voltage generated in the heart muscle. It has an amplitude of over 5 millivolts and becomes quite annoying when one is trying to measure microvolt signals in the vicinity of the heart. This is especially true since the frequency spectrum of most of the various noise sources in the animal is from DC to about 100 cps. Figure 12-3 is an illustration of the frequency spectrum of a variety of noise sources found in the animal itself.

ESTABLISHING A GOOD GROUND IN THE LABORATORY

Every laboratory using electronic instrumentation should be provided with a *good ground.* This means that electrostatic shields must be provided with an extremely low resistance path to the earth. In this way excess electrons can travel to and from the earth, which acts like an infinitely large source of electrons and also an infinitely large sink into which electrons can flow.

This is not as easy as it sounds. In some laboratories the cold-water plumbing is used because it is supposed to be metallic and it is connected to outside water pipes which are sunk into the earth below the water level. As often as not, cold water pipes, radiators, etc., do not provide a low enough resistance path to ground. The result is that they may even generate additional noise because of hidden loops and pass this noise on to the instrument itself.

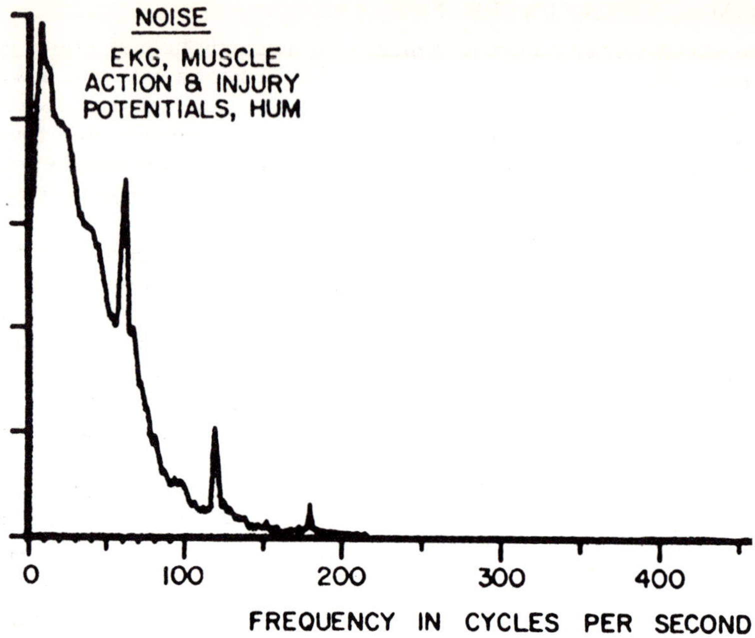

Figure 12-3. Frequency spectrum of noise sources in the animal body. (Courtesy of H. W. Shirer, R. B. Shackelford, and K. E. Jochim, A magnetic flowmeter for recording cardiac output. IRE Transactions on Medical Electronics, vol. ME-6, no. 4, p. 232; published by the Professional Group on Medical Electronics of the IRE, 1959. Used by permission of The Institute of Electrical and Electronics Engineers.)

A location for a *good ground* may be where the main water pipe enters the building. If this is to be used as a ground a large-core coaxial cable, such as RG83U, which is a 14-gauge wire insulated with a thick plastic coating and a closely woven electrostatic shield, is laid from the laboratory to the place where the main water pipe enters the building. At this point both the shield and the core are securely attached to the water pipe by brazing or by drilling and tapping the water pipe.

An ideal way to provide a suitable ground is to dig holes in the earth about 10 feet deep and lay in a number of copper rods about 1 to 2 inches in diameter. Each rod is several feet from the other. The rods are then connected together as shown in Figure 12-4, and a large-core coaxial cable is run to the laboratory. It is sometimes necessary to isolate the 60-cps house current from the outside power transformers to avoid transfer of noise to the instrument. This is done by using a large shielded isolation transformer.

When a good ground has been established in the laboratory, *this point is chosen as a reference.* A sensitive AC voltmeter, 1 millivolt full scale, and preferably battery operated (such as Hewlett-Packard Model 403B), is connected from the reference ground point to the chassis of the instrument being used. The shielded leads from the voltmeter to the chassis must be kept as short as possible. The voltage, with the instrument off or on, should be no more than 1 millivolt. It may, however, be over 1 volt if an improper ground, like a water pipe, is used, in which

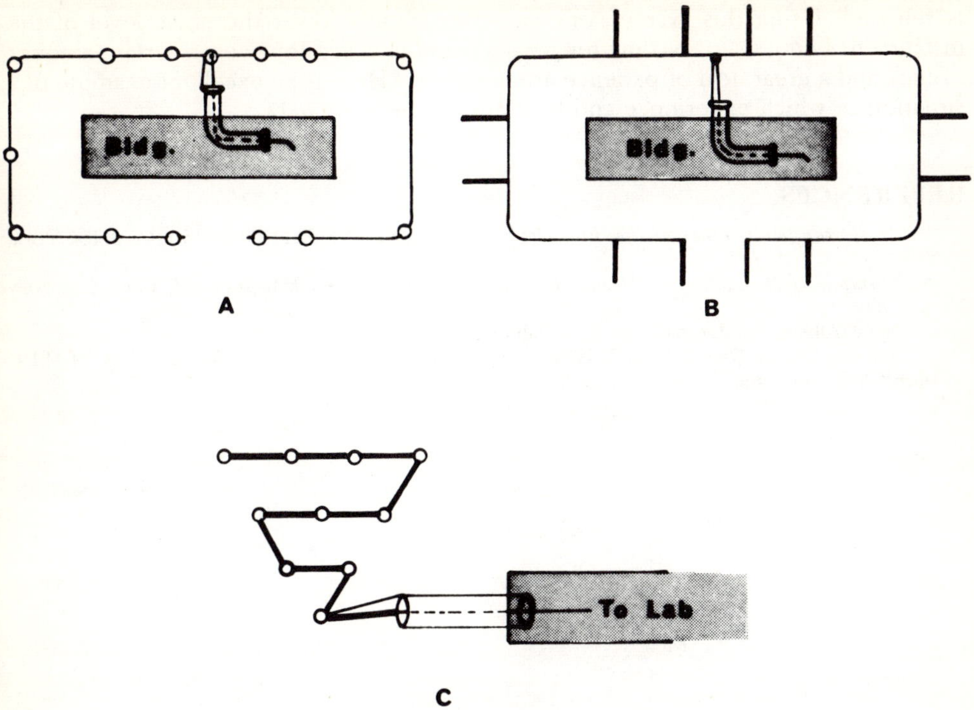

Figure 12-4. Patterns for establishing a good earth ground.

case a new reference point is chosen and the procedure repeated. The best reference point is the one which yields the lowest voltmeter reading.

Every piece of equipment is then grounded to this *reference ground point,* care being taken to avoid an excessive number of ground leads. It is often helpful to place a sheet of tinfoil (if the animal table is not metal) under the animal. The tinfoil is then insulated from the animal and carefully grounded to the reference point. It *may* also be helpful to ground the animal itself. The animal table and the instrument must not be grounded to the reference point and the table then grounded to the instrument, however. If this were done a *ground loop* would be established and the noise would increase considerably. All ground wires from the reference ground to instruments, shields, etc., should be arranged as symmetrically as possible and should be of at least 12-gauge insulated wire.

If the above directions are carefully followed and the circuit is still noisy, the voltage from the reference ground to each of the grounds from various pieces of equipment and appliances should be measured with the AC voltmeter.* Sometimes it is best to leave the voltmeter connected to the reference ground and the chassis of the instrument and watch the dial while each of the other ground leads

*Use an AC voltmeter such as the Hewlett-Packard Model 403B. Connect one lead to the ground reference point and the other lead to the piece of equipment (or other point) the ground of which is being tested. Keep the leads as short as possible and shield if necessary.

is released. Using this system, one can considerably lower the noise level of the instrument being used. No matter what system is used, however, a certain amount of luck and a great deal of patience are necessary. Here is an excellent example of a situation in which experience and luck are the best teachers.

REFERENCES:

An Introduction to Electronics for Physiological Workers, I.C. Whitfield (Macmillan Co., New York, 1959).

Essentials of Biological and Medical Electronics, Ralph W. Stacy (McGraw-Hill Book Co., New York, 1960).

Noise, Albert Van der Ziel (Prentice-Hall, New York, 1959).

Noise in Electron Devices, Louis D. Smullin, and Herman A. Haus, ed. (Technology Press of M.I.T. and John Wiley and Sons, New York, 1959).

APPENDICES

Appendix I. Some Physical Units and Conversion Factors

Name of Unit in System Named*

Quantity	*Defined by*	*Electrostatic*	*MKS*	*Electro-magnetic*	*Equivalence*
Electric charge	$F = \frac{kq_1q_2}{d^2}$	ESU of charge (stat-coulomb)	coulomb		1 coul = 3×10^9 esu
Electric field intensity	F/q	dynes/esu	$\frac{\text{newtons}}{\text{coulomb}}$		$1 \frac{\text{newt}}{\text{coul}} = \frac{1}{3 \times 10^4} \frac{\text{dynes}}{\text{esu}}$
Potential difference (voltage)	Work/q	ergs/esu	joules/coul		$1 \frac{\text{joule}}{\text{coul}} = \frac{1}{300} \frac{\text{ergs}}{\text{esu}}$
Magnetic field intensity	F/m		$\frac{\text{webers}}{\text{meter}^2}$	oersted	$1 \frac{\text{weber}}{\text{meter}^2} = 10^4$ oersted
Magnetic flux			webers	maxwells	1 weber = 10^8 maxwells

*From R. W. Kenworthy, College Physics, F. A. Davis Co.

Appendix II. Conventional Electronic Symbols

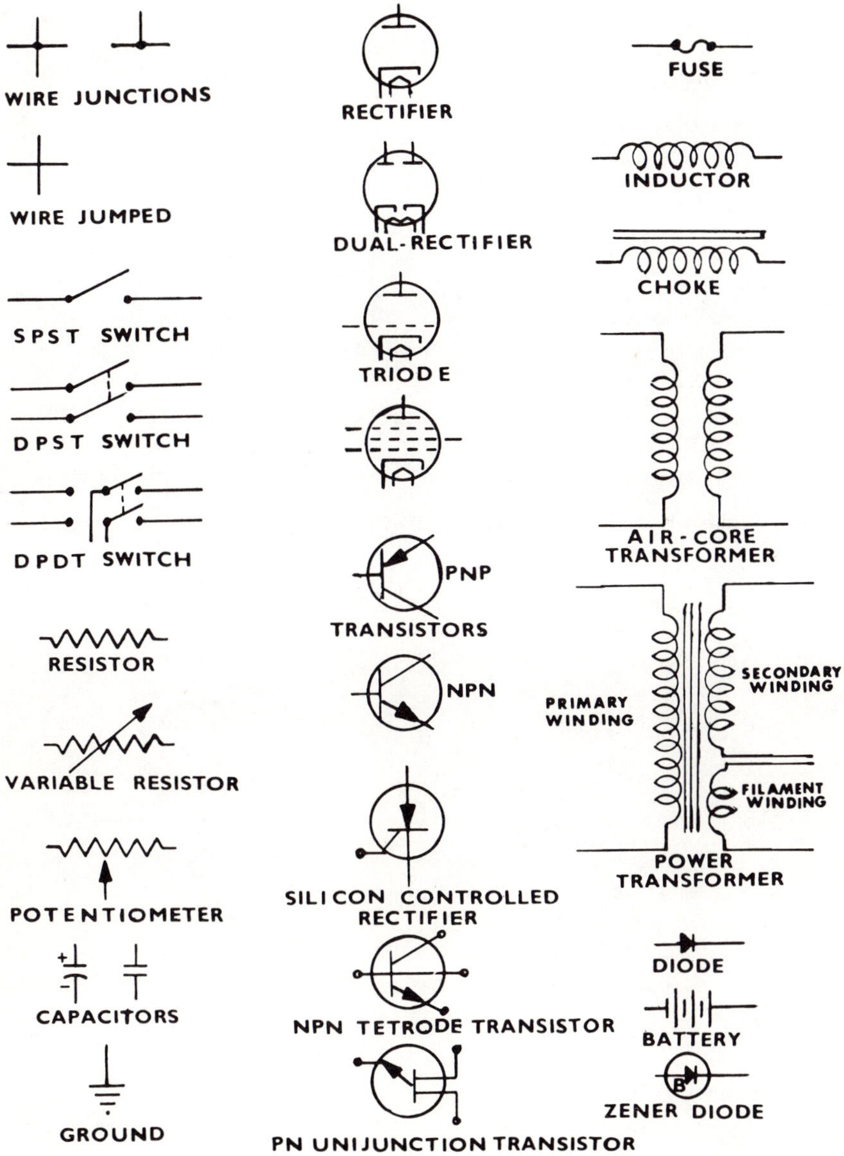

Appendix III. Identification of Components (Color Codes, etc.)

THE FIXED RESISTOR

The commonly used carbon resistor with axial leads is coded by a system of color bands. These color bands provide the user with the magnitude of the resistance in ohms and the manufacturing tolerance of the resistance, which may be 5 per cent, 10 per cent, or 20 per cent. These resistors are available in five increasing physical sizes according to their power capacity: 0.1, 0.25, 0.5, 1, and 2 watts. Figure III-1 shows a carbon axial lead resistor, with the color bands as A, B, C, and D. The A and B bands indicate the significant figures in the magnitude of the resistance according to the values listed in Table III-1. The C band defines the decimal multiplier. The D band indicates the permissible variation of the resistance in terms of the percentage of the resistance in ohms. It may be a silver band, which indicates 10 per cent tolerance, or a gold band, which indicates 5 per cent tolerance. If there is no D band, tolerance is 20 per cent. These resistors may be obtained with values listed in Table III-2. Larger values, up to 100,000 times as great, may also be easily obtained.

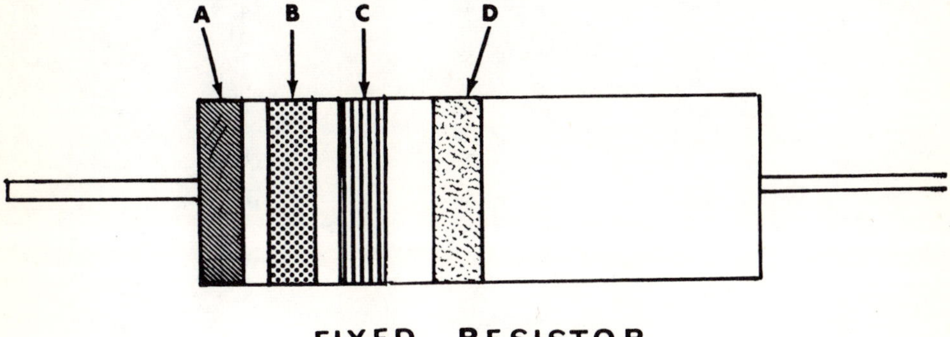

Fig. III-1

TABLE III-1
Resistor-Capacitor Color Code

Color	*Significant Figure*	*Decimal Multiplier*	*Tolerance (%)*	*Voltage Rating**
Black	0	1	—	—
Brown	1	10	1*	100
Red	2	100	2*	200
Orange	3	1,000	3*	300
Yellow	4	10,000	4*	400
Green	5	100,000	5*	500
Blue	6	1,000,000	6*	600
Violet	7	10,000,000	7*	700
Gray	8	100,000,000	8*	800
White	9	1,000,000,000	9*	900
Gold	—	0.1	5	1000
Silver	—	0.01	10	2000
No Color	—	—	20	500

*Applies to capacitors only.

TABLE III-2
Standard Component Values for the First 100 ohms*

20% Tolerance	*10% Tolerance*	*5% Tolerance*
10	10	10
		11
	12	12
		13
15	15	15
		16
	18	18
		20
22	22	22
		24
	27	27
		30
33	33	33
		36
	39	39
		43
47	47	47
		51
	56	56
		62
68	68	68
		75
	82	82
		91
100	100	100

*For values between 1 KΩ and 10 KΩ, multiply each value above by 10; between 10 KΩ and 100 KΩ, multiply each value by 100; between 100 KΩ and 1 MΩ, multiply each value by 1000.

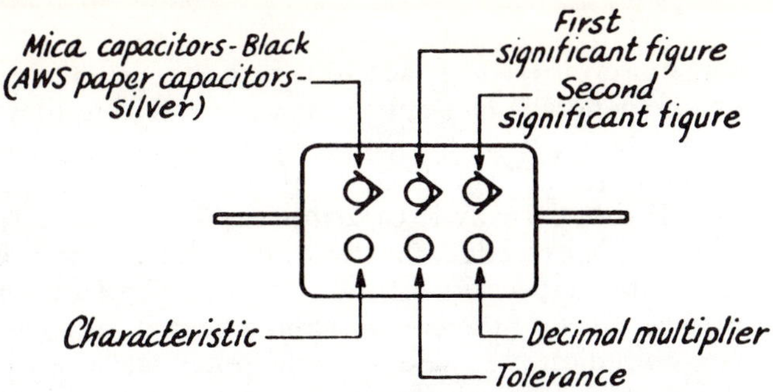

AWS and JAN fixed capacitors

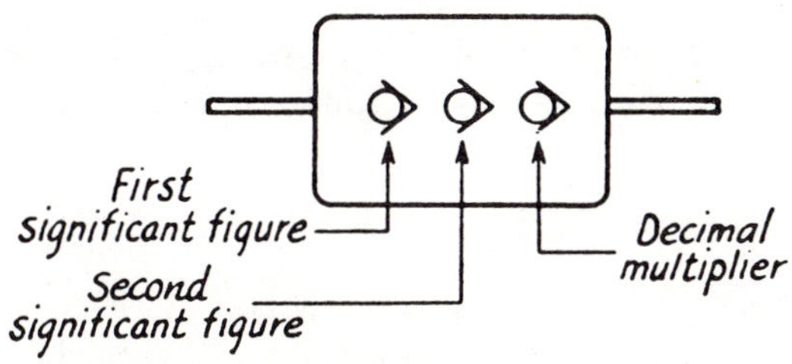

RETMA 3-dot 500-volt, ±20% tolerance only

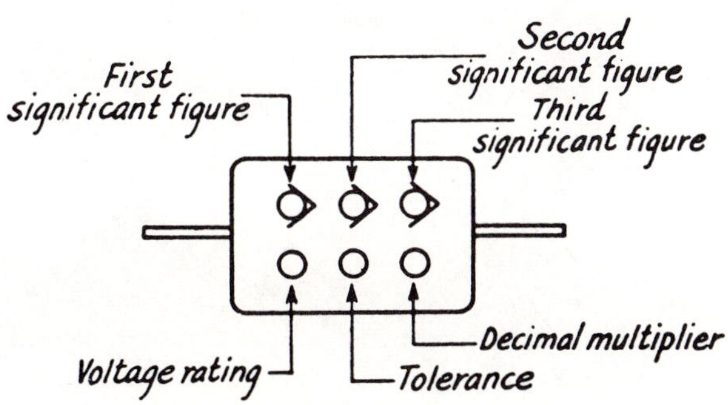

RETMA 6-dot

Fig. III-2

There are many other resistors in use which do not follow the above color-code scheme. These are, for the most part, power resistors (over 2 watts), wire-wound resistors, and special purpose resistors with a 1 per cent tolerance such as metal film resistors. These resistors generally are imprinted with the magnitude of the resistance. Values between those listed for the 5 per cent tolerance in Table III-2 are available as 1 per cent resistors.

CAPACITORS

There are many more varieties of capacitors than there are resistors. Capacitors over 2 microfarads are generally electrolytic capacitors, that is, they are polarized. These capacitors have the value of the capacitance and a voltage, which must not be exceeded, written on them. This voltage should be at least two times the voltage across the capacitor. If the electrolytic capacitor is put into the circuit in reverse or if the working voltage is exceeded, the capacitor will pop and it is then useless.

There are many other capacitors from 0.001 to 2 microfarads which are non-electrolytic and may be put into the circuit without regard for sign. Most of these capacitors, however, do indicate the side to be the potential closest to ground.

Capacitors of less than 1000 picofarads (equivalent to micromicrofarads) are generally either ceramic or mica components. These capacitors are coded according to the 6-dot RETMA (Radio-Electronics-Television-Manufacturers Association) code or the JAN code (Joint Army-Navy specifications). There is also a 3-dot RETMA code but these components are all 500 volt, ± 20 per cent tolerance. Figure III-2 shows the various forms of postage-stamp capacitors and identifies the various dots. The value of the dots may be found in Table III-1.

Appendix IV. Typical Vacuum-Tube Characteristic Curves

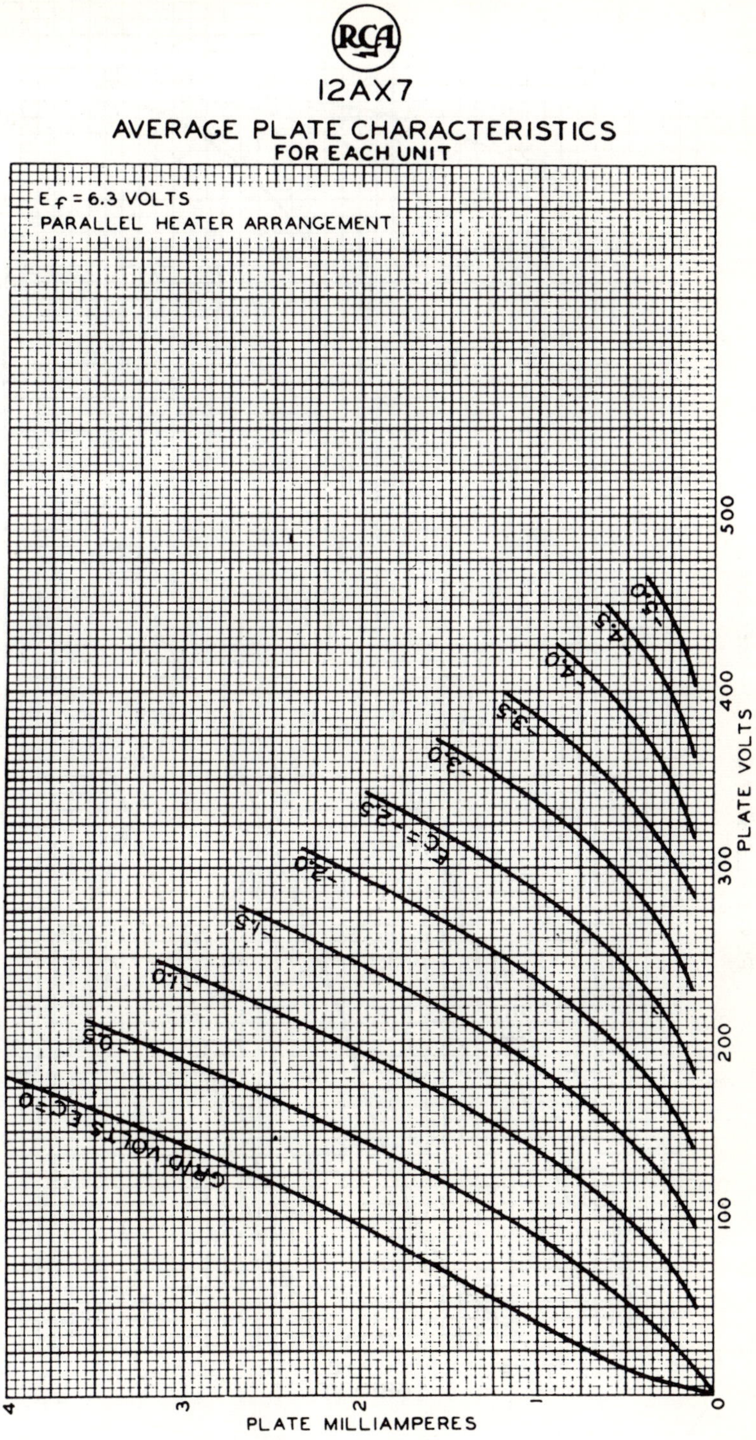

Fig. IV-1

Note: In RCA characteristic curves E stands for voltage; E_f is the filament voltage.

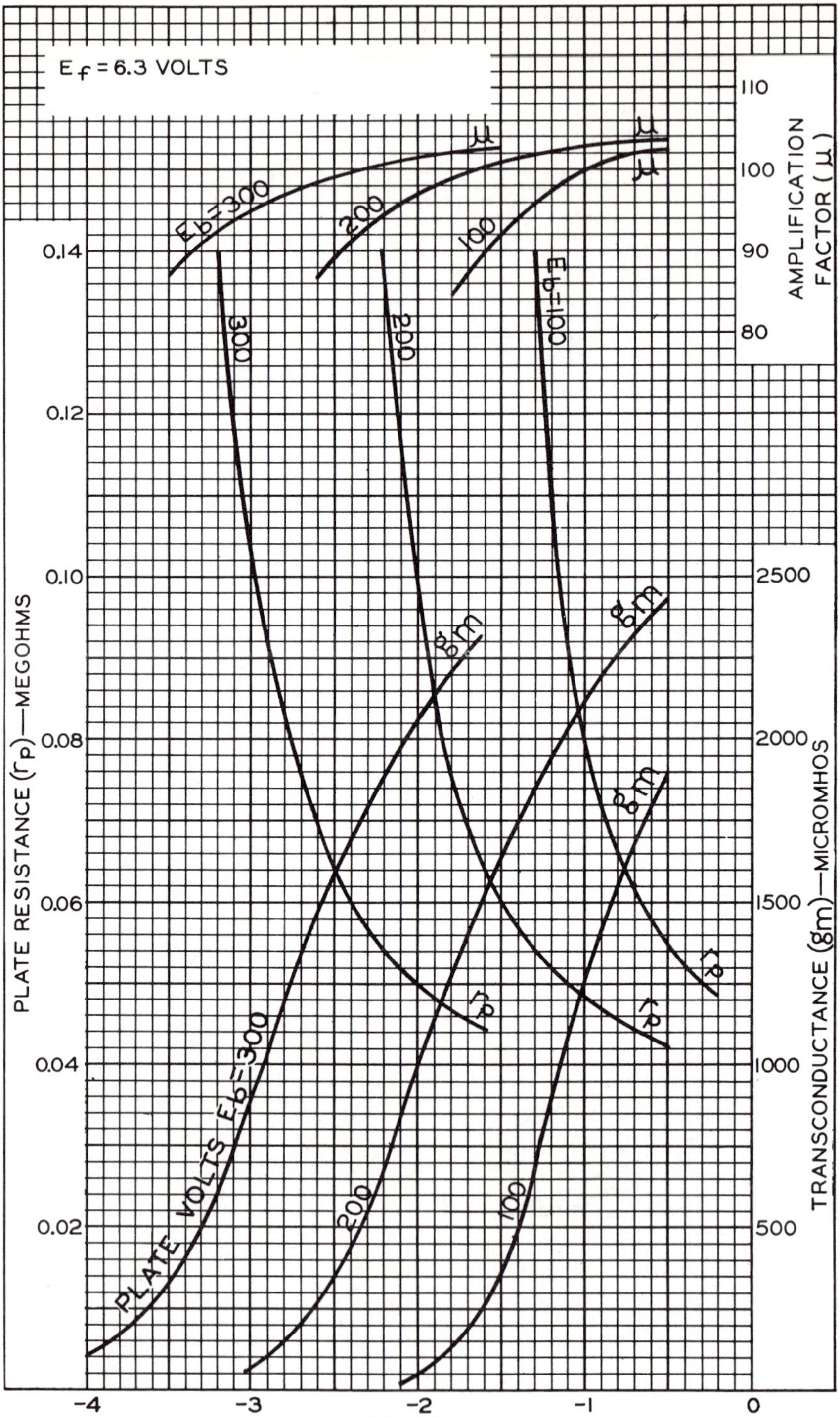
E f = 6.3 VOLTS
AMPLIFICATION FACTOR (μ)
110
100
90
80
μ
μ
μ
Eb=300
200
100
Eb=100
300
200
PLATE RESISTANCE (rp)—MEGOHMS
0.14
0.12
0.10
0.08
0.06
0.04
0.02
TRANSCONDUCTANCE (gm)—MICROMHOS
2500
2000
1500
1000
500
gm
gm
gm
rp
rp
rp
PLATE VOLTS Eb=300
200
100
−4
−3
−2
−1
0
GRID VOLTS

Fig. IV-2

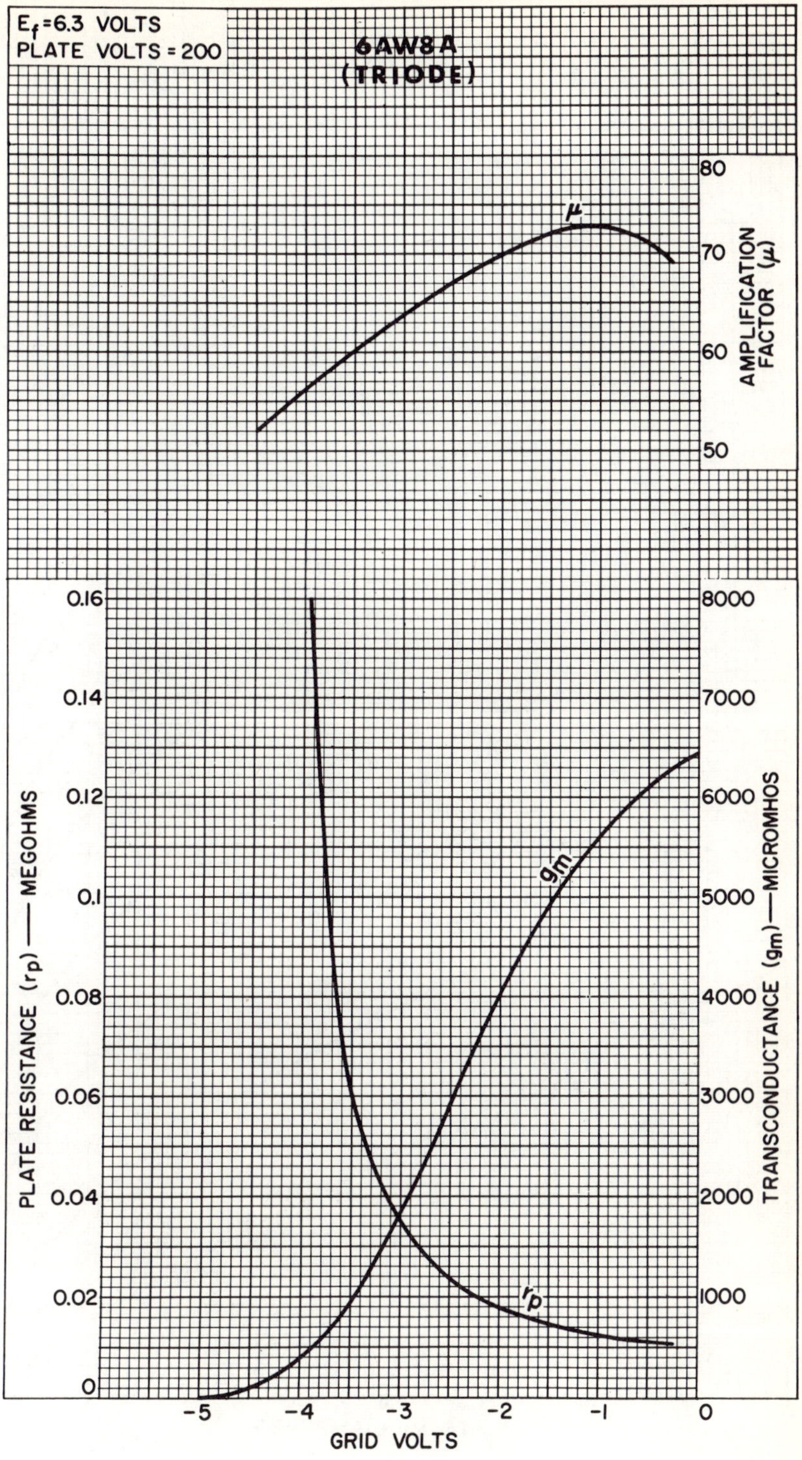

Ef=6.3 VOLTS
PLATE VOLTS = 200
6AW8A
(TRIODE)
μ
AMPLIFICATION FACTOR (μ)
80
70
60
50
PLATE RESISTANCE (rp) —— MEGOHMS
0.16
0.14
0.12
0.1
0.08
0.06
0.04
0.02
0
TRANSCONDUCTANCE (gm)—MICROMHOS
8000
7000
6000
5000
4000
3000
2000
1000
gm
rp
-5
-4
-3
-2
-1
0
GRID VOLTS

Fig. IV-3

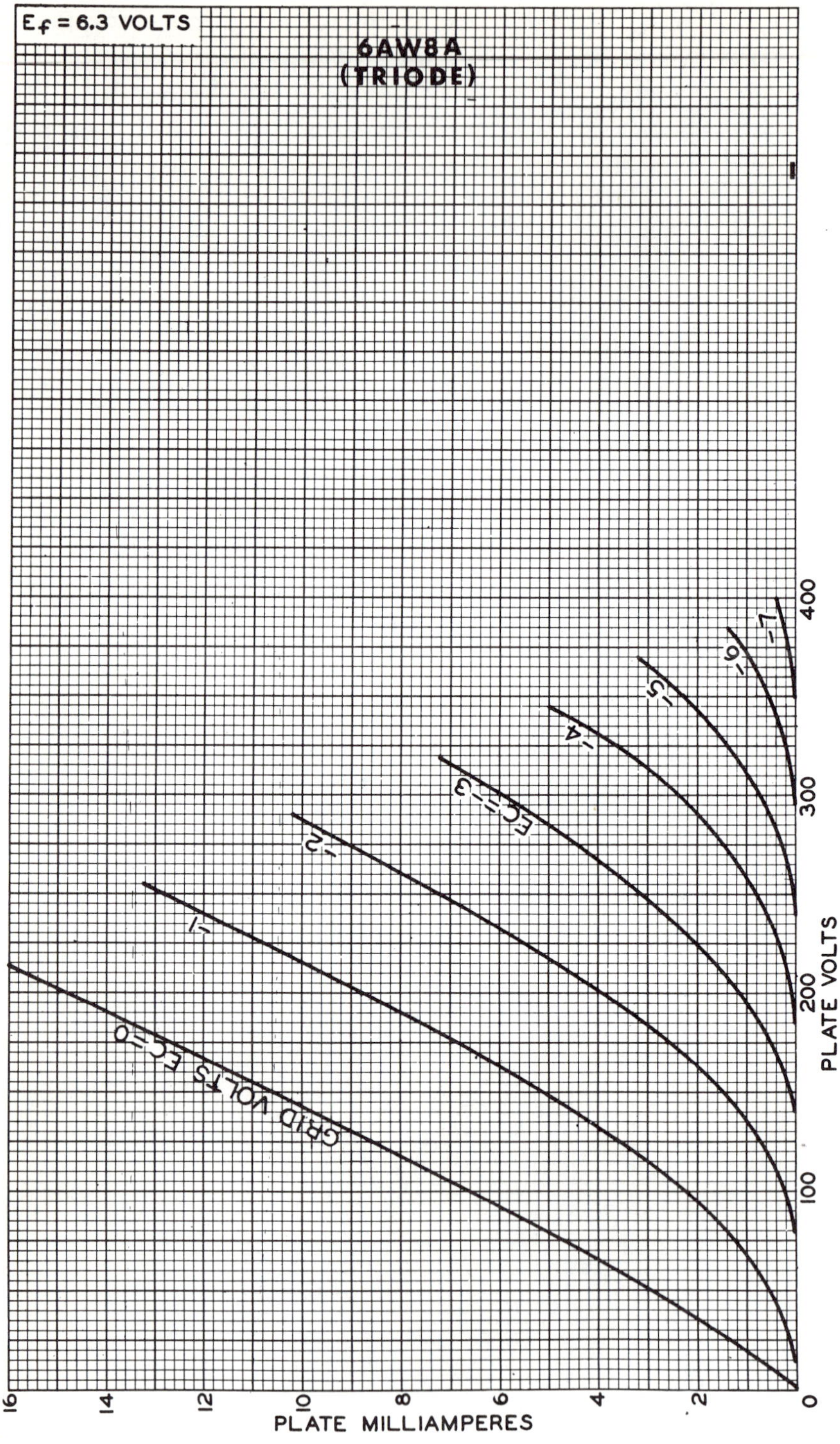
E_f = 6.3 VOLTS
6AW8A
(TRIODE)
GRID VOLTS EC=0
-1
-2
EC=-3
-4
-5
-6
-7
PLATE MILLIAMPERES
16
14
12
10
8
6
4
2
0
PLATE VOLTS
100
200
300
400

Fig. IV-4

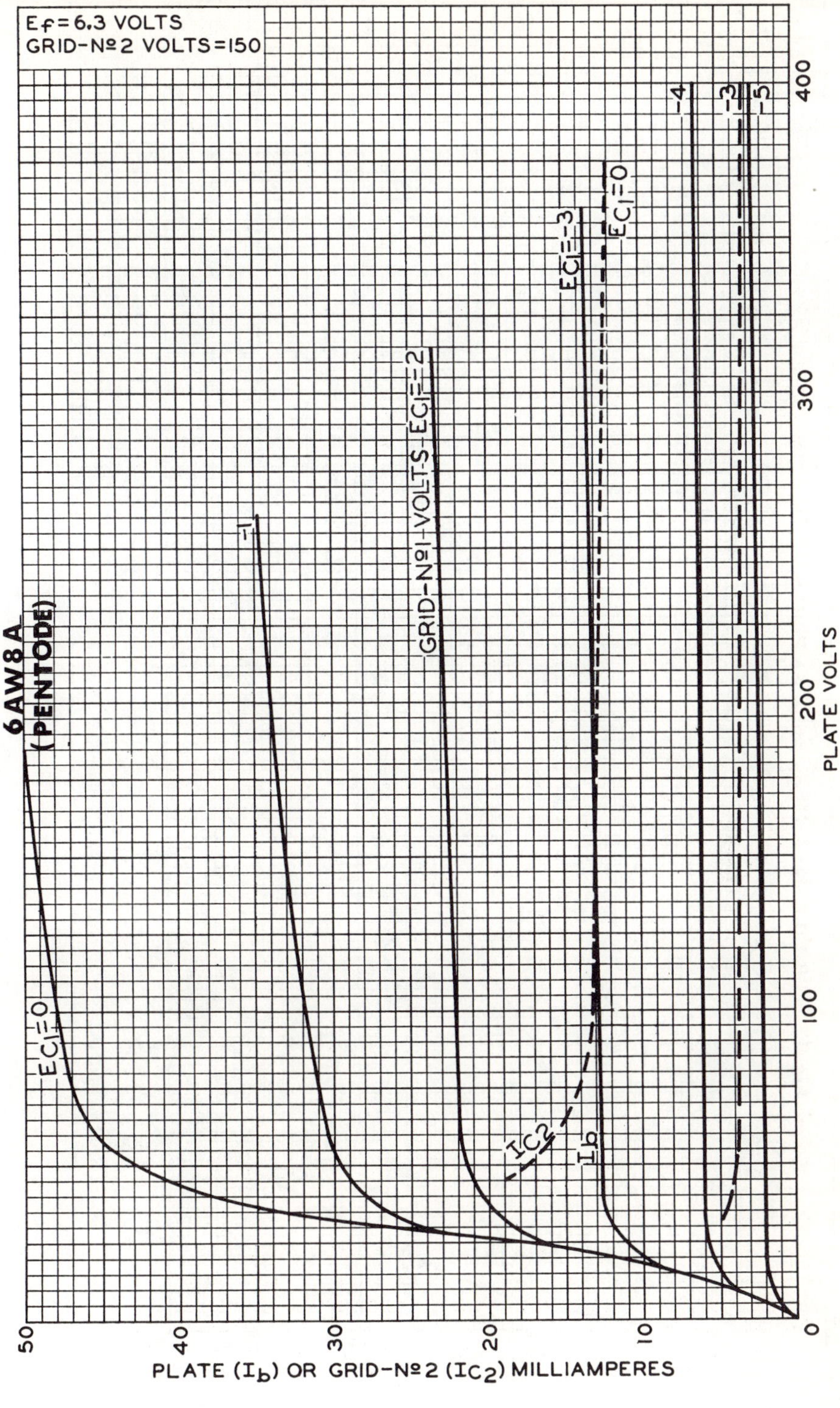
Ef = 6.3 VOLTS
GRID-№2 VOLTS=150
6AW8A
(PENTODE)
EC1=0
-1
GRID-№1-VOLTS-EC1=-2
EC1=-3
EC1=0
IC2
Ib
-4
-3
-5
0
100
200
300
400
PLATE VOLTS
50
40
30
20
10
0
PLATE (Ib) OR GRID-№2 (IC2) MILLIAMPERES

Fig. IV-5

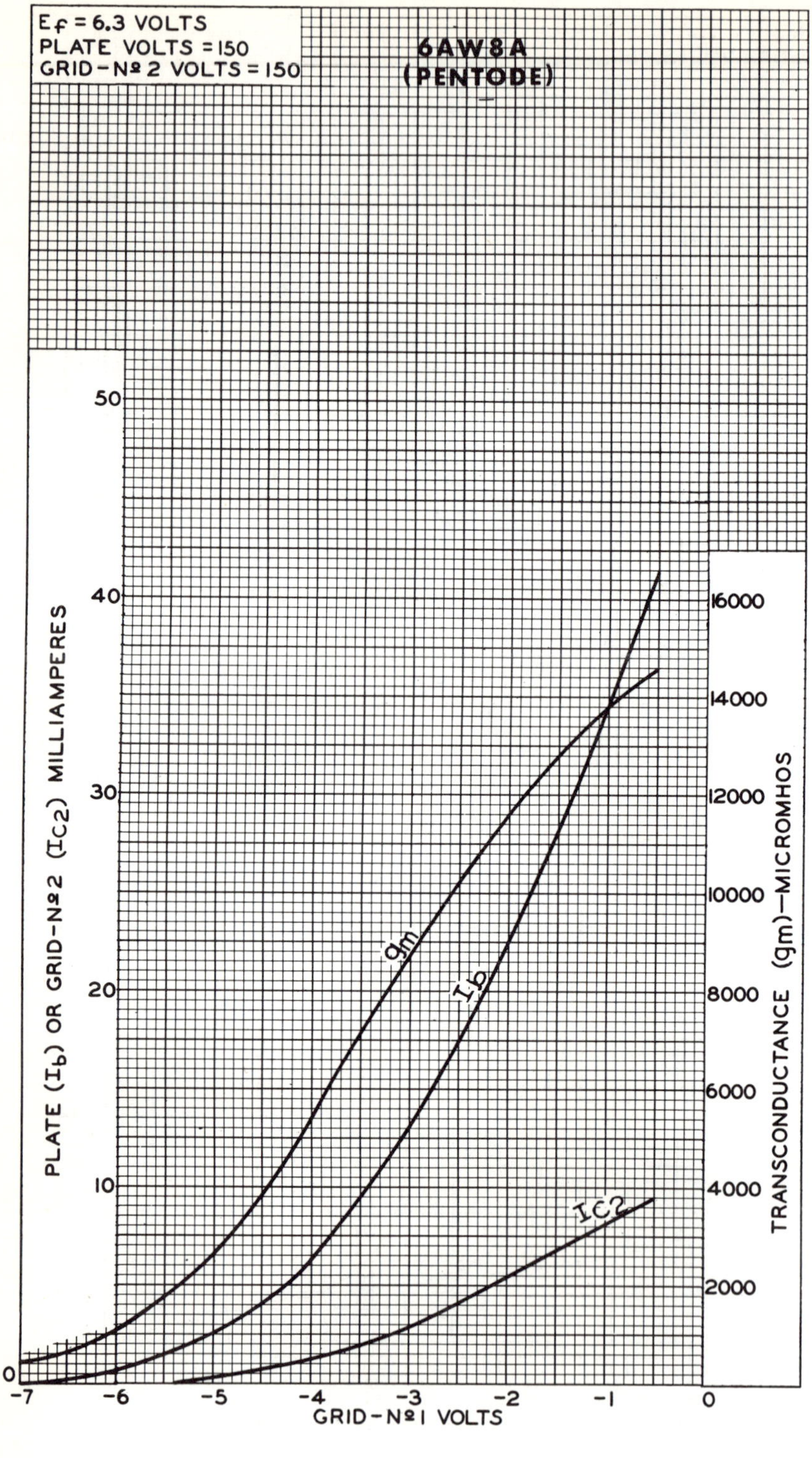
Ef = 6.3 VOLTS
PLATE VOLTS = 150
GRID-№ 2 VOLTS = 150
6AW8A
(PENTODE)
50
40
30
20
10
0
PLATE (Ib) OR GRID-№2 (Ic2) MILLIAMPERES
16000
14000
12000
10000
8000
6000
4000
2000
TRANSCONDUCTANCE (gm)—MICROMHOS
gm
Ib
Ic2
-7
-6
-5
-4
-3
-2
-1
0
GRID-№1 VOLTS

Fig. IV-6

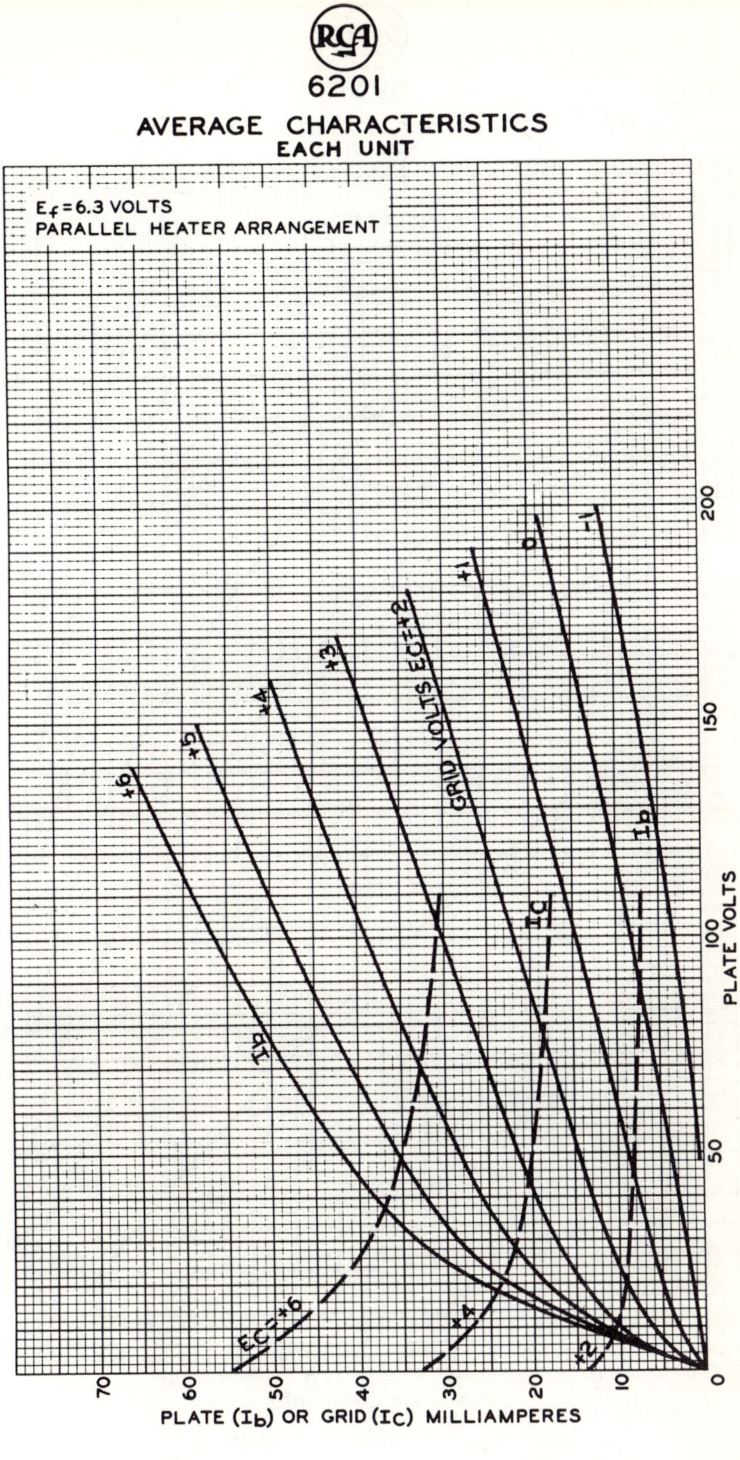
RCA
6201
AVERAGE CHARACTERISTICS
EACH UNIT
Ef = 6.3 VOLTS
PARALLEL HEATER ARRANGEMENT
GRID VOLTS EC=+2
+6
+5
+4
+3
+1
0
-1
Ib
Ic
EC=+6
+4
+2
PLATE VOLTS
0
50
100
150
200
70
60
50
40
30
20
10
0
PLATE (Ib) OR GRID (IC) MILLIAMPERES

Fig. IV-7

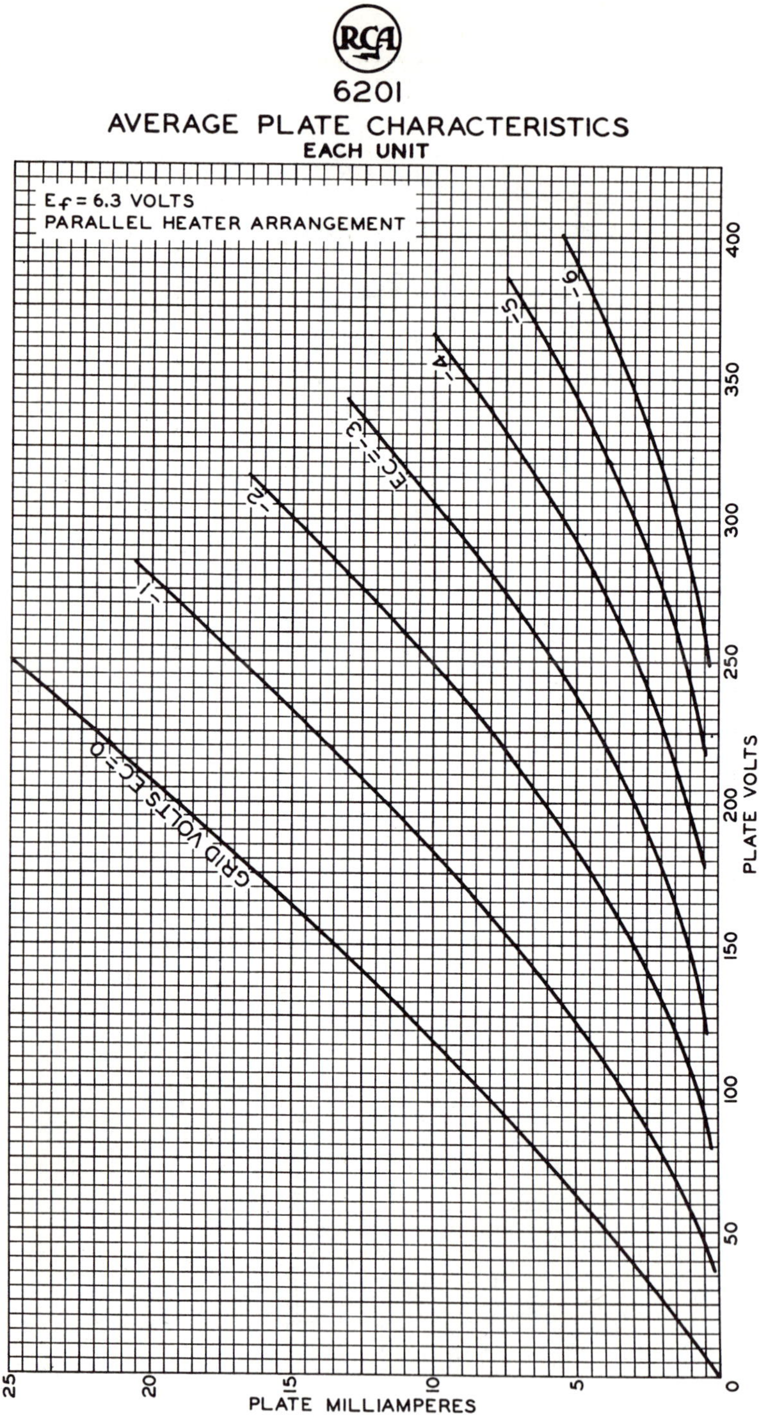
RCA
6201
AVERAGE PLATE CHARACTERISTICS
EACH UNIT
Ef = 6.3 VOLTS
PARALLEL HEATER ARRANGEMENT
GRID VOLTS Ec = 0
-1
-2
Ec = -3
-4
-5
-6
PLATE VOLTS
0
50
100
150
200
250
300
350
400
PLATE MILLIAMPERES
25
20
15
10
5
0

Fig. IV-8

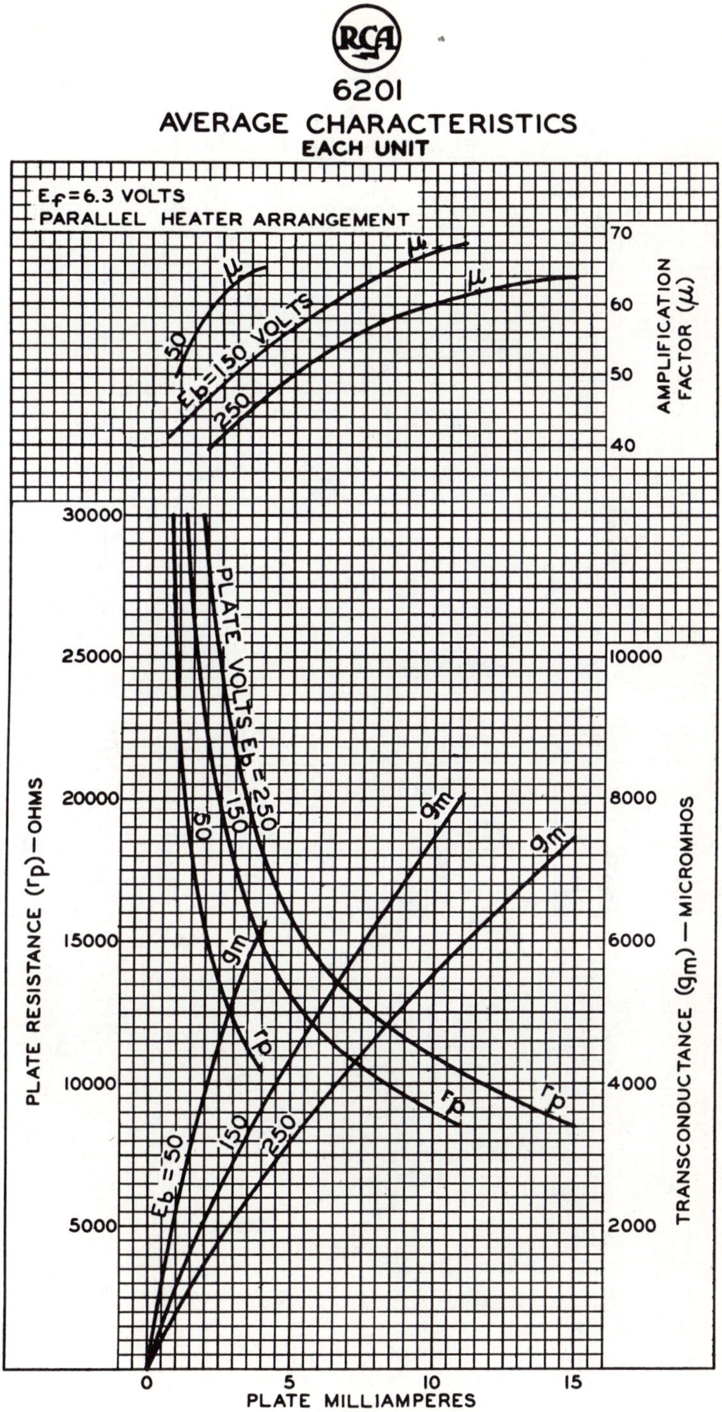
RCA
6201
AVERAGE CHARACTERISTICS
EACH UNIT
Ef = 6.3 VOLTS
PARALLEL HEATER ARRANGEMENT
μ
50
Eb=150 VOLTS
250
70
60
50
40
AMPLIFICATION FACTOR (μ)
30000
25000
20000
15000
10000
5000
PLATE RESISTANCE (rp)—OHMS
PLATE VOLTS Eb=250
150
50
rp
gm
Eb=50
150
250
10000
8000
6000
4000
2000
TRANSCONDUCTANCE (gm)—MICROMHOS
0
5
10
15
PLATE MILLIAMPERES

Fig. IV-9

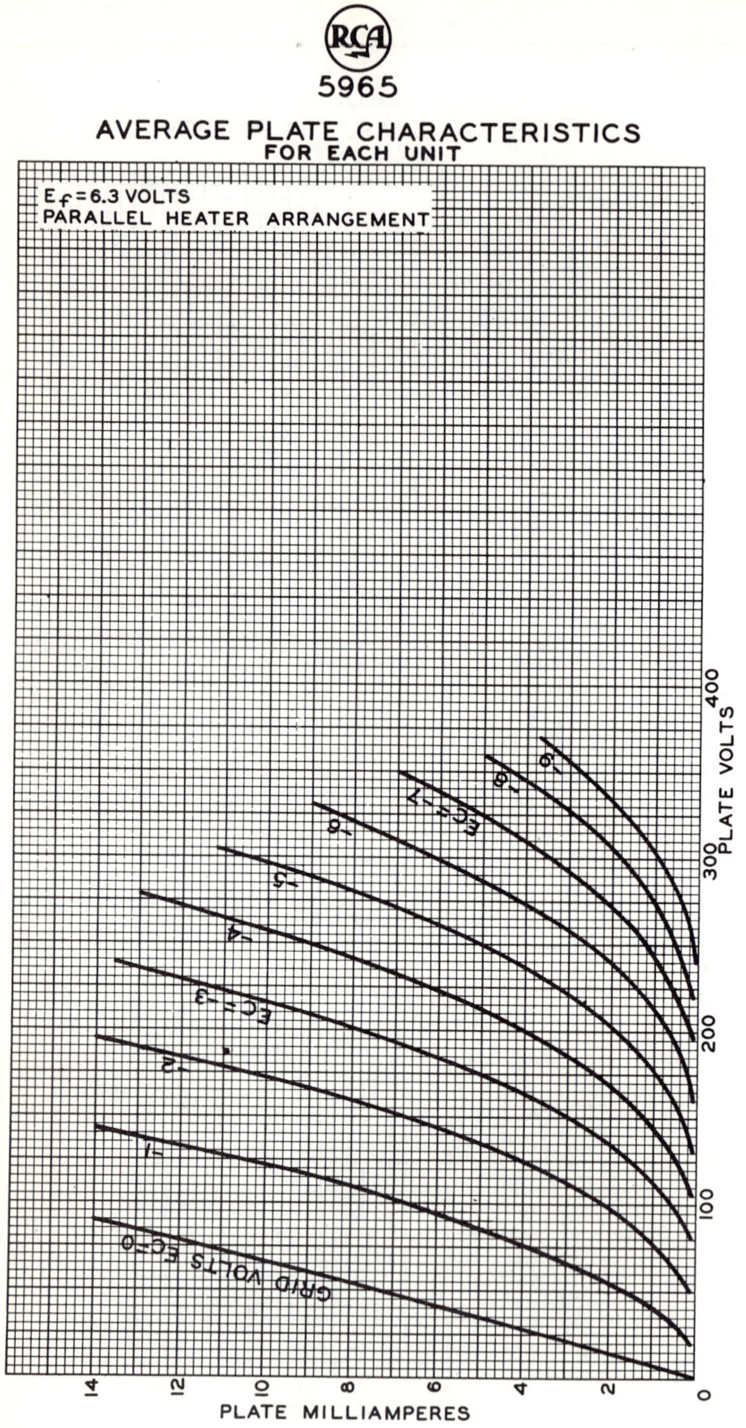
RCA
5965
AVERAGE PLATE CHARACTERISTICS
FOR EACH UNIT
Ef = 6.3 VOLTS
PARALLEL HEATER ARRANGEMENT
GRID VOLTS EC=0
-1
-2
EC=-3
-4
-5
-6
EC=-7
-8
-9
PLATE VOLTS
0
100
200
300
400
PLATE MILLIAMPERES
14
12
10
8
6
4
2
0

Fig. IV-10

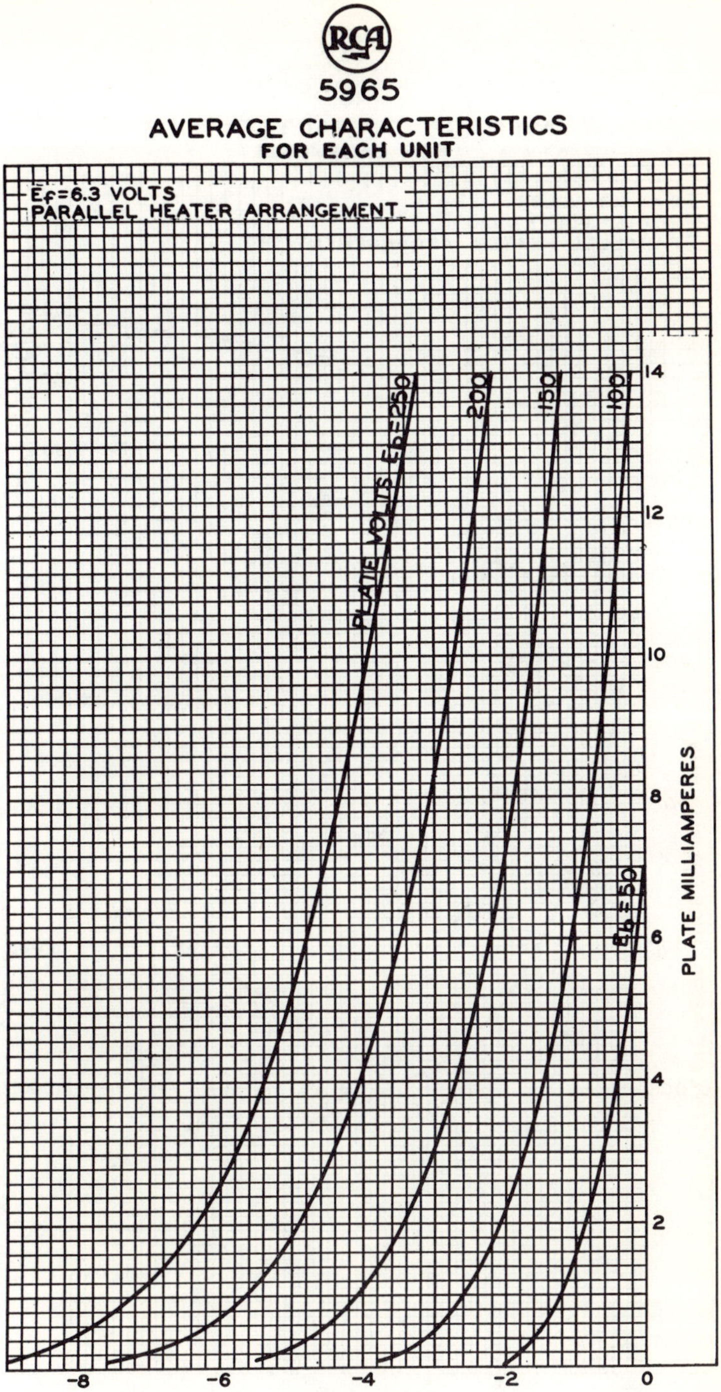
RCA
5965
AVERAGE CHARACTERISTICS
FOR EACH UNIT
Ef=6.3 VOLTS
PARALLEL HEATER ARRANGEMENT
PLATE VOLTS Eb=250
200
150
100
Eb=50
PLATE MILLIAMPERES
14
12
10
8
6
4
2
-8
-6
-4
-2
0
GRID VOLTS

Fig. IV-11

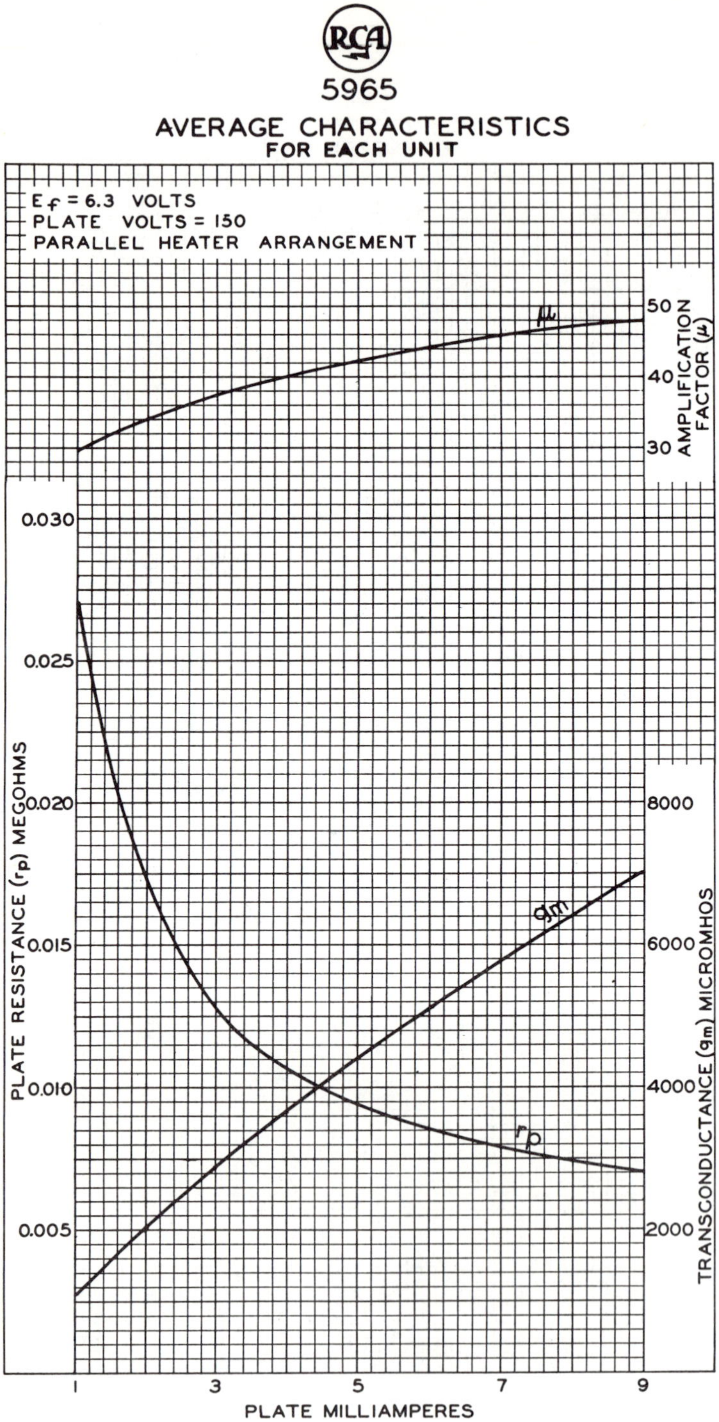
5965
AVERAGE CHARACTERISTICS
FOR EACH UNIT
Ef = 6.3 VOLTS
PLATE VOLTS = 150
PARALLEL HEATER ARRANGEMENT
μ
50
40
30
AMPLIFICATION FACTOR (μ)
0.030
0.025
0.020
0.015
0.010
0.005
PLATE RESISTANCE (rp) MEGOHMS
8000
6000
4000
2000
TRANSCONDUCTANCE (gm) MICROMHOS
gm
rp
1
3
5
7
9
PLATE MILLIAMPERES

Fig. IV-12

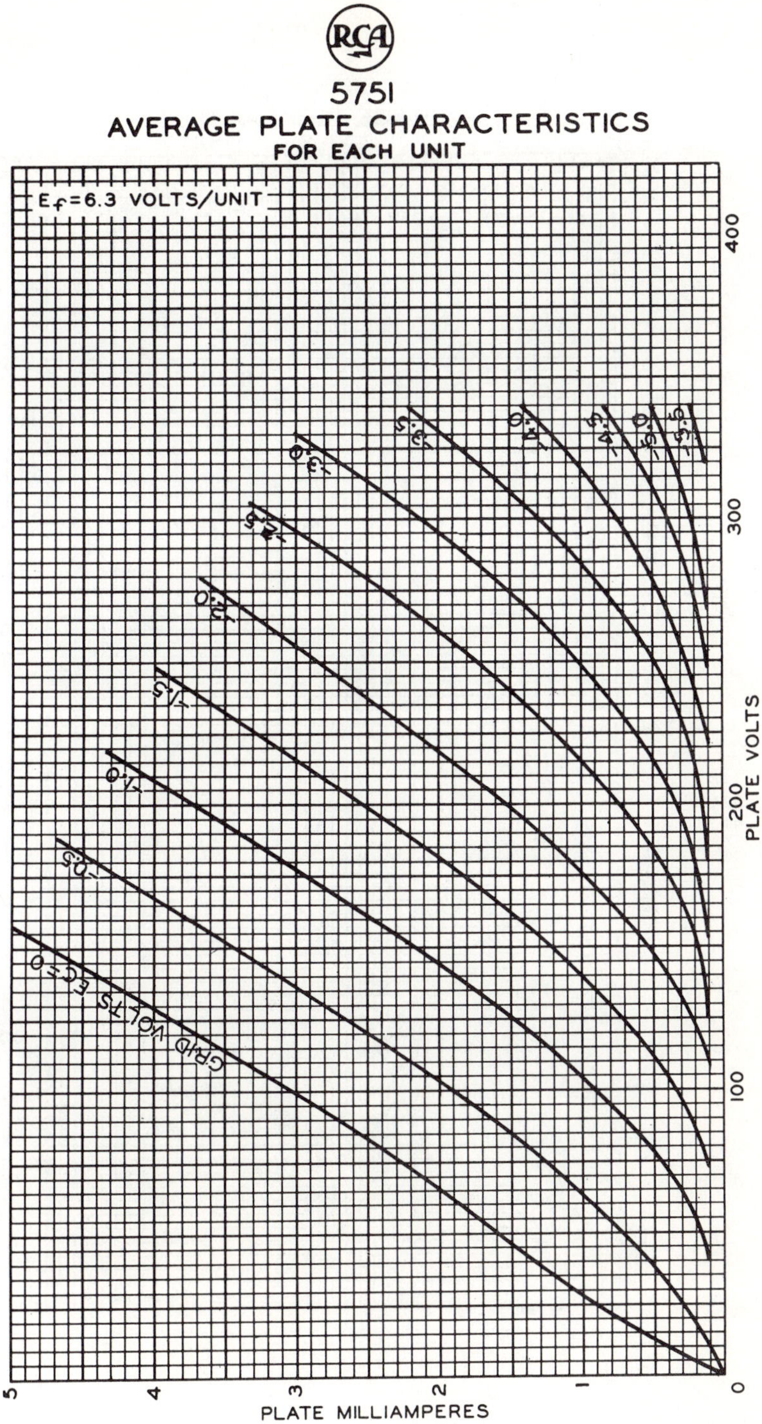
RCA
5751
AVERAGE PLATE CHARACTERISTICS
FOR EACH UNIT
Ef=6.3 VOLTS/UNIT
PLATE VOLTS
PLATE MILLIAMPERES
GRID VOLTS Ec=0

Fig. IV-13

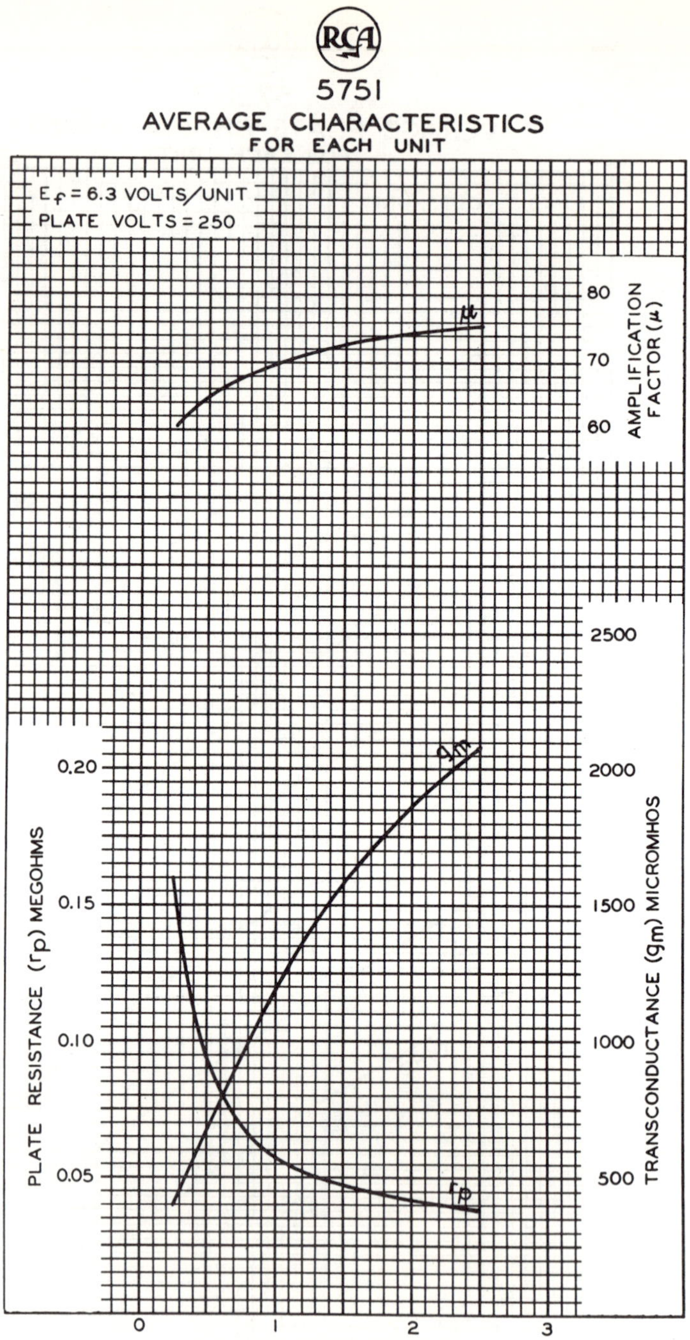
RCA
5751
AVERAGE CHARACTERISTICS
FOR EACH UNIT
Ef = 6.3 VOLTS/UNIT
PLATE VOLTS = 250
80
70
60
AMPLIFICATION FACTOR (μ)
μ
2500
2000
1500
1000
500
TRANSCONDUCTANCE (gm) MICROMHOS
gm
0.20
0.15
0.10
0.05
PLATE RESISTANCE (rp) MEGOHMS
rp
0
1
2
3

Fig. IV-14

Appendix V. Typical Transistor Characteristic Curves

Fig. V-1

TYPICAL COLLECTOR CHARACTERISTICS
FOR TYPES 2N3241 AND 2N3242

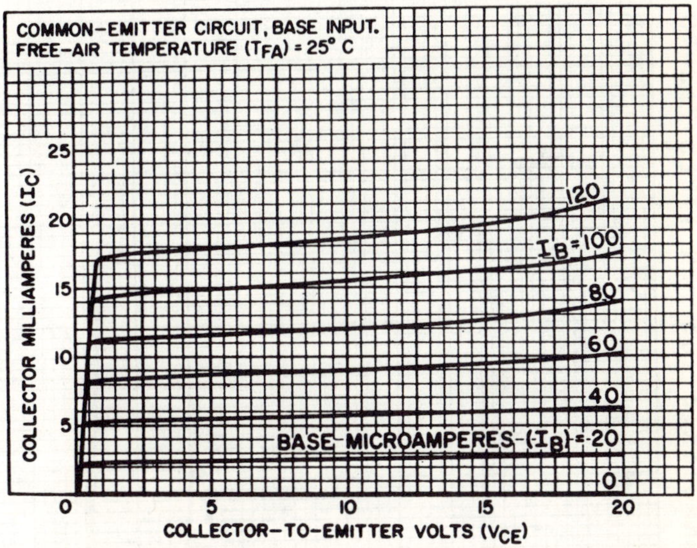

Fig. V-2

TYPICAL COLLECTOR CHARACTERISTICS
FOR TYPES 2N3241 AND 2N3242

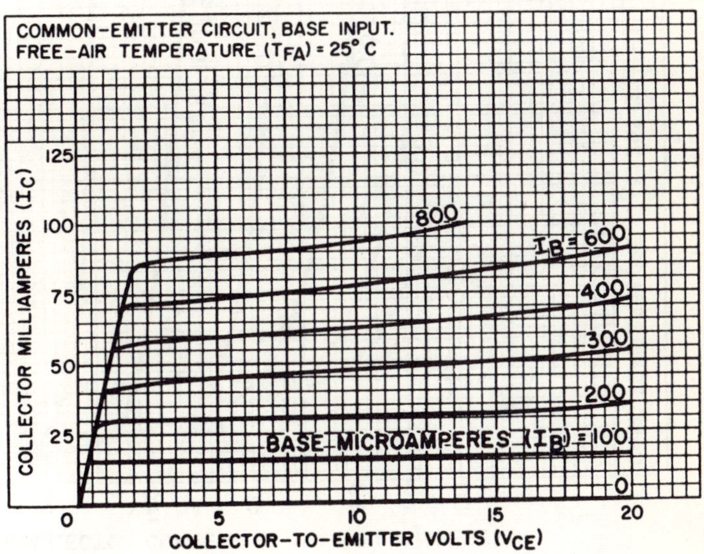

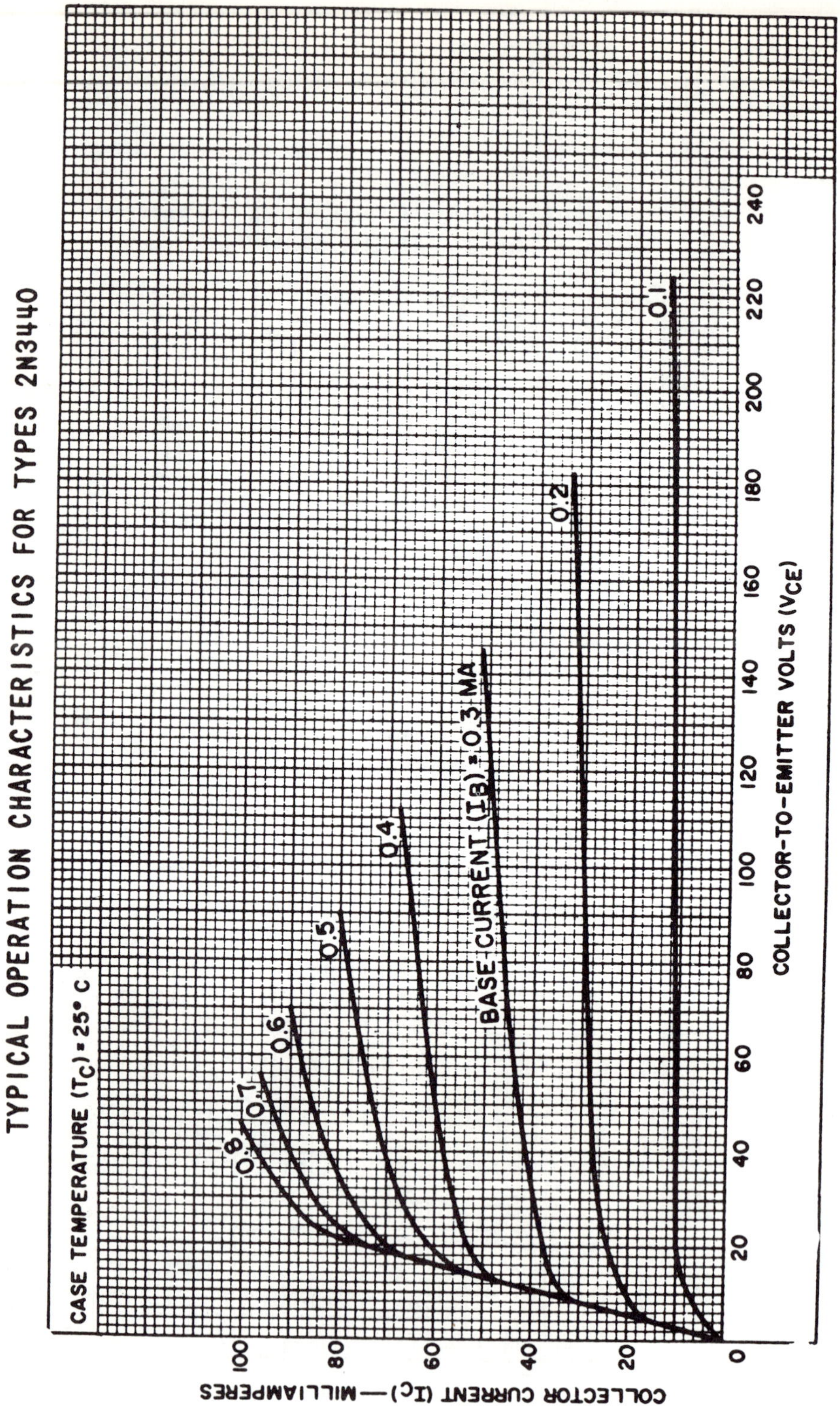
TYPICAL OPERATION CHARACTERISTICS FOR TYPES 2N3440
CASE TEMPERATURE (T_C) = 25° C
0.8
0.7
0.6
0.5
0.4
BASE CURRENT (I_B) = 0.3 MA
0.2
0.1
0
20
40
60
80
100
120
140
160
180
200
220
240
COLLECTOR-TO-EMITTER VOLTS (V_{CE})
COLLECTOR CURRENT (I_C)—MILLIAMPERES

Fig. V-3

TYPICAL OPERATION CHARACTERISTICS FOR TYPE 2N3441

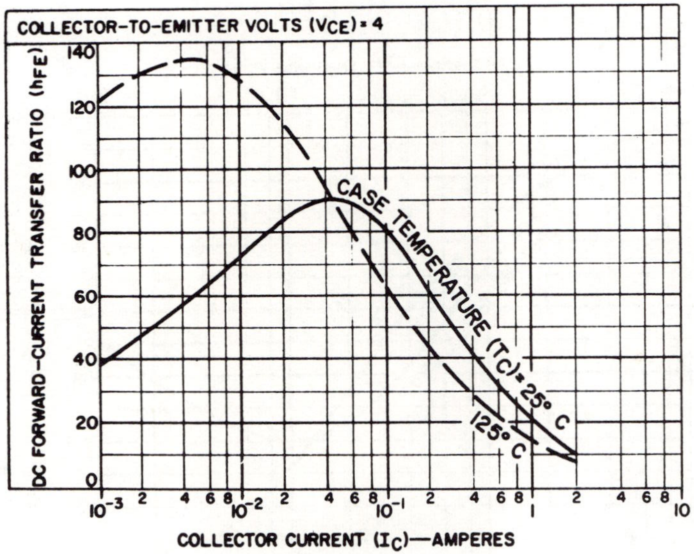

Fig. V-4A

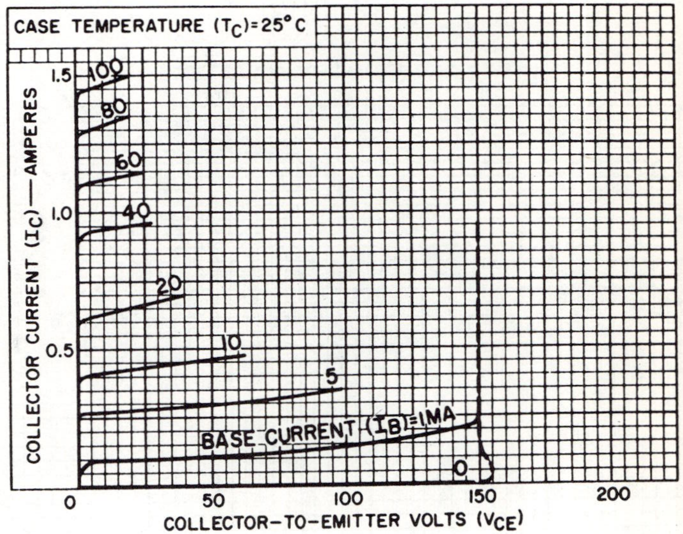

Fig. V-4B

TYPICAL OPERATION CHARACTERISTICS FOR TYPE 2N3442

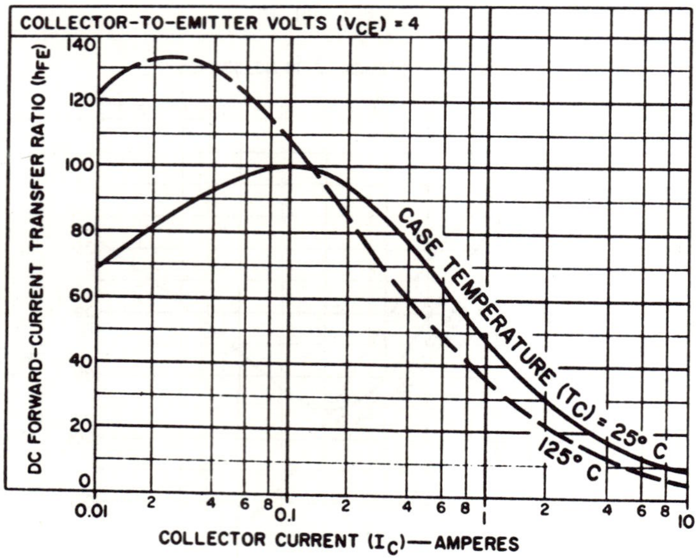

Fig. V-5A

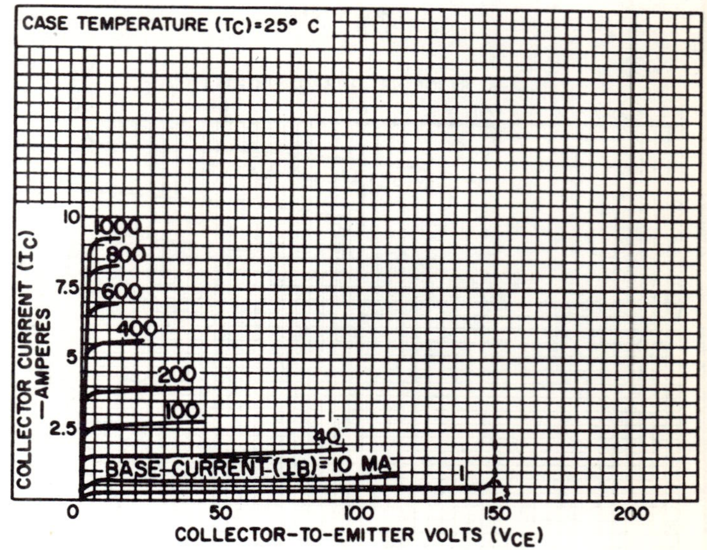

Fig. V-5B

Appendix VI. Operational Amplifier Applications

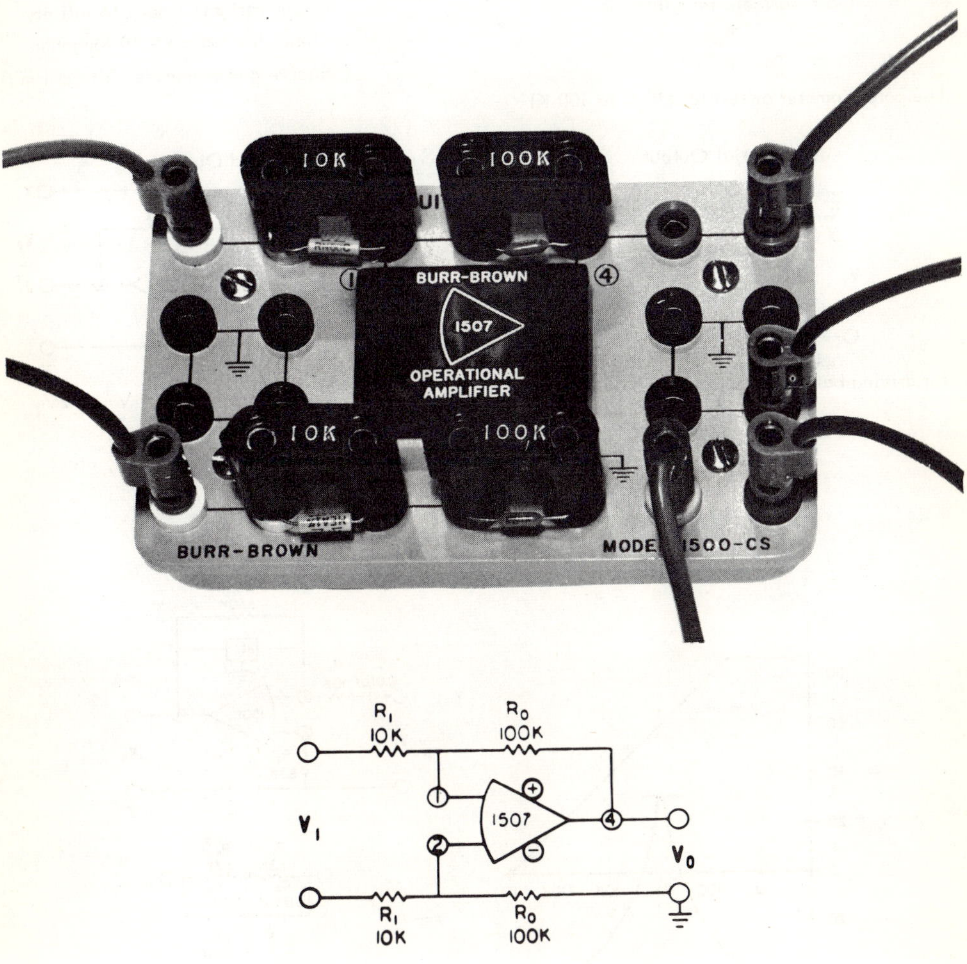

$$V_0 = \frac{R_0}{R_1} V_1 = 10\,V_1$$

Fig. VI-1. Differential amplifier with a gain of 10. (Courtesy of Burr-Brown Research Corporation, Tucson, Ariz.)

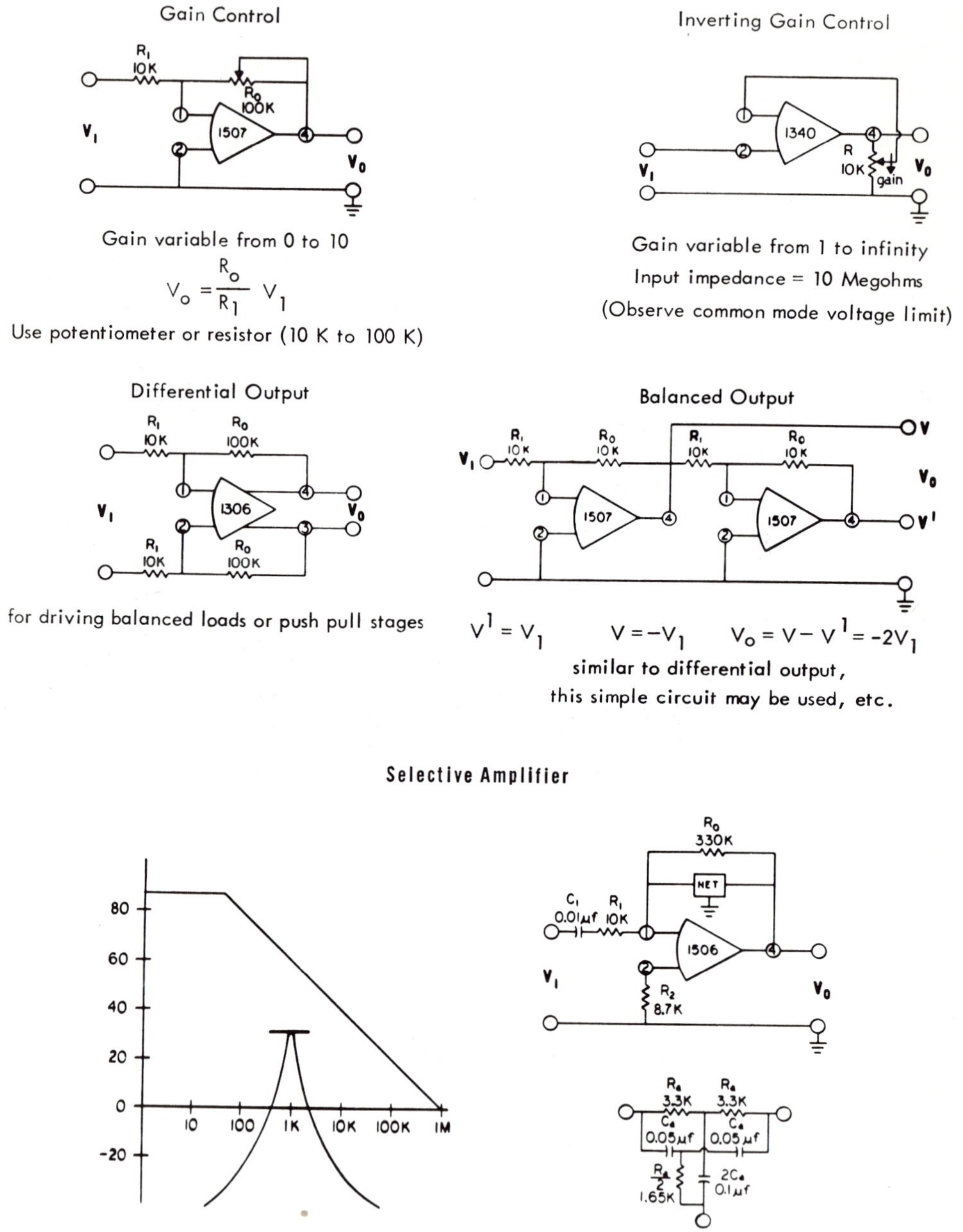

Fig. VI-2. Applications for the operational amplifier module. (Courtesy of Burr-Brown Research Corporation, Tucson, Ariz.)

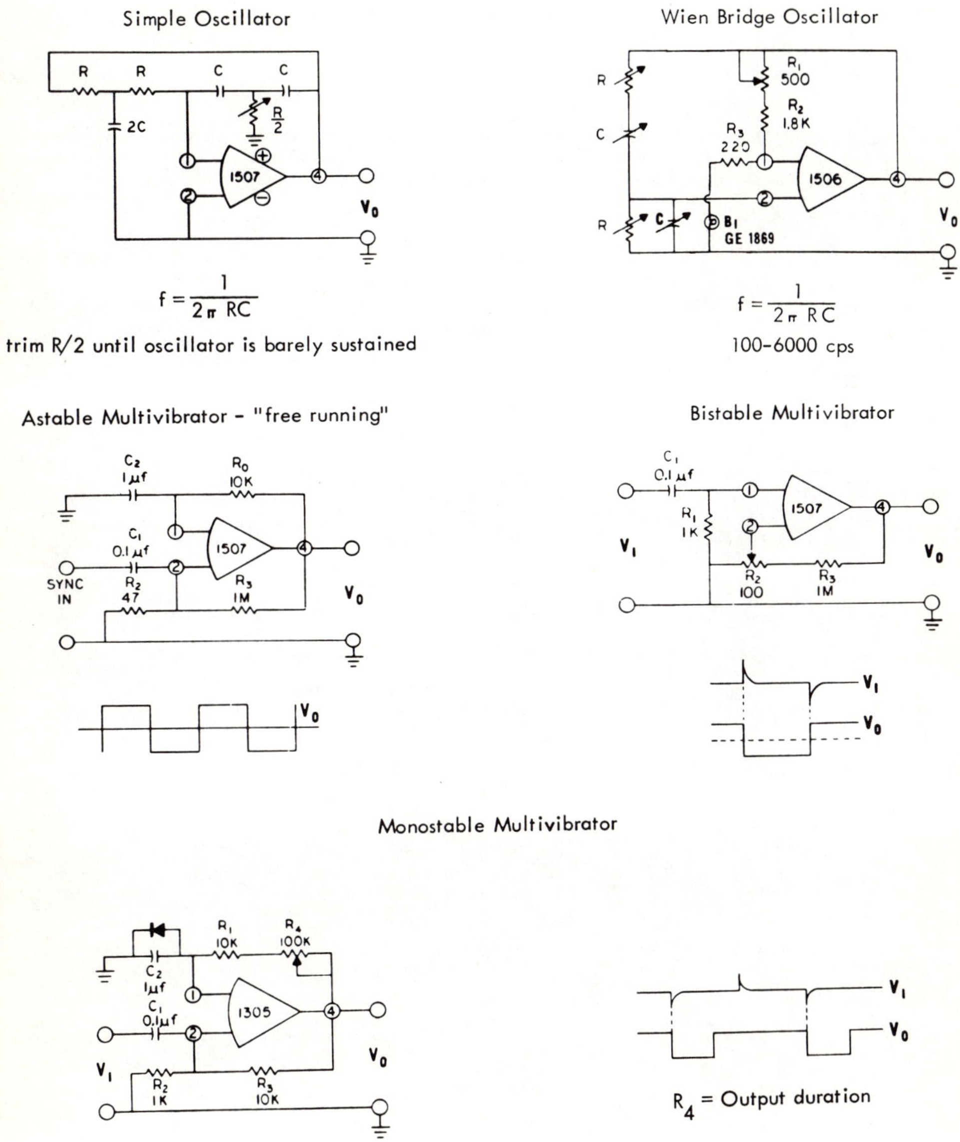

Fig. VI-3. Additional applications for the operational amplifier module. (Courtesy of Burr-Brown Research Corporation, Tucson, Ariz.)

INDEX

AC
basic concepts of, 25-26
definition of, 25
AC circuit theory, introduction to, 25-47
Accuracy of measuring instruments, 51-52
Action potential, definition of, 252
Air flow rate, measurement of, 183-184
Alpha cut-off of transistors, 114
Alternating current
basic concepts of, 25-26
definition of, 25
Alternating current circuit theory, introduction to, 25-47
Ampere. See also *Current.*
definition of, 7
Ampex Corp., 247, 248
Amplification, 83-127
Amplifier
chopper, 124-125
common-base, 116
common-collector, 116
common-emitter, 116
differential, 151, 345
error, 153
gated, 296
grounded-base, 116
grounded-emitter, 116
notch, 105
characteristics of, 105-107
operational, 106-109, 346
slide-wire, 231
transistor as, 111-117
Amplifier circuits
examples of, 97-101
multiple stages of, 97-101
Amplitude, relative, of manometer, 278
Analysis, transient, of catheter-manometer system, 276
Animal, noise in, 313
d'Arsonval galvanometer, 56
Associated Specialties Co., 153, 154, 155
Astable multivibrator, 138-139, 347
Avalanche voltage, 74

BACK-emf, 21
definition of, 21
Back resistance, 73
Ballistocardiogram, 166
Bar, forming, 227
Beckman Instruments, Inc., 205, 206, 207, 208, 209, 210, 211, 222, 224, 225, 226, 227, 228, 229, 230, 231, 262, 263
Bellows pneumograph, 183-184
Beta cut-off of transistors, 114
Bias, 86, 91-92, 94, 121
fixed, 87, 118
forward, 74, 78, 112, 201
grid, 83
of transistor circuit, 118-119
Bias (*continued*)
reverse, 74, 78, 112, 200
self-, 88, 119
Bias voltage, 87
Bioelectric potential(s), 251-265
definition of, 251
Biomedical instrument(s)
calibration of, 164-169, 195, 199, 271, 274-284, 304
classification of, 162-163
frequency response of, 164-165, 192, 271, 274-284, 304
definition of, 50
determination of, 165
ideal, 163
parameters measured by, 169-170. See also *Measurements.*
parts of, 163-164
Biomedical instrumentation, examples of, 251-305
Biomedical measurements. See *Measurements.*
Biomedical parameters, characteristics of, 169-170
Bistable multivibrator, 140-144, 347
Bleeder resistor, 147
Blocking capacitor, 97
Blood flow, measurement of, 192, 201, 284-305
Blood flow transducers, 286, 289-290, 293-294, 297, 300, 302
Blood pressure
diastolic, 265
measurement of, 64, 165, 173-177, 178, 182, 183, 265-284
direct, 284
indirect, 284
pulsatile, 265
systolic, 265
Blood pressure transducer(s), 268, 269-270, 272, 273
catheter tip, 269, 273
glass bubble, 266-269
stretched rubber membrane, 267
Bonded strain gauge, 174-177
Bootstrap, definition of, 102, 142
Bourdon tube pressure transducer, 178, 182-183
Branch point, definition of, 40
Breadboard, definition of, 85
Breakdown voltage, 114, 115
Breathing rate, measurement of, 183-184, 185-186
Bridge
impedance, 63
wheatstone, 62-64
Bristle flowmeter, 289-290
Burr-Brown Research Corp., 345, 346, 347

CADMIUM sulfide photocell, 199
Calibration of instruments, 164-169, 195, 199, 217, 218-219, 235-237, 271, 274-284, 280-281, 303-304. See also individual instruments.
Camera, kymograph, 244

Capacitance, 13-15
  interelectrode, of triode, 96
  stray, 97, 296
Capacitance manometer, 268
Capacitance multiplier, 147
Capacitative displacement transducer, 190
Capacitor(s), 326
  blocking, 97
  definition of, 13
  identification of, 326
  in parallel combination, 15-16
  in series combination, 15-16
Capacitor-input filter, 146, 153
Capacitor-resistor color code, 324
Carbon dioxide tension, measurement of, 205, 210-211
Cardiac pacemaker, definition of, 257
Cardiovascular system, measurement of, 166
Cascade shunt regulator, 149
Catheter-manometer system, transient analysis of, 276
Catheter tip manometer, 269, 273
Cathode, 64, 71
Cathode follower, 95-96, 117
Cathode-ray oscilloscope, 64
Cathode-ray vacuum tube, 64
Caton, Richard, 161
Cell
  muscle, stimulation of, 251
  nerve, stimulation of, 251
  photovoltaic, 200
Cell membrane, polarized, 251
Charge, 3-5
Choke-input filter, 146, 147
Chopper amplifier, 124-125
Circuit(s)
  AC, theory of, introduction to, 25-47
  amplifier
    examples of, 97-101
    multiple stages of, 97-101
  clipping, 77, 78
  gate, 296
  grounded-collector, 116
  hybrid, 111
  integrated, 123
  multivibrator, 138-139, 346, 347
    astable, 138-139, 347
    bistable, 142-143, 347
  squaring, 140-141, 299
  sweep, 64
  tank, 131
Clapp oscillator, 135-136
Clipping circuit, 77, 78
Codes, color, for identification of components, 323
Coil
  inductance of, 21
  primary, definition of, 22
  secondary, definition of, 22
Collector-characteristic curve for transistors, 117
Color codes for identification of components, 323
Colpitts oscillator, 134-135
Common-base amplifier, 116
Common-collector amplifier, 116
Common-emitter amplifier, 116
Common-emitter configuration, characteristic curves of, 116
Common mode signals, 90
Component(s)
  identification of, 323-326
  transformer, 296
Conductance
  definition of, 9
  of triode, 94
Conduction, 7
Conduction velocity of impulse, measurement of, 254, 264, 265
Conductor, definition of, 8
Constant, spring, of instrument, 50
Control grid, 83
Conversion factors, physical units and, 319
Core transformer, 24
Coulomb, definition of, 5
Coulomb's law, 5
Coupling of transistor stages, 100-101, 119, 121
Critical response in RLC circuit, 38
CRO, 64
CRT, 64
Crystal oscillator, 137-138
Current, 7-8
  alternating
    basic concept of, 25-26
    circuit theory, introduction to, 25-47
    definition of, 25
  capacity of wire for carrying, 14
  conduction of, 7
  dark, definition of, 201
  direct, 25
    definition of, 25
  eddy, 24
  grid, 83
  house, 145
  measurement of, 49-67
  peak, reading of, meter for, 126
  plate, of rectifier, 72
Current gain of transistor, 113
Curve(s)
  characteristic
    common-emitter configuration, 116
    for cadmium sulfide photocell, 199
    for double triode vacuum tubes, 89, 327, 336-340
    for gas-filled phototube, 195, 197
    for junction photodiode, 200
    for multiplier phototube, 198
    for pentode vacuum tubes, 331-332
    for PN junction diode, 75, 79, 111
    for transistors, 116-117, 341-344
    for triode vacuum tubes, 329-330
    for tunnel diode, 79-80
    for vacuum tube, 327-340
    for vacuum tube rectifier, 72, 73
    for zener diode, 78-79
  plate-characteristic, 84, 88
  static characteristic, of rectifier, 72

Curve(s) (*continued*)
  transfer-characteristic
    for transistors, 117
    for vacuum-tube triode, 84, 87
    typical, 84, 87
Curvilinear recording
  definition of, 217
  versus rectilinear recording, 218
Cut-off
  alpha, of transistors, 114
  beta, of transistors, 114
Cut-off frequency, 44

DAMPING
  in blood pressure measurement, 277-279
  in RLC circuit, 37-39
Damping resistor, 216
Dark current, definition of, 201
d'Arsonval, 56
d'Arsonval galvanometer, 56
Data Device Corporation, 106
DC, 25
  definition of, 25
Deflection plates, 64
DeForest, Lee, 83
Degeneration, definition of, 102
Demodulator, 296
Department of the Army, 71, 72, 73, 75, 84, 86, 91, 92, 93, 94, 100, 101, 102, 103, 105, 110, 112, 130, 131, 132, 133, 134, 135, 136, 137, 139, 143
Deprez, 56
Detector, phase, 299
Diastolic pressure, 265
Dielectric, 13
Differential amplifier, 151
Differential force transducer, 189
Differential pressure, 268
  measurement of, 268-272
Differential transformer, 183, 184
Differentiator, 109
Digital voltmeter, 60
Diode(s), 69-81
  definition of, 69
  dual, 74
  optoelectronic, 201
  PN junction, characteristic curve of, 75, 79, 111
  semiconductor, 74-80
  solid state, applications of, 76
  tunnel, 78
    characteristic curve for, 79-80
  zener, 78, 148
    characteristic curve for, 78-79
    reference, 149
Direct current, 25
  definition of, 25
Displacement
  measurement of, 172-193
  volume, of pressure transducers, 267
Displacement transducer, 177-178, 180-181
  rheostatic, 173
Doppler ultrasonic flowmeter, 292-294
Double cone pneumotach, 183
Drop counter, measurement of, 168
Dynamic errors in instruments, 52
Dynamometer, 57
Dynode, 195

E & M INSTRUMENT CO., INC., 180, 181, 184, 185, 192, 203, 212, 218, 238, 239, 240, 252, 254, 255, 260, 262, 274, 275, 285, 286, 287, 288
Earth. See *Ground.*
ECG, 254, 258, 260-262
Edison effect, 69
Eddy currents, 24
EEG, 259, 263
Efficiency
  plate, 87
    definition of, 87
  quantum, of photodiode, 200
Einstein, Albert, 193
Einthoven, Willem, 161
EKG, 254, 258, 260-262
Electrical power, 12-13
Electrocardiogram, 254, 258, 260-262
Electrocardiography, 254
  definition of, 252
Electrode(s)
  fluid-filled cup, 257, 261
  glass, 208
  needle, 259
  of moist nerve chamber for recording action potential, 259
  pin, 252
  plate, 260
  silver-silver chloride surface, 258-262
  wick, 253
  wire cloth, skin, 257, 261
Electrode system, 258
Electroencephalogram, 259, 263
Electroencephalography, 259
  definition of, 252
Electromagnetic flowmeter. See *Flowmeter, electromagnetic.*
Electromagnetic interference, 313
Electrometer, 208
Electromotive force, 21
  back, 21
  induction of, 21, 23
Electromyogram, 180-181, 264-265
Electromyography, 264
  definition of, 252
Electron(s), source of, 70-72
Electron cloud, 70, 71
Electronic Associates, Inc., 60
Electronic filter, 148
Electronic noise, definition of, 307
Electronic switch of oscilloscope, 66
Electronic symbols, 321
Electroretinogram, 167, 252
Electroretinography, definition of, 252
Electrostatic field, 4-7
Electrostatic interference, 90, 310-311
Electroturbinometer, 288-289

Emf, 21
back-, 21
induction of, 21, 23
EMG, 264-265
Emission, thermionic, 70-72
Emission noise, secondary, 309
Engineered Electronics Co., 138, 140, 141, 144
Environmental errors in instruments, 52
ERG, 167, 252
Error(s) in biomedical measurements, 51-54
dynamic, 52, 218-219, 236-237
environmental, 52
in curvilinear direct-writing pen, 218-219
in rectilinear direct-writing pen, 221, 234-237
mechanism, 52
random, 54
reading, 53
scale, 52
systematic, 54
Error amplifier, 153
Excitation voltage, 165

FACTOR(S)
conversion, physical units and, 319
gauge, 173
Farad, 15. See also *Capacitance.*
Faraday, Sir Michael, 15
Feedback, 101-106, 132
definition of, 101
inverse, definition of, 102
negative, definition of, 102
positive, definition of, 102
Feedback signal, positional, 231
Field
electrostatic, 4-7
magnetic, 16-24
Figure of merit, 51
Filter(s)
capacitive multiplier, 147
capacitor-input, 146, 153
choke-input, 147
definition of, 74
electronic, 148
high-pass, 44-45
low-pass, 43-44
parallel-tee. See *Twin-tee.*
ripple, for power supplies, 146, 153
special purpose, 42-47
twin-tee. See *Twin-tee.*
types of, 146
Filtering networks. See also *Filters.*
special, 42-47
Fixed bias, 87, 118
Flicker noise, 309
Flip-flop, 140-144
Flowmeter(s), 284-305
bristle, 289-290
Doppler, 292-294
electromagnetic
calibration of, 303
sine-wave, 293-297
Flowmeter(s)
electromagnetic (*continued*)
square-wave, 298-299
trapezoidal wave, 300-305
isothermal, 290, 291
photoelectric, 203-204
radar, 292-294
rotameter, 286, 288
ultrasonic, 290-294
Flow rate of air, measurement of, 183-184
Flow-through oximeter, 212
Fluid-filled cup electrode, 257, 261
Follower, cathode, 95-96
Force
lines of. See *Lines of force.*
measurement of, 175
Force transducer, 175-177, 189
differential, 189
Forming bar, 227
Forward bias, 74, 78, 112, 201
Forward current transfer ratio, 113
Forward resistance, 73
Fourier analysis, 40, 41, 265-267
Frequency
cut-off, 44
definition of, 26
null, 46
of wave, definition of, 26
rejection, 46
Frequency limit, 216
Frequency response of biomedical instruments, 164-165, 192, 271, 274-284, 304
definition of, 50
determination of, 165
Frequency tripler, 76
Full-wave rectifier, 74
Full-wave bridge rectifier, 74

GAIN, voltage, 85, 87, 113-114, 116-117. See also *Amplifiers.*
Gallium arsenide photodiodes, 201
Galvani, Luigi, 161
Galvanometer, 54-57, 243
d'Arsonval, 56
pen motor, 231-236
pivoted coil, 56
Galvanometer coil, series resistor and, 58-60
Gas analysis, physiological, 205-213
Gas-filled phototube, 195-196
characteristic curve of, 195-197
Gaseous regulator tube, 151
Gate circuit, 296
Gated amplifier, 296
Gauge, strain. See *Strain gauge.*
Gauge factor, 173
Gauss, 18. See also *Magnetic field.*
General Electric Co., 75, 79, 80, 111, 124, 125
General Radio Co., 63
Glass bubble manometer, 267
Glass electrode, 208
Glow tube, 151
Grass Instrument Co., 232, 244, 256

Grid
  control, 83
  screen, 96
  suppressor, 96
Grid bias, 83
Grid current, 83
Grid noise, induced, 309
Ground, 7, 310-311, 313-316
  good, establishing in laboratory, 313-316
  virtual, 108
Ground loop, 315
Ground point, reference, 314, 315
Grounded-base amplifier, 116
Grounded-collector circuit, 116
Grounded-emitter amplifier, 116
GSR, measurement of, 168

HALLWACHS, Wilhelm, 193
Hansen "pop" technique for transient analysis of pressure system, 276
Harmonics, Fourier, 40, 41, 265-267
Hartley oscillator, 134-135
Heart muscle, properties of, study of, 178, 180
Heart rate, measurement of, 192, 203-204
Heart sounds, measurement of, 192-193
Heart wall, movement of, measurement of, 180-181
Heat sink, 76
Heat-tracing recorders, 236
Heath Co., 257
Henry, Joseph, 21
Henrys, 21
Hertz, Heinrich, 193
High-frequency recorders, 240-245
High-pressure network, 44-45
Hole, 8, 198
Houston Instrument Corp., 220
Hybrid circuit, 111
Hydrogen ion concentration, measurement of, 205

IMPEDANCE, 11, 27
  concept of, 27-29
  fluid, 29
  in RC circuit, 32
  in RL circuit, 34-35
  in RLC circuit, 39
  input, 217
  output, 217
  source, 296
Impedance bridge, 63
IMPScope, 204-205, 254, 256-257
  accessories for, 174, 204, 253, 259
Impulse, definition of, 252
Induced grid noise, 309
Inductance, 21-24
  definition of, 21
Inductance manometer, 268
Inductor(s)
  in parallel combination, 22
  in series combination, 22
In-phase signals, 90
Input, definition of, 42
Input impedance, 217
Instrument(s), biomedical. See *Biomedical instruments.*
Instrumentation, biomedical, examples of, 251-305
Insulator, 8-10
  definition of, 8
Integrated circuits, 123
Interelectrode capacitance of triode, 96
Interference
  electromagnetic, 313
  electrostatic, 90, 310-311
Inverter, phase, 88-90
Isothermal flowmeter, 290, 291
Isothermal velocity probe, 290

JOHNSON, F. B., 310
Johnson's noise, 310
Junction, definition of, 74
Junction photocell, 200

KELVIN, Lord, 54
Kirchhoff's laws, 40-42
Kymograph camera, 244

LABORATORY, establishing good ground in, 313-316
Lag, 27
Lawrence Radiation Laboratory, 126, 127, 151, 152, 156, 295, 296, 299, 300, 303
Lead, 27
Lenz's law, 22-23
Light sensors, 193-205
Limit, frequency, 216
Lines of force
  electrostatic, 6
  of bar magnet, 16-18
  of solenoid, 18, 20
  of straight conductor, 18, 19
  of toroid, 21
Load, 83
  determination of, 146
Load resistance, 46, 84
Load resistor, 73
Loop
  definition of, 40
  ground, 315
Low-frequency recorders, 219-223
Low-pass network, 43-44

MAGNETIC field, 16-24
Magnetic tape recorders, 245-250
Manometer
  capacitance, 268
  catheter tip, 269, 273
  glass bubble, 266
  inductance, 268
  optical, 268
  relative amplitude of, 278
  strain-gauge, 272-273, 274
  stretched rubber membrane, 266, 267
  Wiggers', 266, 267

Maxwell, James C., 193
Measurement(s)
errors in. See *Errors in biomedical measurements.*
of acceleration, 168
of action potential, 254, 258-259
of air flow rate, 183-184
of bioelectric potentials, 167, 251-265. See also *Electrodes* and *Electrocardiogram.*
of blood flow, 192, 201, 284-305
of blood pressure, 64, 165, 173-177, 178, 182, 183, 265-285
direct, 284
indirect, 284
of breathing rate, 183-184, 185-186
of carbon dioxide tension, 205, 210-211
of cardiovascular system, 166
of conduction velocity, 254, 264, 265
of current, 49-67
of displacement, 175, 177
of drop rate, 168
of flow rate of air, 183-184
of force, 175
of heart rate, 192, 203-204
of heart sounds, 192-193, 203
of heart wall movement, 175
of oxygen tension, 205-213
of pH, 205-211
of resistance, 60-62
of respiratory system, 167
of skin resistance, 168
of temperature, 168
of velocity, 168
of voice, 192-193
of voltage, 49-67
principles of, 49
Mechanism errors in instruments, 52
Mechanoelectric transducer, 188, 191, 290
Medium-frequency recorders, 223-240
Membrane, cell, polarized, 251
Merit, figure of, 51, 101
Metal, work function of, 70
Minneapolis-Honeywell Regulator Co., 176, 242, 243, 261
Monostable multivibrator, 347
Motorola Semiconductor Products, Inc., 118, 147, 148
Müller, Johannes, 161
Multigrid tubes, 96
Multiplier, capacitance, 147
Multiplier phototube, 195-198
characteristic curve of, 198
Multivibrator, 138
astable, 138-139, 347
bistable, 140-144, 347
monostable, 347
Multivibrator circuits, 138-139, 345
Muscle, cardiac, properties of, study of, 178, 180
Muscle cell, stimulation of, 251
Muscle potential, recording of, 264-265
Myocardium, properties of, study of, 178, 180
Myograph, 177-178, 180-181

N-TYPE semiconductors, 9
Needle electrode, 252
Nerve cell, stimulation of, 251
Nerve chamber, 254, 259
Network
filtering, special, 42-47. See also *Filters.*
high-pressure, 44-45
low-pass, 43-44
passive, 42
twin-tee as, 46
RC, low-pass, 43
voltage-divider, 119, 120
Node, definition of, 40
Noise, 307-316
definition of, 46
electronic, definition of, 307
elimination of, 307-316
emission, secondary, 309
flicker, 309
generation of, in animal, 313
grid, induced, 309
in amplifier, 96
in components in electronic circuitry, 310
in semiconductors, 310
in vacuum tube, 308-309
induced grid, 309
Johnson's, 310
partition, 309
secondary emission, 309
shorted-input, 165, 308
shot, 309
source, 308
thermionic, 72, 90
Non-sinusoidal oscillators, 138-144
Non-sinusoidal wave, 40, 138-144
Notch amplifier, 105
characteristics of, 105-107
Null frequency, 46

OERSTED, H.C., 18, 54
Ohm, 9
Ohm, George Simon, 9, 11
Ohm's law, 11-13
Ohm's law triangle, 12
Ohmmeter, 60-64
impedance bridge and, 63
parallel, 61
series, 61
On-line operation, 164
One-shot, 142
Operational amplifiers, 106-109, 345
Optical manometer, 268
Optoelectronic diode, 201
Optoelectronics, 201
Oscillation
parasitic, 155
sinusoidal, frequencies of, 129
Oscillator(s), 129-144
Clapp, 135-136
Colpitts, 134-135
crystal, 137-138

Oscillator(s) (*continued*)
Hartley, 134-135
non-sinusoidal, 138-144
phase-shift, 136-137
simple, 347
sinusoidal, 130-137
split-tank, 134
tuned-base, 132
tuned-grid, 131
tuned-grid-tuned-plate, 133
Wien bridge, 347
Oscillator circuits, 124
Oscillatory response in RLC circuit, 37
Oscilloscope, 64-67, 248-249, 256-257, 258
cathode-ray, 64
dual-beam, 65
Out-of-phase signals, 90
Output, definition of, 42
Output impedance, 217
Output resistance, 84
Output signal, 215-249
Overdamped response in RLC circuit, 38
Oximeter, flow-through, 212
Oximetry, 212
Oxygen, partial pressure of, measurement of, 205-213

P-TYPE semiconductors, 9
Pacemaker, cardiac, definition of, 257
Parallax, reading errors due to, 53, 54
Parallel combination
capacitors in, 15-16
inductors in, 22
resistors in, 10-11
Parallel-tee, 45-47
characteristics of, 46
typical values of resistance and capacitance in, 47
Parameters, biomedical, characteristics of, 169-170
Parasitic oscillation, 155
Partition noise, 309
Passive network, 42
twin-tee as, 46
Peak of tunnel diode, characteristic, 79
Peak current, reading of, meter for, 126
Pentode, 96
Permeability
magnetic field holding and, 20
of core material, 20
Persistence on oscilloscope screen, 65
pH, measurement of, 205
Phase, 27
definition of, 27
in RC circuit, 30
in resistive branch, 30
in RL circuit, 34
in RLC circuit, 39
Phase detector, 299
Phase inverter, 88-90
Phase-shift oscillator, 136-137
Phase shifter, 108
Photocell(s), 198-204
cadmium sulfide, 199
definition of, 198
junction, 200
spectral response of, 195, 202
types of, comparison of characteristics of, 202
Photoconductive material, 198
Photodiodes
gallium arsenide, 201
junction, characteristic curve for, 200
quantum efficiency of, 200
Photoelectric effect, 193
Photoelectric flowmeter, 203-204
Photoelectric plethysmograph, 203, 204
Photoelectric transducers, 177-178, 180-184, 199, 201, 203-204, 212
Phototube
gas-filled, 195-196
characteristic curve of, 195, 197
multiplier, 195-198
characteristic curve of, 198
vacuum, 193-195
spectral response of, 193-195, 202
Photovoltaic cell, 200
Physical units, conversion factors and, 319
Physiograph, 238
Physiological gas analysis, 205-213
Piezoelectric pulse transducers, 190, 192-193
Pin electrodes, 252
Pivoted coil galvanometer, 56
Plate(s), deflection, 64
Plate-characteristic curve, 84, 88
Plate current of rectifier, 72
Plate cut-off voltage, 85
Plate efficiency, 87
definition of, 87
Plate electrode, 253
Plate resistance, 84
Plate voltage of rectifier, 72
Plethysmograph, photoelectric, 203-204
PN-junction diode, characteristic curve of, 75, 79, 111
Pneumograph, bellows, 183-184
Pneumotach, double cone, 183
Point
branch, definition of, 40
reference, ground, 314, 315
Polarization
of cell membrane, 251
of electrode, 257-258
Polygraph, 232
Potential(s). See also *Voltage.*
action, definition of, 252
bioelectric, 74-75, 87, 91-92, 94, 112, 121, 251-265
in brain tissue, measurement of, 259, 263
resting, 251-252
definition of, 251
Potentiometer, 58, 59-60
Potentiometric strip-chart recorder, 223
Potting material, 170

Power, electrical, 12-13
Power amplifier, 126-127
Power supply, 145-157
  elements of, 153-157
  regulation of, 148-157
Preamplifier, definition of, 124
Precision Instrument Co., 245
Pressure
  blood. See *Blood pressure.*
  diastolic, 265
  differential, 268
    measurement of, 268-272
  pulse, 265
  systolic, 265
Pressure gradient, definition of, 266
Pressure transducer, 178, 182-183, 266-273
  Bourdon tube, 178, 182-183
  capacitance manometer, 268
  differential transformer manometer, 269
  glass bubble manometer, 267
  strain-gauge manometer, 272-273, 274
  Wiggers' manometer, 266-267
Primary coil, definition of, 22
Pulse pickup
  photoelectric, 201, 203-204
  piezoelectric, 192, 193
Pulse pressure, 265
Pump, pressure and flow calibration with, 280-281
Push-pull operation, 90-95

Q, 38
Quadrature voltage, 296
Quantum efficiency of photodiode, 200

RADAR flowmeter, 292-294
Radio Corporation of America, 85, 87, 88, 89, 90, 96, 97, 98, 99, 109, 113, 114, 115, 116, 118, 119, 120, 121, 122, 123, 124, 132, 142, 143, 149, 150, 191, 194, 195, 196, 197, 198, 199, 200, 202, 327, 333, 334, 335, 336, 337, 338, 339, 340
Range of measuring instruments, 49
Ratio
  forward current transfer, 113
  signal-to-noise, 307
RC circuit, 30-32
  time constant of, 31-32
RC filter network
  high-pass, 44
  low-pass, 43
Reactance, 27
  definition of, 11
Read-out devices, 163, 215
  characteristics of, 215
Reading errors in biomedical measurements, 53-54
Recorder(s)
  curvilinear tracing of, 218, 229
  errors in, 215-219, 237
  heat tracing, 236
  high-frequency, 240-245
  low-frequency, 219-223
  magnetic tape, 245-250
Recorder(s) (*continued*)
  medium-frequency, 223-240
  optical, 241-245
  potentiometric strip-chart, 223
  rectilinear recording of, 218, 221, 229, 231, 237
  strip-chart, 220
    potentiometric, 223
  tape, 245-248
Recordings, types of, 217
Rectification
  full-wave, 74
  half-wave, 73
Rectifier(s), 69-81
  curve for, typical, 72, 73
  definition of, 69
  full-wave, 74
  full-wave bridge, 74
  operation of, 72-74
  silicon-controlled, 201
  vacuum-tube, 69-70
Rectilinear recording
  definition of, 217
  versus curvilinear recording, 218
Reference ground point, 314, 315
Reference voltage, 153
Regeneration, definition of, 102
Regulation
  of power supply, 148-157
  of voltage, 148
Regulator
  cascade shunt, 149
  series, 153
Regulator tube, gaseous, 151
Rejection frequency, 46
Resistance, 9-10, 27
  back, 73
  definition of, 9
  effective, 45
  forward, 73
  in AC circuits, 27, 29-30
  load, 46, 84
  measurement of, 60-64
  output, 84
  plate, 84
  source, 45
Resistive branch of AC circuit, 29-30
Resistivity
  definition of, 9
  of materials at room temperature, 10
Resistor(s)
  bleeder, 147
  damping, 216
  fixed, identification of, 323
  in parallel combination, 10-11
  in series combination, 10-11
  load, 73
  series, galvanometer coil and, 58-60
  shunt, 57
Resistor-capacitor color code, 324
Resonance, series, quad, 38
Respiratory system, measurement of, 167

Response
critical, in RLC circuit, 38
frequency
of biomedical instruments, 164-165, 192, 271, 274-284, 304
definition of, 50
determination of, 165
spectral, 193-195, 202
steady state. See *Steady state response.*
transient. See *Transient response.*
Resting potential, 251-252
definition of, 251
Reverse bias, 74, 78, 112, 200
Rheostatic displacement transducer, 173
Ripple, 146-148
definition of, 145
RL circuit, 32-35
time constant of, 33
RLC circuit, 35-37
series, steady state response in, 39-40
time constant of, 38
transient response in, 37-38
RMS value, 29
Rotameter, 286, 288
Runaway, thermal, 111, 119

SANBORN Co., 186, 187, 188, 189, 233, 234, 235, 236, 241, 246, 249, 270, 271
Scale errors in instruments, 52
Scalp potentials, measurement of, 248-250, 270
Schweigger, Johannes S.C., 161
Screen grid, 96
Secondary coil, definition of, 22
Secondary emission noise, 309
Self-bias, 88, 119
Semiconductor(s), 8-10. See also *Transistor, Diode, Zener diode, Photodiode, Optoelectronic diode.*
N-type, 9
noise in, 310
P-type, 9
Semiconductor diode, 74-80
Sensitivity, 50-51
definition of, 50
of measuring instruments, 50-51
Sensors, light, 193-205
Series combinations
capacitors in, 15-16
inductors in, 22
resistors in, 10-11
Series regulator, 149, 153
Series resistor, galvanometer coil and, 58-60
Series resonance, Q and, 38
Shifter, phase, 108
Short, virtual, 108
Shorted-input noise, 165, 308
Shot noise, 309
Shunt-fed Hartley oscillator, 134
Shunt regulator, 149
Shunt resistor, 57
Signal(s)
common mode, 90
Signal(s) (*continued*)
feedback, positional, 231
in-phase, 90
out-of-phase, 90
output, 215-249
Signal-to-noise ratio, 307
Silver-silver chloride surface electrodes, 258-262
Silicon-controlled rectifier, 201
Simpson Electric Co., 55, 59
Sine wave
frequency of, 26
generation of, 8, 26
Sink, heat, 76
Sinusoidal oscillation, frequencies of, 129
Sinusoidal oscillator, 130-137
Skin electrode, wire cloth, 257, 261
Skin resistance, measurement of, 168
Slide-wire amplifier, 231
Solenoid, generation of magnetic field in, 20
Solid state diode, applications of, 76
Solid State Electronics Corporation, 126
Source impedance, 296
Source noise, 308
Source resistance, 45
Spectral response, 193-195, 202
Split-tank oscillator, 134
Spring constant of instrument, 50
Squaring circuit, 140-141, 299
Statham Physiological Transducers, Inc., 177, 178, 179, 272, 273, 274
Static characteristic curve of rectifier, 72
Steady state response, 29, 32-33, 39-40
definition of, 29
in RC circuit, 32
in RL Circuit, 33
in RLC Circuit, 37
in series RLC circuit, 39-40
Stimulation of nerve or muscle cell, 251
Stimulator, 256
of nerve or muscle, 254, 257
Strain gauge, 64, 173, 175-177
bonded, 174-177
unbonded, 64, 175
Strain gauge manometer, catheter tip, 273
Stray capacitance, 97, 296
Strip-chart recorder(s), 220
potentiometric, 223
Stromuhr, 288
thermal, 289
Super-conductivity, 9
Suppressor grid, 96
Sweep circuit, 64
Switch, electronic, of oscilloscope, 66
Symbols, electronic, 321
Synapse, definition of, 252
Systematic errors in measurements, 54
Systolic pressure, 265

TANK circuit, 131
Tape recorder(s), 245-248
magnetic, 245-250

Tektronix, Inc., 66, 103, 104
Temperature, measurement of, 168
Tetrode, 96
Texas Instruments, Inc., 221
Thermal runaway, 111, 119
Thermal stromuhr, 289
Thermionic emission, 70-72
Thermionic noise, 72, 90
Thermostromuhrs, 289
Thomson, J. J., 193
Threshold, definition of, 251
Time constant
  of circuit, 31
  of RC circuit, 31-32
  of RL circuit, 33
  of RLC circuit, 38
Toroid, generation of magnetic field in, 21
Tracings, types of, 217
Transconductance of triode, 84
Transducer(s), 171-213
  blood flow, 286, 289-290, 293-294, 297, 300, 302
  blood pressure. See *Blood pressure transducer.*
  Bourdon tube, 178, 182-183
  calibration of, 172
  capacitance displacement, 187-188
  definition of, 171
  differential transformer, 183, 185-189
  displacement, 173, 177-178, 180-181
    photoelectric, 177-178, 180-181
    rheostatic, 173-174
  force, 175-177, 189
    differential, 189
  mechanoelectronic, 188, 191, 290
  photoelectric, 177-184, 199, 201, 203-204, 212
  piezoelectric, 190, 192-193
  pressure, 178, 182-183
  strain-gauge. See *Strain gauge.*
Transfer-characteristic curve
  for transistors, 117
  for vacuum-tube triode, 84, 87
  typical, 84, 87
Transformer(s), 22-24
  core, 24
    material of, 24
  definition of, 22
  differential, 183, 184
  primary coil of, 22
  secondary coil of, 22
Transformer component, 296
Transient analysis of catheter-manometer system, 276
Transient response, 29-38
  definition of, 29
  in RC circuit, 31-32
  in RL circuit, 33
  in RLC circuit, 37-38
  types of, 37
Transistor(s), 109-111
  basic circuit arrangements of, 116
  breakdown voltage of, 114
Transistor(s) (*continued*)
  characteristic curves for, 116-117, 341-344
  circuit arrangements in, 116
    biasing of, 118-119
Transistor amplifier, 111-117
  circuits arrangements in, examples of, 120-126
  classes of, 118
Transistor stages, coupling of, 119
Trapezoidal-wave flowmeter, 300
Trigger level of sweep, 65
Triode(s), vacuum tube, 83-88
Tube(s)
  gaseous regulator, 151
  glow, 151
  multigrid, 96
  regulator, gaseous, 151
  vacuum. See *Vacuum tubes.*
Tuned-base oscillator, 132
Tuned-grid oscillator, 131
Tuned-grid-tuned-plate oscillator, 133
Tunnel diode, 78, 79-80
  characteristic curve for, 79-80
Twin-tee, 45-47
  characteristics of, 46
  typical values of resistance and capacitance in, 47

ULTRASONIC flowmeter, 290-294
Unbonded strain gauge, 64, 175
Underdamped response in RLC circuit, 38
Units, physical, conversion factors and, 319

VACUUM phototube, 193-195
  spectral response of, 193-195, 202
Vacuum tube(s)
  cathode-ray, 64
  characteristic curves for, 326-340
  double triode, characteristic curves for, 89, 327, 336-340
  noise in, 308-309
Vacuum tube pentode, characteristic curves for, 331-332
Vacuum tube rectifier, 69-70
Vacuum tube triode, 83-88
  characteristic curves for, 329-330
Valley point of tunnel diode, characteristic, 79
Velocity
  conduction, of impulse, measurement of, 254, 264, 265
  measurement of, 168
Virtual ground, 108
Visicorder, 242-243
Voice, measurement of, 192-193
Volt, definition of, 7
Volt-ohmmeter, 58, 59
Volta, Allesandro, 7
Voltage(s), 5-7
  avalanche, 74
  bias, 87
  breakdown, 114, 115
  clipping circuit and, 77, 78
  definition of, 5

Voltage(s) (*continued*)
excitation, 165
measurement of, 49-67
plate, of rectifier, 72
plate cut-off, 85
quadrature, 296
reference, 153
regulation of, 148
Voltage-divider network, 119, 120
Voltage doubler, 76, 78
Voltage gain, 85-87, 113-114, 116-117. See also *Amplifiers.*
Voltage quadrupler, 76, 78
Voltage tripler, 76, 78
Voltmeter, 58, 60
digital, 60
Volume displacement of pressure transducers, 267
VOM, 58, 59

WALLER, Augustus D., 161
Wave(s)
non-sinusoidal, 40, 138-144
sine
frequency of, 26
generation of, 26
Waveforms, 40
Wheatstone bridge, 62-64. See also *Strain gauge.*
White Instrument Laboratories, 45, 46, 106, 107
Wick electrode, 253
Wien bridge oscillator, 347
Wiggers' manometer, 266, 267
Wire cloth skin electrode, 257, 261
Work function of metal, 70

X-Y RECORDER, 220

Z. See *Impedance.*
Zener diode, 78, 148
characteristic curve for, 78-79
reference, 149